PROTEIN BIOCHEMISTRY, SYNTHESIS, STRUCTURE AND CELLULAR FUNCTIONS

HISTONES

CLASS, STRUCTURE AND FUNCTION

PROTEIN BIOCHEMISTRY, SYNTHESIS, STRUCTURE AND CELLULAR FUNCTIONS

Additional books in this series can be found on Nova's website under the Series tab.

Additional E-books in this series can be found on Nova's website under the E-book tab.

PROTEIN BIOCHEMISTRY, SYNTHESIS, STRUCTURE AND CELLULAR FUNCTIONS

HISTONES

CLASS, STRUCTURE AND FUNCTION

CHANG-HUI SHEN
EDITOR

nova publishers
New York

For permission to use material from this book please contact us:
Telephone 631-231-7269; Fax 631-231-8175
Web Site: http://www.novapublishers.com

NOTICE TO THE READER

Additional color graphics may be available in the e-book version of this book.

LIBRARY OF CONGRESS CATALOGING-IN-PUBLICATION DATA

Histones : class, structure, and function / editor, Chang-Hui Shen.
p. ; cm.
Includes bibliographical references and index.
ISBN 978-1-62100-274-1 (hardcover)
I. Shen, Chang-Hui, 1967-
[DNLM: 1. Histones--chemistry. 2. Histones--metabolism. 3. Histones--physiology. QU 56]
LC classification not assigned
572'.66--dc23

2011033097

Published by Nova Science Publishers, Inc. † New York

CONTENTS

PREFACE

Histones are DNA-binding proteins that help in the packaging of DNA to fit into the cell nucleus in a hierarchy of organization. During the packaging and unfolding process, histones are involved in crucial biological activities at the molecular level. This book presents research in the study of histone class, structure and function, including the roles of linker histones in chromatin structure, the post-translational modifications of histones and the importance of chromatin structure and epigenetics in gene regulation and carcinogenesis.

Chapter 1 - The H1 linker histone family of proteins is the most divergent heterogeneous class of histone proteins. In mammals, there are eleven different subtypes. Histone H1 binds to the DNA entering and exiting the nucleosomal core particle and has an important role in the establishment and maintenance of higher order chromatin structure. As such, this histone class was traditionally regarded as being the architectural proteins of chromatin involved in chromatin condensation. This led to the speculation that this class of histone proteins functions as global repressors of transcription. However, their evolutionary conserved heterogeneity, as well as more recent data, has led to the hypothesis that the H1 subtypes are also involved in chromatin remodeling and genomic integrity and function not only as negative regulators, but also as positive regulators of certain genes. The major epigenetic modification (post-translational modification) of the H1 histones is phosphorylation. The results of many investigations have also implicated phosphorylation of H1 in the regulation of chromatin remodeling. Changes in the H1 subtype composition of chromatin have been extensively studied in many cell and tissue systems during various biological processes, such as development and differentiation. However, whether there is a functional relevance in these subtype switches had not been addressed in these classic studies. More recent results now indicate that specific subtypes may have unexpected and discrete functional roles, not only during development and differentiation, but also during DNA replication, DNA repair, apoptosis and aging. In this light, the aim of this review is to present the highlights of the accumulated available data which strongly support that specific H1 subtypes may have functionally significant roles in certain biological processes.

Chapter 2 - The manner in which linker histone globular domains associate with the nucleosome is a key issue in understanding the location of the linker histone tails, and thus their functional roles in regulating gene expression. Understanding of the location of the linker histone globular domain within the nucleosome can provide insight into the nucleosome architecture in the context of the location relative to the dyad on a statistically significant nucleosome. It can also indicate any possible association between linker histones and different nucleosomes in the higher order chromatin structure. Therefore, it would contribute to the implication of gene regulation and higher-order structure models. However, studies on the binding of the linker histone globular domain to the nucleosome have not yielded a unified picture. In this chapter, I will discuss the current findings in this field including the linker histone variants, the structure of the linker histone globular domain, the location of the linker histone within a chromatosome, the functional domains of the linker histone tail, the contribution of linker histones to higher order chromatin structure and its connection to gene expression.

Chapter 3 - The advantages of using *Saccharomyces cerevisiae* as a model for chromatin studies imply its relatively small number of genes, about 6000, which enables uncomplicated genome-wide study. On the other hand, its short cell cycle - around two hours for genome duplication is another benefit for research. It has been shown that *S. cerevisiae* cells possess about 28% homology of genes with human. Interestingly 20% of genes known to take part in the development and progression of certain human diseases have their homologs in yeast cells. Furthermore, yeast genome possesses a full complement of histones and other chromatin proteins which express vast homology with the respective proteins in human cells. These characteristics of *S. cerevisiae* allow obtained results on them to be easily approximated for human cells. Chromatin has always been a matter of extensive scientific research. A good part of the knowledge of the authors on its structure and function has been gathered by experimenting on *S. cerevisiae*. Strangely enough, among the plethora of proteins which build up yeast chromatin the linker histone catches a specific attention with its odd features. *S. cerevisiae* linker histone, Hho1p, is the only histone known to possess two globular domains instead of one. Previous hypotheses suggested that it is responsible for the organization of yeast chromatin mostly in euchromatin. Nonetheless, recent data, including ours, have shown that Hho1p functions in yeast chromatin are far more diverse. It has been proved to be involved in the building and maintenance of the higher-order chromatin organization in yeast, from the 30 nm fiber, chromatin loop organization up to further levels of chromatin compaction. Altogether these data brings this histone protein closer to linker histones of higher eukaryotic cells and points out that linker histones and higher-order chromatin structures are more evolutionary conserved than it is generally believed.

Chapter 4 - Recent quantitative proteome analyses of human metaphase chromosomes have revealed various types of histones and their amounts involved in metaphase chromosomes. Core histones, and their variants such as H2A.Z and MacroH2A account for 70 % of the total amount of chromosome proteins. Significant amounts of ubiquinized H2A are also found in chromosomes. In addition, different types of linker histone H1 and its variants, including H1.X, are found in total 6 % of chromosome proteins. Unlike authentic linker histones that localize over the entire chromatin region, histone H1.X accumulates in the nucleolus during interphase and is at the chromosome periphery during mitosis, suggesting a distinct structural role for H1.X. Knockdown analysis of H1.X results in the failure of alignment and/or segregation of mitotic chromosomes. HP1-BP74 which has an H1 globular domain at the central region has been identified from proteome analysis. The HP1-BP74 globular domain binds to the linker DNA between nucleosomes, facilitating the formation of the higher-order structure, chromatosome. Proteome analyses of metaphase chromosomes shed light on the importance of histones in the formation of higher-order chromosome structure.

Chapter 5 - Histones are subjected to post translational modifications such as acetylation, deacetylation, methylation, phosphorylation, ubiquitination, sumoylation, ADP ribosylation, glycosylation, biotinylation, and carbonylation. The first four post-translational modifications (acetylation, deacetylation, methylation, and phosphorylation) have been studied extensively in comparison with the other types of modification. Histone modification activity has been shown to participate in a variety of cellular processes including, but not limited to, gene activity. In this chapter, the authors will discuss current findings of histone modifications and its connection to gene activity.

Chapter 6 - Histone modifications play a crucial role in the epigenetic regulation of various aspects of eukaryotic development. Histone modifications such as methylation and acetylation of histone tails are regulated by histone modifiers which are specific enzymatic complexes including histone methyltransferases, histone demethylases, histone acetyltransferases and histone deacetylases. Histone modifications result in the modulation of chromatin structure, a process which leads to the optimal gene expression programmes that eukaryotic organisms require during development and in response to changing environments. The action of histone modifiers leads to open (active) or closed (inactive) forms of chromatin, which are associated with activation or silencing of genes, respectively. Whereas histone modifiers have been well investigated and reviewed in the model plant *Arabidopsis*, comprehensive reports regarding the study of histone modifiers in plants of high economic value, such as cereals, are limited. This review will present the current knowledge on epigenetic histone modifiers in cereal model plants and crops such as *Brachypodium*, rice, maize, and barley.

Chapter 7 - Chromatin condensation and subsequent decondensation are processes required for proper execution of various cellular events including cell division and DNA replication. During mitosis, chromatin compaction is at its highest, whereas relaxation of chromatin is necessary for DNA replication, repair, recombination, and gene transcription. Since histone proteins are directly complexed with DNA in the form of a nucleosome, great emphasis is put on deciphering histone post-translational modifications that control the chromatin condensation state. Histone H3 phosphorylation is a mark present in mitosis, where chromatin condensation is necessary, and in transcriptional activation of genes, when chromatin needs to be decondensed. There are four characterized phospho residues within the H3 N-terminal tail during mitosis: Thr3, Ser10, Thr11, and Ser28. Interestingly, H3 phosphorylated at Ser10, Thr11, and Ser28 has been observed on genomic regions of transcriptionally active genes. Therefore, H3 phosphorylation is involved in processes requiring opposing chromatin states. The level of H3 phosphorylation is mediated by opposing actions of specific kinases and phosphatases during mitosis and gene transcription. The cellular contexts under which specific residues on H3 are phosphorylated in mitosis and interphase are known to some extent. However, the functional consequences of H3 phosphorylation are still unclear. Likewise the role of epigenetic signaling events, regarding methylation and acetylation of histone proteins, that regulate histone modification during mitosis is still actively being explored. Several converging and reinforcing signals, including transcription factors, noncoding RNAs, DNA methylation, and histone modifications orchestrate complex epigenetic states. Although all of these pathways modulate transcription from chromatin in vivo, the mechanism by which epigenetic information is transmitted through cell division remains unclear.

Chapter 8 - A nuclear genome physically exists as a nuclear DNA molecules or a set of chromosomes. The structural parameters of a nucleus or chromosome would determine the functional basis of a genome, i.e. regulation of gene expression and transmission of the genome to the next generation. In plants, there are distinct upper limits of structural parameters for maintaining genome functions. The genomic DNA content in a nucleus has an upper limit of 3% (v/v, DNA/nucleus) for construction and maintenance of a functional nucleus. The length of a chromosome arm is also restricted to an upper limit that is less than a half of the length between the two spindle poles. Moreover each chromosome functionally differentiates region from to region which is characterized with specific combinations of histone modifications. The three-dimensional distribution of the each chromosomal region with specific patterns of histone modification and the functional differentiation of the nuclear space are also closely related to the plants with large-sized genomes. This fact clearly indicates that the higher structure of the nucleus and the chromosomes with different histone modification set play essential roles in genome functions.

Chapter 9 - Oxidative stress causes DNA and protein damage, which may compromise genomic integrity. While oxidant-induced DNA damage has been extensively studied, less is known about nuclear protein damage, particularly of histones, which support chromatin structure and influence gene expression. Among other mechanisms, protein damage may be caused via non-enzymatic glycation initiated by reducing sugars (e.g., glucose and ADP-ribose) or reactive dicarbonyl compounds (e.g., glyoxal). Glycation is a non-enzymatic reaction between the reducing sugar carbonyl and protein amino groups, which undergo rearrangements to stable ketoamines and advanced glycation end products (AGEs) such as fluorescent (e.g., argpyrimidine) and non-fluorescent (e.g., N^{ϵ}-carboxymethyllysine, CML) protein adducts and crosslinks. Protein glycation, induced by hyperglycemia, has been implicated in diabetic complications, although the contribution of histone glycation to these processes is unknown. High local concentrations of ADP-ribose are produced in the chromatin microenvironment during DNA repair, potentially contributing to histone glycation. This chapter describes the authors' detection of non-enzymatic modifications occurring *in vivo* on histone H1 from calf thymus (Part I) and the basis for inhibition of these processes. In particular, the authors will discuss the roles of rutin metabolites as alternative, effective natural product AGE inhibitors for prevention of histone H1 modification and resulting disease complications (Part II).

Chapter 10 - The organisation of DNA and histone proteins into chromatin plays a central role in the regulation of eukaryotic gene expression. The coordinated regulation of gene expression programs is critical in directing developmental programs and in enabling cells to respond to extracellular signals. The combination of transcription factor binding sites found in gene promoters and enhancers, and the availability of these transcription factors, is important in co-ordinating these gene expression programs. However transcription factors must operate within the chromatin landscape, which regulates their ability to access their binding sites and drive transcription. The chromatin architecture of a gene can vary dramatically in different cell types, with differences in histone density, composition, modification and the presence of chromatin remodelers. While there would appear to be almost unlimited permutations of chromatin architectures, recent data generated using genome-wide technologies has begun to define chromatin signatures that are associated with particular gene states, and provide insight into how these architectures interact with extracellular signals. In this chapter the authors review current information on the role of chromatin in regulating inducible gene expression, using the immune system as a model.

Chapter 11 - The production of ribosomal RNA (rRNA) accounts for more than half the transcriptional activity of an actively growing eukaryotic cell. Not surprisingly given its importance to the cell's well-being, the entire system of rRNA production is subject to several levels of regulation that often distinguish it from those of the protein-coding genes. The regulation of rRNA gene expression occurs at many levels, including that of chromatin structure. The authors' current understanding of rRNA gene chromatin (r-chromatin) in higher eukaryotes is that it exists in several forms ranging from a nucleosomal and silenced chromatin in which promoter DNA is methylated at several positions to a non-nucleosomal, methylation-poor chromatin associated with the ribosomal transcription factor UBF. The roughly equal ratio of silenced to active rDNA copies usually changes very little between resting and actively growing cells, and most of the expression regulation is due to the rate of transcription initiation and elongation at the active rDNA copies. This peculiar observation suggests that the organization of r-chromatin plays an important role in more than gene expression regulation, and is likely to involve the stability of the recombination-prone ribosomal RNA genes. In this chapter, the authors describe different structural aspects of the r-chromatin and their possible relation to the regulation of its expression.

Chapter 12 - Cancer development is a multi-factorial event, involving nuclear processes that promote the expression of genes linked to cell survival and proliferation. A clear connection has been established between epigenetic events and cancer biology. However, the nature of this relationship as it relates to disease progression and treatment remains unclear. With a particular emphasis on breast cancer, here the authors review advancements in the fields of DNA methylation, histone acetylation and histone methylation as they relate to tumor development, progression, and treatment.

In: Histones: Class, Structure and Function
Editor: Chang-Hui Shen
ISBN: 978-1-62100-274-1

Chapter 1

THE FUNCTIONALLY DISCRETE BIOLOGICAL ROLES OF THE H1 LINKER HISTONE SUBTYPES

Thomae G. Sourlingas*
Institute of Biology, Laboratory of Nuclear Proteins and Chromatin Function,
National Centre for Scientific Research, "Demokritos"
Aghia Paraskevi, Attiki, Greece

ABSTRACT

The H1 linker histone family of proteins is the most divergent heterogeneous class of histone proteins. In mammals, there are eleven different subtypes. Histone H1 binds to the DNA entering and exiting the nucleosomal core particle and has an important role in the establishment and maintenance of higher order chromatin structure. As such, this histone class was traditionally regarded as being the architectural proteins of chromatin involved in chromatin condensation. This led to the speculation that this class of histone proteins functions as global repressors of transcription. However, their evolutionary conserved heterogeneity, as well as more recent data, has led to the hypothesis that the H1 subtypes are also involved in chromatin remodeling and genomic integrity and function not only as negative regulators, but also as positive regulators of certain genes. The major epigenetic modification (post-translational modification) of the H1 histones is phosphorylation. The results of many investigations have also implicated phosphorylation of H1 in the regulation of chromatin remodeling. Changes in the H1 subtype composition of chromatin have been extensively studied in many cell and tissue systems during various biological processes, such as development and differentiation. However, whether there is a functional relevance in these subtype switches had not been addressed in these classic studies. More recent results now indicate that specific subtypes may have unexpected and discrete functional roles, not only during development and differentiation, but also during DNA replication, DNA repair, apoptosis and aging. In this light, the aim of this review is to present the highlights of the accumulated available data which strongly support that specific H1 subtypes may have functionally significant roles in certain biological processes.

* E-mail: sourlin@bio.demokritos.gr.

1. INTRODUCTION

The histone family of proteins is comprised of five classes, H2A, H2B, H3, H4 (core histones) and H1. Two molecules each of the first four classes form the core histone particle (histone core octamer) around which are wrapped 146 or 147 nucleotide base pairs (bp). This is the fundamental repeat unit of eukaryotic chromatin, i.e., the nucleosome [1, 2, 3, 4, 5] The fifth histone class, H1, also termed linker histone (LH), associates with the nucleosomal core particles and its globular region constrains an additional ~ 20bp and the resulting chromatosomes comprised of approximately 167bp [6] are connected by stretches of linker DNA with lengths varying between 0 and ~ 70bp [reviewed in 7]. H1 seals and organizes the DNA entering and exiting the nucleosomal core particle [8, 9]. This is the first order of compaction, the 11 nm fiber (also known as beads-on-a-string). Histone H1 stabilizes the interaction of nucleosomes in compacted chromatin [8, 9]. In this respect, the basic role of H1 is architectural in that it is involved in the second order of DNA compaction (condensation), the formation of the 30 nm fiber, the chromatin fiber. Classic *in vitro* studies showed that the H1 histones had a fundamental structural role in the generation of the chromatin fiber [10, 11, 12, 13, 14]. This extensive *in vitro* line of work led to the conclusion that the H1 histones were static structural proteins. As such, these *in vitro* properties also led to the speculation that these proteins function as global repressors of transcription.

All histones are known to be conserved in evolution. However, although the H1 histone proteins also have conserved regions, unlike the core histones, which demonstrate remarkable sequence conservation, the H1 linker histones diverge significantly in sequence and structure through evolution [9]. Moreover, all histone classes have interspecies non-allelic subtypes and all, except H4, have intraspecies subtypes. However, the H1 histone family is the most heterogeneous class of histones. To date, eleven different H1 subtypes have been identified in the human genome. These subtypes are classified according to the cell-type and/or according to the phase of the cell cycle in which they are expressed [15]. The H1 subtypes that are expressed in somatic cells during the S phase are five: H1.1, H1.2, H1.3, H1.4 and H1.5 (somatic subtypes that are cell cycle-dependent or DNA replication-dependent) [16, 17, 18, 19, 20]. There are also two subtypes expressed in somatic cells that are expressed throughout the cell cycle (replication-independent subtypes). These are the replacement H1 subtype, H1.0 (also known as $H1^{o}$ (degrees) and H1 zero) [21, 22, 23] and the ubiquitous subtype, H1x [24, 25]. Four H1 subtypes are expressed in germ cells, i.e., the testicular isoforms, H1t, H1T2 and H1LS1 [26, 27, 28, 29, 30] and an oocyte-specific subtype, H1oo [31] [also reviewed in 15, 32].

Aside from the large body of *in vitro* work that led to the traditional delineation of their role in chromatin formation and maintenance, three decades of research also investigated the H1 subtype composition (constitution) of chromatin in many cell types and tissues. Among others, some of these early studies included lymphocytes, fibroblasts, spermatocytes, non-dividing and/or quiescent (Go) cell/tissue systems, dividing somatic cell types, during tissue-specific development [reviewed in 33], rat brain cortical neurons [34], differentiating cell systems [34], leukemic cell types [35], etc. These studies revealed the extensive diversity of the H1 subtypes that are present in varying proportions depending on the species and/or the cell/tissue system.

Moreover, aside from their subtype heterogeneity, the H1 subtypes are also covalently modified (post translational modifications). Core histone modifications have been, and are, extensively studied and have given rise to the hypothesis of the "histone code" [36, 37, 38, 39]. However, little attention has been given to H1 modifications and much less is known as to their roles. The best studied H1 modification currently is phosphorylation. Traditionally, H1 phosphorylation levels were associated with the cell cycle. Observations showed that the levels of H1 phosphorylation were lowest in the G1 phase and increases continuously during S and G2, reaching a maximum in late G2 [40]. This led to the assumption that high levels of H1 phosphorylation were related to, or played a role in, chromosome condensation. For many years, it had also been known that H1 is also ADP-ribosylated [41, 42], but little is known about the function of this modification. More recently, other H1 modifications have been identified. For example, H1 can be ubiquinated *in vivo* [43] and sites of H1 acetylation and methylation have been identified by mass spectrometry (MS) [44]. These findings further extend the spectrum of H1 heterogeneity.

Having delineated the basic, but in actuality, complex aspects/properties of the H1 family of proteins, some questions come to mind. Why is this histone class so heterogeneous? Why are there subtype switches and changes in the proportion of the H1 subtypes during biological processes such as development and differentiation? Is there a functional relevance to all this heterogeneity? In fact, more recent data have shown that (1) the H1 histones are not only global repressors but may also act as positive regulators of gene transcription, (2) H1 phosphorylation can in fact play a role in chromatin remodeling and transcriptional regulation, (3) *in vivo* studies have shown that the H1 histone subtypes are not just static structural proteins, but have preferential dynamic binding mobilities/properties and localizations on eu- and heterochromatin regions, which further predisposes us to look towards functionally discrete roles for certain subtypes and, (4) certain H1 subtypes may have discrete functional roles in other biological processes, aside from those of development and differentiation, such as DNA repair, replication, apoptosis and aging.

There is no doubt as to H1's role as a chromatin architectural protein. However, these more recent findings warrant a reappraisal of the traditional concept of H1. It may in fact be that the H1 subtypes have dual roles, both structural and functional. Indeed, all recent data involving the H1 subtypes point towards that direction. In this light, it is the aim of this chapter to present the highlights of the accumulated recent data that support the idea that specific H1 subtypes have discrete functional roles in certain biological processes.

2. Structure of H1 and its Subtypes

In higher eukaryote organisms, the linker histones have a three-domain (tripartite) structure consisting of a conserved central globular domain, flanked by relatively unstructured (in solution) N- and C- terminal domains (CTD). The central domain is comprised of approximately 80 amino acid residues which are mostly hydrophobic. Its structure has been determined by X-ray crystallography [12, 45, 46]. This domain consists of a three helix bundle containing a winged-helix motif similar to that found in prokaryotic and eukaryotic DNA-binding proteins i.e., transcription factors [45, 47]. The central globular domain is believed to bind at or near the dyad axis of symmetry of the nucleosomal core particle, where the entering and exiting DNA duplexes cross [12, 48, 49]. This binding serves to seal two full turns of DNA around the nucleosome [6]. However, the exact location and binding site of H1 on the nucleosome has yet to be determined. Pruss and Wolffe proposed the off-dyad model [50], whereas Muyldermans and colleagues [8] postulated the dyad-axis binding model [reviewed in 51].

The linker histone molecule is intrinsically asymmetric. The basic N-terminal and C-terminal tails are very different in length. The C-terminal tail comprises about 50 % of the protein [51]. The short N-terminal domain, with a length of about 45 amino acid residues, is enriched in basic amino acids in its second half, adjacent to the central globular domain. The C-terminal domain is extended, has a length of about 100 amino acids and is highly enriched in lysine, serine, proline and alanine [9, 15]. However, despite this description, it must be noted that this domain is highly variable in both length and net charge [9]. The C-terminal domain is required, in addition to the globular domain, for H1-dependent chromatin folding and compaction into higher-order structures, but is not sufficient alone, for the full micrococcal nuclease protection, as ascertained from early micrococcal nuclease digestion studies on reconstituted chromatin [52, reviewed in 47]. On the other hand, these same early studies showed that the shorter N-terminal basic tail was not required for inducing higher order structures [52]. As proposed by Wolffe and colleagues, the C-terminal tail domain binds to the linker DNA, neutralizing its charge and facilitating chromatin condensation [11]. In essence, the Wolffe model postulates that the globular domain binds to nucleosomal DNA and places the C-terminal tail of the linker histone at the edge of the nucleosomal DNA so that it may interact with the linker DNA. However, this model precludes the interaction of the N-terminal tail (the second, unstructured, DNA binding motif) with any part of the DNA contained in the nucleosomal particle [50, also reviewed in 51]. In the more recent model of Muyldermans and colleagues [8] the N-terminal basic tail is in close contact with the linker DNA, while the C-terminal tail is projected towards the entering and exiting nucleosomal DNA, and is available to make additional contacts with either of them [8, 51].

The C-terminal domain region possesses other distinct features. For example, the number of lysine residues and the (S/T)PXK phosphorylation motifs of this domain in each subtype are highly conserved between species. However, these features vary between the individual intraspecies subtypes. Therefore, it is likely that the functional heterogeneity observed among the mammalian subtypes of histone H1, resides in differences in the C-terminal domain [53]. Specifically, the H1 subtypes, H1.1 and H1.2 have the shortest C-terminal regions, whereas H1.4 and H1.5 have the longest C-terminal domains. Although these domains are very lysine-rich, each subtype contains a different amount of lysines and arginines that confer different overall net positive charges on the C termini. The 103-amino acid long C-terminal domain of histone H1.1 has the fewest, with 38 lysines and 2 arginines, whereas histones H1.4 (110 amino acids, containing 43 lysines and 1 arginine) and H1.5 (114 amino acids, containing 46 lysines) have the largest number of positively charged amino acids. Not only is histone H1.1 the shortest subtype, it also differs from the remaining subtypes in having only two of the conserved cdk-dependent S/TPXK phosphorylation sites. Interestly, histone H1.0 is the shortest of the linker histones, with a C-terminal domain consisting of 97 amino acids, but contains the highest density of basic amino acids, with 1 arginine and 41 lysines. It also has one more S/TPXK motif than histone H1.1 [54].

3. STOICHIOMETRY OF H1

Whereas the histones that comprise the core particle have a definite and unchanging stoichiometry per nucleosome, irrespective of cell type, the extension of this concept to the H1 class, i.e., that there is 1 H1 molecule per nucleosome in all cell types and species, has been misleading. For example, Bates and Thomas found that lymphocyte chromatin had almost exactly 1 H1 molecule per nucleosome, whereas chicken erythrocytes had on average 1.3 H1 molecules per nucleosome [55]. In contrast, liver nuclei had only 0.7 H1 molecules per nucleosome. Despite the fact that this early study showed that there were differences in the stoichiometry of H1 amongst different cell types, the generalization of one H1 per nucleosome has persisted. However, this topic has more recently been examined in detail in a number of mouse tissues. Fan *et al.* quantitated the amounts of each H1 subtype in normal mouse hepatocytes, spleenocytes, thymocytes, and embryonic stem (ES) cells and found the following cell-specific total levels of H1 per nucleosome: spleenocytes 0.79, thymocytes 0.83, adult hepatocytes 0.79, neonatal hepatocytes, 0.76, ES cells 0.5 [56, 57]. They also examined the relationship between nucleosomal repeat length (NRL) and H1 content/nucleosome. They found a robust linear relationship between NRL and H1 content, i.e., as the amount of linker histone per nucleosome decreases, a reduction in nucleosomal spacing is also found due to a decrease in NRL [also reviewed in 58 Woodcock *et al.*, 2006]

The *in vivo* H1-depletion studies of Fan *et al.* also showed that loss of H1 resulted in polynucleosomes being significantly less compact [56, 57]. Moreover, deletion of one or two H1 subtypes resulted in a compensatory up-regulation of other subtypes, resulting in a return to the normal level of H1/nucleosome, and no apparent phenotype [59, 60, 56]. Knocking out additional subtypes apparently creates a situation in which the up-regulation process fails to fully compensate, at least in some tissues. Under these circumstances, a reduction in NRL occurs [reviewed in 58]. In this sense, the apparent redundancy of the H1 subtypes appears to be consistent with their structural role. However, triple null mutant mice lacking H1.2, H1.3 and H1.4, which have about 50% of the normal ratio of H1 to nucleosomes, die by mid-gestation with a broad range of developmental defects [56]. In contrast, ES cells can be propagated from early embryos of both wild-type and mutant animals. The wild-type ES cells are already comparatively low in H1 with 0.5 molecules per nucleosome, and in the triple null mutant cells, this is halved, giving the surprisingly low value of 0.25 H1 per nucleosome [57]. Yet, the mutant ES cells show no obvious changes in morphology or growth characteristics. These results suggest that in the mouse, certain developmental programs require more H1 than do the housekeeping functions needed for the continuous cell cycling of ES cells, consistent with a role for H1 in epigenetic regulation. These results suggest a new and unexpected aspect to H1 function, unrelated to chromatin structure. Whether this function causes the embryonic lethal developmental failures in H1-depleted mice remains to be seen [5]. One concept related to transcriptional control that continues to hold is that H1 content increases during terminal differentiation of cells, a period of progressive shut down of groups of genes. Thus, at least for vertebrate systems, the maximum level of total linker histone is found in the fully differentiated and transcriptionally inert avian erythrocyte, and the minimum in pluripotent ES cells [58]. In this respect, and in contrast to the earlier *in vitro* studies, the more recent *in vivo* work of Fan *et al.*, who show that certain subtypes are essential in embryonic development, indicates that a reappraisal of the concept of H1 is necessary [56, 57]. Further *in vivo* studies, to be described below, also substantiate that a re-examination of the functional roles of the H1 subtypes is warranted.

4. Dynamic Binding and Localization of the H1 Subtypes

Fusion of H1 to green fluorescent protein (GFP) was used to monitor histone H1 movement directly by fluorescence recovery after photobleaching (FRAP) in living cells. Results from these initial *in vivo* experiments showed that H1 was highly mobile with a residence time of about 1 min [61, 62]. These results confirmed earlier *in vitro* studies indicating that H1 was in a state of free exchange [reviewed in 63]. Relative to the histone proteins that comprise the nucleosomal core particle, H1 histones have lower binding affinities to chromatin and have a shorter residence time. However, *in vivo* results using FRAP to assess differences in the binding affinities of specific H1 subtypes were incomplete or even contradictory. For example, *in vivo* results from Misteli *et al.* indicated that H1.0 and H1.2 had similar binding affinities [61], a finding that came at odds with the more recent work of Th'ng *et al.* [54]. Using the same experimental strategy, Lever *et al.*, showed that the H1.1-GFP fusion protein exchanged rapidly in both condensed and decondensed chromatin, throughout the cell cycle [62].

These findings suggested that modulations in the transcriptional competence of a gene could be achieved by the binding of activator molecules to control sequences during temporary dissociation of H1 molecules [reviewed in 47, 64]. H1 mobility, combined with an H1 content of less than one per nucleosome, may likely increase the potential for chromatin remodeling and access to chromatin by regulatory complexes [58].

Furthermore, more recent work carried out by Hendzel and colleagues gave a more in depth profile of the differences in binding mobilities amongst the individual mammalian H1 subtypes and the domain of the H1 molecule that contributes to the control of, and regulation of, H1 function with respect to binding and residency [54]. Moreover, in the same study, the sub-localizations of the H1 subtypes on chromatin were analyzed. In a previous study, the same group found that the binding of histone H1.1 to chromatin becomes destabilized with a single amino acid substitution at position 152 (in the C-terminal domain). This result was consistent with the contribution of the C-terminal domain in H1-DNA binding *in vitro*. Based on this, Hendzel and colleagues reasoned that the small but significant differences (or variations) in primary sequence of the CTD among the six human somatic cell H1 histones will result in quite distinct binding properties *in vivo* [54]. Their FRAP experiments showed that there are in fact significant differences in the binding properties between the H1 histones. Histones H1.1 and H1.2, which are distinguished by having the shortest C-terminal tails, rapidly undergo binding and dissociation events, requiring only 1-2 min to reach equilibrium. H1.0 has diverged in amino acid sequence, and the highly divergent CTD is the shortest among all of the H1 histones. However, unlike H1.1 and H1.2, H1.0 takes longer to recover, requiring several minutes to equilibrate. It is more closely aligned with the remaining H1 histones that have longer C-terminal domains, which exchange much more slowly. These histones, H1.3, H1.4 and H1.5, all require as much as 15 min to equilibrate after the initial photobleaching event. These results suggest that the histone H1 subtypes bind to chromatin with different affinities *in vivo*. Both H1.1 and H1.2 bound to chromatin with the lowest affinity, and this affinity increased sequentially with increasing CTD lengths of histones H1.3, H1.4 and H1.5. Specifically, H1.1 and H1.2 each had similar, to each other, but distinctly different binding properties from all other histone H1 subtypes. H1.0 and H1.3 showed only modest differences that were not statistically significant from each other, but were significantly different from the remaining histone H1 subtypes. Similarly, H1.4 and H1.5 had recoveries that were similar to each other but differed significantly from H1.0 and H1.3. These findings suggest that the lengths of the CTD of histone H1.1–H1.5 have a major influence on their binding to chromatin. Although histone H1.0 has the shortest CTD, the binding studies suggest that other variations in the rest of the molecule may contribute to its increased binding affinity, such as the higher density of lysine content, making this subtype the most basic of all, and also the density and distribution of the DNA-binding S/TP*X*K motifs [54].

To further investigate the extent that the C-terminal domain defines the *in vivo* binding properties of the histone H1 subtypes, the same group performed domain switches between histone H1.1, the histone H1 subtype with the lowest binding affinity, and histones H1.4 and H1.5, the subtypes with the highest binding affinity. The replacement of the C-terminal domain of either H1.4 or H1.5 with that of H1.1 significantly reduced the duration of time required for the protein to equilibrate following photobleaching, down to that typical for histone H1.1. Conversely, when the C-terminal domain of histone H1.1 was replaced with either the C-terminal domain of histone H1.4 or histone H1.5, the duration of time required for the protein to equilibrate following photobleaching was significantly increased. These results showed that the C-terminal domain defines the binding properties of the H1 subtypes. Moreover, to further evaluate the role of the length of the C-terminal domain in binding affinity, Th'ng *et al.* truncated histone H1.5. The C-terminal truncation of H1.5 resulted in a significant reduction of the time required for H1.5 to equilibrate in the photobleached region [54].

Using GFP-tagged H1 histone subtypes (N-terminally tagged) and Hoechst staining (stains DNA) in living cell nuclei, Hendzel and colleagues also investigated the localization and distribution of H1 histones in eu- and heterochromatin regions (open, transcriptionally active or poised for transcription, and condensed, transcriptionally inactive, respectively)[54]. Histones H1.1-H1.3 were more commonly found in euchromatin regions, whereas histones H1.4 and H1.5 are preferentially located in heterochromatin regions of the genome. Despite the fact that H1.0 is known to accumulate during differentiation where heterochromatin is increased (see section 9 of this chapter), the green fluorescent H1.0 tended to localize in the euchromatin regions of the nucleus. Interestingly, only a single heterochromatin domain was found to be enriched in histone H1.0, whereas pericentromeric heterochromatin tended to be depleted of histone H1.0 [54].

The binding affinities of each subtype correlate well with the C-terminal domain, with the exception of H1.0. The fact that H1.0 has a strong binding affinity, similar to the subtypes with the longer CTDs, though it has the shortest CTD, can be attributed to the fact that it has the most basic amino acid composition in the CTD region (more basic than even H1.4 and H1.5). This gives H1.0 a 20% higher positive charge with respect to the other linker subtypes [65], which may consequently confer upon this subtype the observed higher binding affinity.

Histone H1.0 has been the most actively studied single histone H1 subtype from mammalian species. Its sequence divergence and accumulation in terminally differentiated and aging tissues (see section 9) has led to the speculation that it is important in the terminal differentiation process. A previous study showed that H1.0 depletion had no phenotype [60]. However, it was more recently demonstrated that histone H1.0 is genetically required in the terminal differentiation of dendritic cells and H1.0 depletion causes a specific defect in dendritic cell differentiation [66]. H1.0 knock-out mice are specifically deficient in this cell type of the immune system. The fact histone H1.0 binds to chromatin with an intermediate affinity and has a preference for euchromatic regions of the genome, while it is found to be depleted in heterochromatin, suggests a gene regulatory role. Enrichment in euchromatic regions of the nucleus imply that it is preferentially involved in the regulation of expressed and expression-competent regions of the genome. Its association with non-dividing cells suggests that it may function to repress expression of genes that are no longer required, such as those associated with cell cycle progression.

In summary, the C-terminal domain distinguishes the linker histones from each other and plays a major role in the control of their binding association with chromatin. More importantly, their differential bindings, mobilities and localizations most probably are major determinants of their functionally distinct roles in different biological processes.

5. Post Translational Modifications

The H1 histones are, just as the core histones, targets of post translational modifications. The most well-studied H1 modification is phosphorylation. However, modern MS has revealed quite a number of other modifications of novel modified residues [reviewed in 15]. Modifications of the linker histones can occur on all three domains. These modifications include the extensively-studied H1 phosphorylation, poly(ADP-ribosyl)ation, acetylation, methylation and the more recently discovered H1 modifications, ubiquitination and formylation [44, 15].

It is now well accepted that acetylation and methylation of the core histones are key regulators of transcription. However, mass spectrometric analyses have shown that acetylation and methylation are also key modifications of the linker histones [reviewed in 67]. Acetylation of the Nα-terminus involves the co-translational cleavage of a methionine, followed by acetylation of the second residue of the N-terminal tail. Initial, non-MS work showed that only H1.0 in human cells and H5 in avian erythrocytes had the Nα-terminus both acetylated and unacetylated. However, it has now been shown that the former modification not only occurs on all linker histone subtypes in human and chicken, but that it is also the most abundant type of acetylation. Linker histones do exist with their Nα-terminus unacetylated but are significantly fewer in number. Although the function of Nα-terminus acetylation is unclear, it was noted in earlier work that whereas the ratio of acetylated to non-acetylated Nα-termini remained constant in avian H5 erythrocytes from newly hatched and adult chickens, it increased for H1.0 in aging rat tissues [68, 69]. Given that H1.0 is most abundant and accumulates in terminally differentiated cells, there may be a link between N-terminus acetylation and differentiation [44, 67]. Moreover, up to nine acetylation sites were mapped to a single H1 subtype, H1.4. Most of the observed lysine acetylation sites of H1.4 were found in the globular domain (GH1) on residues that have been considered to be directly involved in DNA binding [44, 67]. These residues are Lys52, Lys64, Lys85 and Lys97. As acetylation can interfere with DNA binding, it is possible that this modification can significantly change the binding properties of H1, conferring specific properties to some H1 subtypes. A second group of acetylation sites were identified for a single subtype in its N- and C-terminal tails by MS [44].

Although there are reports of methylation of lysines in linker histone proteins in HeLa cells [70, 44] there is a difference in the number of identified sites. Garcia *et al.* found that in H1.4, Lys26 and Ser27 were simultaneously methylated and phosphorylated, respectively [70]. This is similar to the well-known adjacent methylation and phosphorylation of sites at Lys9-Ser10 and Lys27-Ser28 of core histone H3 that are involved in transcription [67]. However, Wisniewski *et al.* found no methylations in this region of H1.4 or of the other subtypes in cultured cells, but they were able to identify lysine methylation in mouse tissues, including mono-, di- and tri-methylated residues (e.g., mono- and dimethylated Lys63) [44]. The inability to identify methylation sites in cultured cells may reflect a lower extent of this modification in cultured cells rather than complete absence of H1 methylation. The observed differences in methylation do not reflect differences between mouse and human because the same group found a number of methylated lysine residues of subtypes H1.2-H1.4 isolated from human tissues.

Ubiquitination of core histones has been well-documented. Observations suggest that ubiquitinated core histones may have multiple functions and structural effects that may generally be associated with increased gene expression. This implies that it may prevent chromatin folding or help maintain an open conformation [reviewed in 43]. For the first time, ubiquitinations of linker histone proteins were identified by MS in human cell cultures and mouse tissues by Wisniewski *et al.* [44]. Another putative novel modification found is that of formylation. H1.1-H1.5 were all found to be formylated. Formylation of lysines has been shown to arise as a result of oxidative damage to DNA [71, 44, also reviewed in 67, 15]. Wisniewski *et al.* mapped thirteen formylated lysine residues, distributed on all three structural domains of H1 [72]. Among the modified sites were lysine residues considered important for DNA binding and those where acetylation or methylation had been observed.

Another modification found to occur in the linker histones H1.2, H1.5 and H1.0 was poly(ADP-ribosyl)ation when mouse fibroblasts were undergoing poly(ADP-ribosyl)ation-mediated DNA repair, implying that this linker histone modification associates with less condensed regions of chromatin [reviewed in 73].

Basically, the recent work of Wisniewski *et al.*, using proteomic MS analyses clearly demonstrates that linker histones are extensively modified, not only by phosphorylation, but also by acetylation, methylation, and ubiquitination [44, 71]. Moreover, the same group also showed that some lysine residues are formylated. Wisniewski *et al.* identified far more post translational modifications than all the previously published work over the last 3 decades [44]. This, of course, is due to the use of modern MS technology which has been shown to be a powerful tool in contemporary proteomic-oriented research. As concluded by these investigators, the extent and diversity of H1 modifications seems to be similar to those described for core histones, suggesting that functions of linker histones are regulated with a complexity similar to that of the core histones. Moreover, against the established view that linker histones are mainly modified within the C-terminal domain, Wisniewski *et al.* showed than many of their newly-found modifications and/or novel residues are located within the globular domain and also in N-terminal part of the proteins [44].

Left for last in this section, is the most intensively studied H1 modification, phosphorylation. Reversible H1 phosphorylation was discovered more than three decades ago [e.g., 74] and since its discovery, a tremendous amount of research work has been devoted to this topic. Needless to say, it is impossible to review all this body of work in this chapter; however an attempt will be made to present some of the highlights regarding H1 phosphorylation.

In human cells, H1 is phoshorylated at up to five sites with the consensus motif (phosphorylation motif, (S/T)PXK) of cyclin-dependent-kinase (CDK) [75]. Up to four of these sites are located in the C-terminal tails of H1 and one in the N-terminal portion of the histone. In addition to these sites, phosphorylation of non-CDK1-sites within the N-terminal tails was reported. These sites are Ser1, Thr3 of H1.4 [70, 76], Thr10 of H1.5 [75], Ser26 of H1.4 [70] and Ser30 of H1.X [70]. Moreover, phosphorylation of Ser35, which is located N-terminally to the globular domain of the H1.4, was also reported [70]. In the study of Wisniewski *et al.* nearly all these sites were identified in tissues and human cells [44] and they also mapped several previously only predicted CDK-1 sites [75]. In human cells, these are H1.2 (pThr146), H1.3 (pThr18; pThr147), and H1.4 (pThr18). They also found novel non-CDK1 consensus sites in all analyzed histones, except for H1t. Apart from the pThr165 site in H1.2, other non-CDK1 sites are located in the N-terminal tails and the globular domain (GH1) of the subtypes. Histone H1.5 is also phosphorylated at the acetylated Ser1 residue of the N-terminal tail. In histones H1.0 and H1.1, only non-CDK1 phosphorylation sites were identified. In the mouse histone subtypes, H1.1-H1.4, a non-cdk1 site, pSer41, that is located within the helix 1 of the globular domain was found. This modification, together with the other modification sites (see above), within the globular domain may potentially regulate interaction of linker histones with DNA.

The functional significance of post-translational modifications of core histones, the "histone code" has clearly been demonstrated. Deciphering how distinct modifications of core histones sequentially, or in combination, regulate chromatin structure and function (chromatin remodeling) is already advanced. The potential involvement of histone H1 in the code has largely been neglected until recently. The data presented in this chapter clearly demonstrate that H1 histones also significantly contribute to this code. Further studies are necessary in order to understand the linker histone code, its relationship and association with the core histone code and their relationship with cellular mechanisms.

6. Functional Aspects of H1 Phosphorylation and Dephosphorylation

Phosphorylation of the H1 histones was observed to be cell-cycle dependent. It progressively increases during the cell cycle, being lowest in G1 and becoming maximal during late G2 and mitosis. In telophase, phosphorylation decreases sharply, thus resetting these histones to a phosphate-free state at the beginning of the next cell cycle. These results were originally from the early work of Gurley and co-workers [77, 78, 79] where they observed that H1 from CHO cells is singly phosphorylated in late G1, triply phophorylated in S and G2, and carried up to six phosphates per molecule in the M phase. These observations led to the original correlation that high H1 phoshorylation levels were related to the profound structural changes that take place during chromosome condensation and that low levels of H1 phosphorylation is associated with decondensed chromatin. However, other findings reveal more complex mechanisms related to H1 phosphorylation and cellular function.

During the S phase, H1 histones exist both as unphosphorylated and phosphorylated forms and there is evidence that this phosphorylation is important for the process of DNA replication [80, 81, 82]. Alexandrow and Hamlin demonstrated, more specifically, that replicating DNA and phosphorylated H1 co-localize *in vivo*, a finding strongly suggesting that this H1 phosphorylation promotes DNA decondensation as a precondition for DNA replication [83]. Moreover, the same study showed that active DNA replication and increasing levels of H1 phosphorylation are tightly linked processes and are Cdk2-dependent (S phase cyclin-dependent kinase) [83]. Previous studies have suggested that phosphorylated H1 is involved in decondensing chromatin which may allow transient access of proteins to DNA [40]. Thus the simplest interpretation of the data of Alexandrow and Hamlin is that recruitment of Cdk2 to replication foci leads to H1 phosphorylation, chromatin decondensation and facilitation of DNA replication [83]. It has been shown that S phase H1 phosphorylation is kinase-dependent through studies where kinase (Cdk2) activity is inhibited resulting in dephosphorylation [40]. Thus the findings of Gurley and colleagues, where H1 phosphorylation increases during the cell cycle, throughout the S phase and culminates during mitosis are reconciled by the prevailing view that increased H1 phosphorylation is necessary for the subsequent decondensation of chromatin which in turn allows access to the factors responsible for the progression of DNA replication [77, 78, 79].

As mentioned above, H1 phosphorylation culminates during mitosis. What, then, is the role of this phosphorylation in the completely condensed chromosome? One proposal is that H1 hyperphosphorylation itself might serve as a mitotic "trigger" and not the actual physical means for condensation [40]. While this may or may not hold true, more recent evidence supports the hypothesis and model originally proposed by Roth and Allis that, rather than playing a direct physical role in the chromosome condensation process, H1 phosphorylation actually leads to a transient decondensation which allows access to specific chromosome condensation factors (for example, topoisomerase II) [40]. This hypothesis is supported by more recent findings from studies which implicate H1 phosphorylation and dephosphorylation in chromatin remodeling events involved in transcriptional regulation and the dynamic exchange of the H1 subtypes between chromatin fibers. These results will be briefly outlined below.

There is mounting evidence that H1 phosphorylation has a functional role in the regulation of gene expression. H1 phosphorylation weakens the binding affinity of H1 to the nucleosome, resulting in either the complete removal of these proteins or, at least its phosphorylated C- and/or N-terminal tails. This was originally thought to induce the profound structural changes that accompany chromosome condensation [84, 85]. However, it is now known that loss of H1 results in significantly less compact polynucleosomes [57]. Moreover, it has been shown in some biological systems where H1 phosphorylation is uncoupled from mitosis and chromosome condensation, that condensed chromatin is enriched in dephosphorylated forms of H1 [40]. For example, in *Tetrahymena*, changes in H1 phosphorylation appear to occur whenever there are marked changes in gene expression due to altered physiological conditions, supporting a role for H1 phosphorylation in transcriptional regulation [86, 87]. Indeed in *Tetrahymena*, data have confirmed the role of histone H1 phosphorylation in the transcriptional regulation of certain genes [88].

Furthermore, using antibodies selective for phosphorylated H1 histones, enrichment of phosphorylated H1 histones in transcriptionally active chromatin has been observed [89, 90].

More recent *in vitro* studies have also implicated phosphorylation of H1 in the regulation of chromatin remodeling [91, 92]. Phosphorylation seems to stimulate the dynamic exchange of linker histones between chromatin fibers by increasing the rate of dissociation of H1 from chromatin [93, 94]. This may provide a mechanism by which phosphorylation can affect H1 function *in vivo*. In other words, the more recent results have shown that dephosphorylation of H1 is associated with the maintenance of H1 subtypes on chromatin and the formation or stabilization of condensed structures, while H1 phosphorylation is associated with transient removal or dynamic exchange of H1 subtypes and the destabilization of condensed structures. This would allow access to transcriptional DNA binding factors and ATP-dependent chromatin remodeling enzymes. In fact, *in vitro* analysis of events occurring during the activation of the hormone-inducible mouse mammary tumor virus (MMTV) promoter assembled in minichromosomes revealed that incorporation of H1 histones into the MMTV chromatin improves the stability of the nucleosomes and decreases basal transcription. After activation of the transcription, initiated by binding of the progesterone receptor (PR) to the hormone responsive element (HRE), H1 positively regulates transcription due to the improvement in nucleosome positioning. Results from these investigations led to a model describing the steps in PR-mediated activation of the assembled MMTV promoter. It is assumed that H1 is phosphorylated by a kinase that is recruited to a PR which is bound onto the promoter. This phosphorylation of H1 now permits the binding of the ATP-dependent remodeling complex NURF to open the nucleosome structure. During the final steps of the transcriptional activation by formation of the transcription initiation complex, H1 leaves the promoter [95, 92]. This model is supported by the finding that unphosphorylated, in contrast to phosphorylated, H1 inhibits a number of chromatin remodeling complexes *in vitro* [91, also reviewed in 15].

The experimental evidence presented above support that phosphorylation and dephosphorylation of H1 decondenses and condenses chromatin, respectively, either locally or on a more global scale. The proposed mechanism by which this occurs is directly associated with the biophysical properties of DNA. The unphosphorylated H1 tails shield the negative charge of the DNA backbone to allow for chromatin fiber interactions within condensed structures. Upon phosphorylation, the negative phosphate groups weaken the interaction of the tails with DNA to allow for local decondensation [40, 88, 94].

The more general implications of these data point towards the fact that specific H1 subtypes and/or their phophorylation states may have specific functional roles in chromatin remodeling events which are responsible for transcriptional regulation that takes place during numerous cellular/biological processes such as development and differentiation. More interestingly, reports have shown that the linker histones may have, or do have, roles in DNA repair [96], apoptosis [97, 98, 99, 100, 101, 102 103, 104], lifespan [105], aging and senescence [106, 107, 108, 109, 110, 111, 112].

7. Linker Histones and Gene Expression

Linker histones have long been considered as being global repressors of transcription. This was initially surmised from *in vitro* experiments whereby binding of H1 to naked DNA inhibited transcription. Biochemical studies also demonstrated that active genes were partially H1-depleted in a diverse number of organisms [113]. Moreover, the general view of their role in chromatin condensation, from *in vitro* structural studies, gave further support to this point of view, i.e., since H1 compacts chromatin and condensed chromatin has generally been considered inactive chromatin, H1 is a general repressor of gene expression. However, more recent studies showed that H1 may not necessarily be a global repressor, but may in fact specifically repress certain genes and/or conversely, be an activator of other genes. For example, Shen and Gorovsky found that in *Tetrahymena*, H1 does not affect transcription globally, but does repress the basal transcription of at least one gene, *ngoA* (function unknown), while it is required for fully activated expression of the *Cyp* gene (encodes a cysteine protease) [113]. The mode in which this can be accomplished may be that by releasing H1 from linker-associated sequences, local chromatin structures around specific genes are modified by enhancing nucleosome mobility (sliding) or altering nucleosome position, thereby exposing or shielding regulatory regions. These changes may be explained if we consider that the chromatin structure of each gene is unique, established by a combination of factors, including sequence-specific placement of nonhistone binding factors, or particular structural features of the underlying DNA itself. In this view, any global change in histones is translated in different changes in the chromatin architecture of each gene, which results in a gene-specific alteration (either positively or negatively) in the binding of transcription factors (positive or negative). An alternative view is that much of the specificity lies with the transcription factors themselves. In this view, the constant features of chromatin structure such as the nucleosome core, the N-terminal tails of core histones, or the linker-associated histones can (but need not) be used differently by gene-specific transcriptional activators or repressors either to facilitate or inhibit their binding, explaining the apparent gene specificity of mutating or deleting these histones. In this sense, the H1 histones, rather than being nonspecific repressors of transcription, as originally thought, may in fact be co-participants with transcription factors in the specificity of transcription *in vivo* [113].

Results from other *in vivo* studies also further enhanced the point of view that the H1 class may in fact be repressors or activators of specific genes. It has long been proposed that selective binding of histone H1 is, in part, responsible for the differential developmental regulation of the oocyte and somatic 5S rRNA genes in *Xenopus laevis*. Howe *et al.* using reconstituted nucleosomal arrays, showed, for the first time, that H1 is actually bound to the somatic genes without repressing transcription, i.e., that H1 can bind nucleosomes without affecting the assembly of the transcription machinery [114]. This was also shown in other studies, where the RNA polymerase II trans-acting factor, NF-1, can bind linker DNA, regardless of the presence of histone H1 and where the RNA polymerase I basal transcription factor, UBF, is able to displace H1 when binding nucleosomal DNA [115, 116]. Moreover, whereas H1 nucleosomal binding did not affect transcription of the somatic 5S rRNA gene, it did repress, to a large extent, oocyte 5S rRNA gene transcription [114]. These results reinforced the idea that H1 can actually be a selective repressor of transcription initiation [114].

During the same time period, Sera and Wolffe further explored the differential transciptional effects of H1 binding to the oocyte versus somatic 5S rRNA genes [117]. They found that H1 selectively represses oocyte 5S rRNA genes by directing the stable positioning of a nucleosome, such that transcription factors can not bind to the gene. This effect did not occur on the somatic-type genes. Their results further showed that the reason for this differential mode of H1 regulation was that H1 binds to the 5' end of the nucleosome core on the somatic 5S rRNA gene, leaving key regulatory elements in the promoter accessible, while H1 binds to the 3' end of the nucleosome core on the oocyte 5S rRNA gene, specifically blocking access to a key promoter element. Since they found that H1 binds with equivalent affinities to nucleosome containing either gene, the investigators concluded that it is the sequence-selective assembly of a specific repressive chromatin structure on the oocyte 5S rRNA genes that accounts for differential transcriptional repression. A final point to be made from this study is that H1 binds selectively to nucleosomal DNA, compared to naked DNA and that H1 contributes to the organization of a specific repressive nucleosomal architecture on the oocyte 5S rRNA genes. Thus, general components of chromatin can determine the assembly of specific regulatory nucleoprotein complexes [117].

Moreover, Hellauer *et al.* using whole-genome microarray analysis of the *S. cerevisiae* genome, identified genes that were affected by H1 removal [118]. Surprisingly, deletion of the gene encoding yeast histone H1 does not result in increased gene expression. Rather it results in a modest reduction. Specifically, they demonstrated that, in the absence of linker histone, only 27 of over 6000 transcripts were altered by a factor of two or more. Additionally, these transcripts were reduced in the absence of the linker histone, suggesting that the yeast linker histone acts as an activator of transcription rather than as a repressor [also reviewed in 119].

The work cited above, regarding the potential role of H1 as a positive or negative regulator of the expression of specific genes, in combination with, or in contrast to, its role as a global repressor, dealt with the entire linker histone class, without referring specifically to a certain H1 subtype. The next question which arises, in light of the fact that there are, to date, 11 identified mammalian subtypes (see Introduction of this chapter) is whether specific subtypes may have distinct regulatory functions in gene expression. To the author's knowledge, Brown *et al.* were the first to find evidence that a specific subtype may be involved in, or participate in transcriptional regulation of a certain gene [120]. They found that overexpression of H1.0 resulted in transient inhibition of cell cycle progression in the G1 and S phase, a result consistent with the proposed role of this subtype in the stabilization of inert chromatin in differentiated cells (see section 9 of this chapter), whereas overexpression of H1.2 (H1c) had no effect on the cell cycle of mouse 3T3 fibroblasts. In order to ascertain whether this effect involved changes in gene expression, they analyzed transcript levels of cell cycle-related genes and certain housekeeping genes. Their analyses showed that overexpression of H1.0 resulted in the significantly reduced expression of a number of genes, including *c-fos, c-myc, cyclin D2* and *cdc2*. However *cyclophilin* and *GADPH* transcript levels remained relatively unaffected. The fact that the observed overproduction of H1.0 affects expression of *c-fos* and *c-myc* strongly implies that these effects are manifested at the level of chromatin structure, as these are immediate early response genes and it has been demonstrated that transitions in chromatin structure are associated with the expression of these genes. Surprisingly, overexpression of H1.2 had little effect on the expression of some genes, but led to a significant increase in the transcript levels of *GAPDH, c-myc* and *cyclophilin.* A possible explanation is that H1.2 association with chromatin promotes a more open conformation, whereas H1.0 promotes an intermediate or more closed conformation, allowing a greater or lesser access to transcription factors, respectively [120]. That H1.0 overexpression, and to a much lesser extent H1.2, results in a decreased rate of digestion of chromatin by micrococcal nuclease explains in part the differential effects of H1.0 and H1.2 in regulating chromatin structure and function [121]. However this explanation does not completely account for the differential effects that the overproduction of each subtype was found to have on the expression levels of certain genes. Obviously, more in depth work is necessary in order to determine the mode in which specific H1 subtypes act as repressors or activators of specific genes.

In another study which was designed to determine whether the H1 histone subtypes play differential roles in the regulation of gene expression, Alami *et al.* combined mice lacking specific H1 histone subtypes with mice carrying transgenes subject to position effects [68]. Because position effects result from the unique chromatin structure created by the juxtaposition of regulatory elements in the transgene and at the site of integration, transgenes can serve as exquisitely sensitive indicators of chromatin structure. The investigators found that some, but not all, linker histone subtypes can attenuate or accentuate position effects. The effects were observed for three different transgenes integrated at three different sites in the mouse genome. Both positive and negative effects of specific H1 subtypes were observed. For example, they found that absence of H1.3 (H1d) or H1.4 (H1e), but not H1.1 (H1a), H1.2 (H1c) or H1.0 attenuates the rate of age-dependent silencing of a multicopy globin transgene. This shows that the H1 subtype composition of chromatin does indeed have important qualitative and quantitative effects on gene expression. From their results, the investigators point out and suggest that the linker histone subtypes play differential roles in the control of gene expression and that the sequential arrangement of the linker histones on the chromatin fiber might regulate higher-order chromatin structure and fine-tune expression levels [68].

Lastly, in a more recent study, Sancho *et al.* developed an inducible shRNA expressing system for the depletion (knock down) of individual H1 subtypes in the human breast cancer cell line, T47D [122]. They found significant differences among the H1 subtypes with regard to cell cycle progression, cell proliferation and gene expression. Knock down of H1.4 leads to cell death in T47D. This is the first time that it has been noted that one subtype is functionally essential for survival in a human cell type. H1.2 was also found to be functionally important for cell cycle progression. Depletion of H1.2 caused G1 arrest, defects in chromatin structure and changes in the expression of specific genes linked to the cell cycle. Specifically, in H1.2-depleted cells, 11 genes were up-regulated and 59 genes were down-regulated. This indicates that H1.2, rather than being a general repressor, is acting directly or indirectly as a positive regulator of gene expression. Depletion of the different H1 subtypes caused an alteration in the expression of a limited number of genes, different in each H1 subtype knock down. Most of the genes were affected by a single H1 subtype knock down, supporting the theory that H1 isoforms play specific roles in gene expression. Nonetheless, a proportion of genes were altered by more that one H1 subtype, suggesting that redundant roles for H1 subtypes may also exist [122]. Not surprising was the finding that depletion of the differentiation-associated H1 subtype, H1.0, caused activation of approximately 20 genes and repression of only 2. The activated genes were cell cycle-related (see section 9 for more details). The results from the altered expression levels of cell cycle-related genes after knock down of the different H1 subtypes, indicated that each subtype affects the expression of these genes diversely. The basic conclusions of these investigators were that the H1 subtypes are not fully redundant but may have differentiated functions in some of their nuclear roles, including nucleosome spacing and gene expression control, which have an impact on important cellular events, such as proliferation [122].

8. Linker Histones, DNA Repair and Apoptosis

DNA damage is caused by numerous internal (free radicals) and external agents (radiation, chemicals). These genotoxic agents result in various types of DNA lesions (damage) that if not properly repaired, may lead to genome instabilities proving fatal to the organism. Depending on the type of lesion, there are a number of repair pathways which the cell may employ. It is outside the scope of this chapter to review the DNA damage response (DDR). However, it must be mentioned that the DDR involves (1) detection of the damaged DNA (sensing or sensor proteins), which is the first event in the DDR, (2) repair of the lesion(s), which must first be proceeded by chromatin remodeling events and (4) restoration of chromatin to its original state. Double strand breaks (DSB) are the most dangerous in terms of affecting genome stability, as they can result in chromosomal rearrangements which may lead to cancer [123]. If the damage is too great to be repaired, there are numerous signaling pathways that will lead the cell to apoptosis. Some of the proteins involved in all these steps are known, while it is evident from all the work in this field that there are many more, as yet to be discovered. An intriguing and surprising aspect to the proteins now known to be involved in the DDR is the discovery that histones also have roles in this process. Though the core histones and their modifications (e.g, phosphorylation of H2A.X (γH2A.X) [124]) are certainly players in DNA damage response processes, the purpose of this section is to briefly present pertinent highlights with respect to the involvement of the linker histones in the DDR.

In response to DNA damage, chromatin becomes decondensed, presumably to facilitate genome surveillance and access of repair factors. As H1 is involved in chromatin compaction [57], Murga and colleagues examined the effect of reduced levels of H1 on the DNA damage response [125]. They found that triple-knock out mouse embryonic stem (ES) cell lines (with three of the six H1 genes inactivated resulting in a decondensed chromatin phenotype [57]) had a resistance to DNA damaging agents, an enhanced checkpoint response and an increase in the amount of activated DDR proteins in comparison to wildtype cells [125]. It is possible that one of the DNA damage response kinases could phosphorylate H1, resulting in its release from chromatin, thus allowing decondensation of the chromatin and access of the DDR proteins [123]. The phosphorylation status of H1 could act as an indicator to the cell by relating the degree of damage sustained by the genome. At lower levels of damage, H1 becomes phosphorylated, and is released from the nucleosome, allowing chromatin decondensation and access of repair proteins. If the damage is extensive and beyond repair, then more H1 would get phosphorylated and released. This could exceed a critical level at which point it may signal to the mitochondria causing release of cytochrome c and instructing the cell to undergo apoptosis. Alternatively, specific linker histone subtypes could preferentially associate with certain types of chromatin and release of that subtype upon damage could result in a particular cellular response [123].

As described in the review of Chubb and Rea, there also appears to be an interplay between modifications of core and linker histones, as the ubiquitylation and acetylation state of nucleosomes are able to influence H1 dynamics [123]. The carboxy-terminal tails of H2A and H2A.X are able to interact with linker histones [1]. These tails are the sites of modifications involved in the early response to DNA DSBs, suggesting that the modification state of these tails could influence the binding status of H1, either directly through physical interactions or indirectly through the recruitment of other histone modifying or remodeling factors [126, 127]. In support of this, Zhu and colleagues found that deubiquitylation of H2A, at K119, by the enzyme 2A-DUB (H2A-Deubiquitinase) promoted the phosphorylation and subsequent release of H1 from the nucleosome [128]. Indeed, it has been shown that after DNA damage, there is cytoplasmic accumulation of histone H1 that is due to active export from the nucleus, rather than inhibition of its import. Results have also indicated that this export may be p53-dependent. p53 could function in the displacement of the linker histones or, more likely, in the translocation of previously displaced linker histones from chromatin into the cytosol [reviewed in 129].

In fact, Konishi *et al.* demonstrated a specific role for the linker histone, H1.2 in triggering apoptosis in response to DNA damage [97]. These investigators found that in X-ray irradiated rat thymocytes (irradiation induces mainly DSBs, which if extensive, lead to apoptosis) H1.2 translocates to the cytoplasm and induces the release of cytochrome c from mitochrondria, thus triggering the apoptotic signaling cascade. The mode in which H1.2 activates the release of cytochrome c is by inducing conformational activation and oligomerization of Bak (member of the proapoptotic Bcl-2 family associated with the mitochrondrial membrane). The investigators further speculated that H1.2 might activate Bak by disrupting the outer mitochondrial membrane directly [97]. Since DNA damage does not lead to any change in the overall level of H1.2 expression, how then is its apoptotic function activated? H1.2 does not undergo any obvious post translational modification as a result of DNA damage. Presumably, DNA damage dislodges some H1.2 from chromatin. This may occur through the physical remodeling or damage-induced modification of other components. This dislodged population is then available for export from the nucleoplasm by an active process which may itself be stimulated by damage [also reviewed in 130]. p53 is essential for DNA damage-induced apoptosis (and also to X-ray irradiation-induced apoptosis in which the damage is primarily DSBs) in many types of cells, and thymocytes from $p53^{-/-}$ mice are resistant to X-ray irradiation-induced apoptosis. Thus Konishi *et al.* examined the relationship between p53 and histone H1.2 [97]. When $p53^{+/+}$, $^{+/-}$, and $^{-/-}$ mice were irradiated, cytochrome c release and caspase activation were seen in the thymus of $p53^{+/+}$ and $^{+/-}$ mice, but not in that of $p53^{-/-}$ mice. Importantly, histone H1.2 and other subtypes of histone H1 were increased in the cytosolic fraction from the thymus of $p53^{+/+}$ and $^{+/-}$ mice, but not in that of $p53^{-/-}$ mice. These results indicated the p53 is necessary for a cytoplasmic increase of various forms of histone H1, including H1.2. It is important to note that while all linker histones are eventually released from chromatin and translocated to the cytoplasm in a p53-dependent manner after irradiation, only H1.2, but not the other H1 forms, induced cytochrome c release in a Bak-dependent manner [97]. Other data from the same work also support that translocation of H1.2 and other H1 histones are down-stream events from p53 activation. However, it is not clear how p53 effects histone translocation. Perhaps p53-mediated DNA repair is involved in displacing H1.2 from chromatin. Alternatively, given that p53 itself undergoes active nucleo-cytoplasmic shuttling, p53 could conceivably play a more direct role in regulating H1.2

translocation after damage. Regardless of the precise mechanism, the clear implication of this work is that H1.2 is inherently proapoptotic if it reaches the cytoplasm [130]. Another take-home message from the work of Konishi *et al.* is that H1.2 plays an important role in the transduction of genotoxic stress signals from the nucleus to the mitochondria [97]. Another enticing speculation is that H1.2 translocates to the cytoplasm in a manner proportional to the amount of damage sustained. If this is the case, then it could be used as a readout of the condition of the genome. Thus low levels of damage would result in low levels of cytosolic H1.2 and no apoptotic response, but high levels of damage would result in high levels of cytocolic H1.2 and, aside, or in conjunction, with its role in the release of cytochrome c, this could also be a signal to the cell that the genome is sufficiently damaged that apoptosis is necessary [also reviewed in 119]. Needless to say, further work in this area promises to reveal new, as yet undiscovered, roles that the linker histone subtypes and their post translational modifications may have in the DDR processes and apoptosis.

With respect to apoptosis, there are now numerous works that relate the H1 histone class, and perhaps specific H1 subtypes with the final stages of apoptosis. In an investigation by Sourlingas *et al.*, it was found that histone deacetylase inhibitors (i.e., sodium butyrate and trichostatin A) increased the levels of histone acetylation, induced the expression levels of the differentiation-associated H1 subtype, H1.0, concomitant to the induction of apoptosis in human peripheral blood lymphocytes [103]. These observations correlated the induction of apoptosis with increased levels of histone acetylation and the induction of H1.0. The induction of H1.0 is known to be the result of increased acetylation in the promoter region of this gene [131]. Both histone acetylation and linker histones play important roles in chromatin remodeling. Chromatin remodeling is a necessary part of the final stages of apoptosis so as to render chromatin accessible to DNA endonucleases that fragment DNA (see below). H1.0 induction may, or may not be, directly involved in apoptosis. However, since linker histones play significant roles in chromatin conformational changes, it is not unreasonable to speculate that induction or increased expression of this specific subtype may perhaps actually be one of the necessary factors involved in the chromatin reorganization events that take part during the last stages of apoptosis. More work with this H1 subtype and other apoptotic agents and cell types is needed so as to clarify this question.

A major hallmark of the terminal stages of apoptosis is internucleosomal DNA fragmentation. In contrast to non-specific DNA fragmentation, or no DNA fragmentation which are also known to occur during apoptosis [132, 133, 134, 101, 102], internucleosomal fragmentation processes the genome of dying cells into "bite-size" oligonucleosomal pieces for efficient clearance by phagocytosis [102]. The major advantage of this specific mode of DNA fragmentation is that it eliminates the transforming potential of any hazardous oncogenes during DNA degradation of tumor or virally transformed cells [135, 136]. The endonuclease responsible for this type of DNA degradation is the DNA fragmentation factor 40 (DFF40). It is now well-documented that DFF40 is the only nuclease that cuts double-stranded DNA [137] in the DNA internucleosomal linker region [100], thus producing oligonucleosomes. This type of degradation can be detected by agarose gel electrophoresis as a characteristic "DNA ladder" as opposed to the "smear-type" appearance after non-specific DNA fragmentation, when other apoptotic endonucleases are activated which cut single-stranded and double-stranded DNA in both intra- and inter- nucleosomal regions [100, 102, 138]. DFF40 exists as a heterodimer with DFF45, chaperone and inhibitor of the endonuclease. DFF45 has no nuclease activity but helps to correctly fold DFF40 and inhibits its enzymatic activity under non-apoptotic conditions.

Numerous cell-free *in vitro* studies, using recombinant DFF (DFF40-DFF45) have shown that apoptotic activation of caspase 3 results in DFF45 proteolysis and release of the active DFF40 nuclease [e.g., 99, 100, 139]. The same results regarding the proteolysis of DFF45 by caspase 3 have also been obtained in human cell lines [e.g., 140]. DFF45 is therefore a nuclear substrate of caspase 3 that is hydrolyzed as the result of nuclear translocation of active caspases [98]. However, the activity of DFF40 must be further enhanced by specific chromosomal proteins such as histone H1, HMGB1/2 and topoisomerase II [99, 100]. In fact, using naked DNA substrates, it has been shown that histone H1 stimulates the activity of DFF40 more than 20-fold [98, 99, 100]. Moreover, in cell free systems using recombinant DFF, it has been shown that histone H1 and HMGB1/2 recruit or target DFF to the internucleosomal linker regions of chromatin and that histone H1 directly interacts with DFF40 conferring DNA binding ability to DFF and simultaneously stimulating the nuclease activity of DFF40. Thus, the body of *in vitro* work showed that histone H1 was a necessary component in the activation of DFF40 under apoptotic conditions.

In light of the association of histone H1 (total fraction) with DFF40 activation during the *in vitro* activation of apoptosis, Ninios *et al.* chose to focus their attention first on the association of DFF40 with total histone H1 and then specifically with individual H1 linker histone subtypes in an *in vivo* cellular milieu using trichostatin A (TSA) as the apoptotic agent in a promyelocytic leukemic cell line (NB4) [104]. Using co-immunoprecipition (co-IP) of the chromatin-enriched nuclear fraction, they provide evidence that DFF40 interacts with all the H1 subtypes that were used in the study under non-apoptotic conditions, however, it preferentially interacts with specific H1 subtypes after the induction of apoptosis by trichostatin A. Specifically, analysis of the H1 subtype composition of the chromatin-enriched fraction showed that trichostatin A induced the increase in the expression levels of histones H1.5, H1.3 and H1.0. The increase in H1.0 levels was expected since it is known from previous work that this agent induces the upregulation of the *H1.0* gene [131]. However, an unexpected finding in this investigation was the increase of H1.5 and H1.3 in the histone H1 subtype composition of chromatin. In contrast to the co-IP results obtained under non-apoptotic conditions, after trichostatin A-induced apoptosis, DFF40 was found to associate with specific H1 subtypes in a preferential manner. The co-IP results showed that DFF40 interacts more with H1.0 and H1.5. This finding may be explained as being due to the increased abundance of these two subtypes in chromatin in the presence of trichostatin A. However, this simplistic explanation comes at odds with the results obtained with H1.3. This subtype was also found to increase after trichostatin A treatment, but in contrast to H1.5 and H1.0, the association of DFF40 with this specific subtype was significantly decreased.

These results strongly imply that DFF40 interacts preferentially with H1.5 and H1.0 and not with H1.3 despite their similar abundance on chromatin in the NB4 leukemic cell line after the induction of apoptosis by TSA. Future work, in this specific line of research, should be aimed at determining whether this is cell type-specific or whether the specific apoptotic agent used (or both?) is the determining factor in the degree of the association of DFF40 with specific H1 subtypes.

9. Linker Histones, Differentiation, Aging and Senescence

One of the most well-studied linker histones is the replacement (replication-independent) somatic subtype, H1.0. It was first detected in tissues exhibiting low rates of cell division and proliferation [141]. Its accumulation can be induced in cultured cells upon growth inhibition. Conversely, the amount of H1.0 in the chromatin of quiescent (Go) cells is reduced upon resumption of proliferation [142]. H1.0 was found to accumulate in various cell systems that can be induced to differentiate. In essence, H1.0 accumulates in all cell/tissue systems during the process of terminal differentiation when cells are committed to a differentiation program [reviewed in 23]. As such, H1.0 has been termed throughout the literature as the differentiation-associated linker histone subtype. H1.0 was first considered to be enriched in condensed chromatin areas (heterochromatin). The logical assumption from these findings is that during differentiation there is a considerable increase in heterochromatin, since during the differentiation process many genes shut down, while others, necessary for the specialized function of the differentiated cell-type, remain active. All these data have led to the postulation that H1.0 is important in the terminal differentiation process. Importantly, in strong support for this point of view, were the results of Gabrilovich *et al.* where they demonstrated that H1.0 is genetically required in the terminal differentiation of dendritic cells [66]. With respect to H1.0 localization and function, more recent findings showed that this subtype localizes to euchromatin (open chromatin, transcriptionally active, or poised for activation) areas [54]. The hypothetical explanation for this newer finding is that H1.0 may function to repress the expression of genes that are no longer required, such as those associated with cell cycle progression [54]. In other words, H1.0 may be a necessary component of the terminal differentiation process because it positioned in such a manner so as to be poised to negatively regulate gene expression.

Aside from H1.0, there are also other well-documented differences in the H1 subtype protein complement of different tissues during development and differentiation. Lennox and Cohen found that (1) there are no tissue-specific subtypes among the somatic tissues that they examined, (2) H1.1 (H1a) and H1.5 (H1b) are synthesized in large amounts only in dividing cells, whereas H1.2 (H1c), H1.3 (H1d) and H1.4 (H1e) are synthesized in both dividing and non-dividing cells and, (3) H1.4 (H1e) always accumulates in non-dividing somatic cells [143]. The differences that these investigators found, even at that early time, led them to support the idea that the individual H1 subtypes differ from each other in some of their functions. Later studies also examined the mRNA levels of the H1 subtypes in terminally differentiated or Go-arrested cells. For example, Wang *et al.* found that all five genes for the somatic H1 histones are expressed in exponentially growing cells [144]. However, the levels of H1.1 (H1a), H1.5 (H1b), and H1.3 (H1d) mRNAs are greatly reduced in cells that are terminally differentiated or arrested in the Go phase of the cell cycle, while H1.2 (H1c) and H1.4 (H1e) continue to be expressed. There seems to be agreement between the proteins levels found by Lennox and Cohen and the mRNA levels for H1.1, H1.2, H1.5 and H1.4, with respect to dividing versus non-dividing cell-types [143]. However though H1.3 protein levels remain high in adult tissue [143], its mRNA levels decrease dramatically [144]. It may be that this protein is more stable than the other histone H1 proteins, partially accounting for its persistence in non-dividing, terminally differentiated cells. Thus, during development and terminal differentiation an interplay exists amongst transcriptional control and protein stability that is most likely unique for each subtype. Needless to say, though a vast amount of work has been accomplished over the past thirty years, much more is necessary in order to further clarify the functional aspects of each subtype's accumulation or decrease during the differentiation process.

Differentiation and senescence and/or the aging process have numerous similarities. Though the aging process is indeed influenced by stochastic factors, it is generally accepted that both require genetic programming. In both cases, certain cell cycle proliferative genes must shut down whereas other genes must be activated. Moreover, reorganization of the genome, i.e., changes in heterochromatin domains, reorganization of active and inactive chromatin domains and maintenance of repressive chromatin structures are hallmarks of both processes [145, 146, 147, 148, 149, 150, 151] and are the underlying cause of the changes in gene expression that must occur. As has been emphasized throughout this chapter, for such chromatin remodeling events to take place, changes in the histone constitution of chromatin and in their epigenetic modifications, must, and do, occur. In the remaining part of this chapter, an attempt will be made to outline some aspects of what is currently known regarding the association and/or the potential functional roles of the linker histones and certain post translational modifications (i.e., the phosphorylation state) of this histone family with the aging process and/or the senescent phenotype.

Although it is known that during aging and senescence there is a reorganization of euchromatic and heterochromatic regions, models on how this chromatin reorganization occurs during aging have been controversial. Recent evidence supports a model where heterochromatin increases during aging [149, 150, 151] not unlike the differentiation process [152]. As mentioned above, linker histone subtype changes in chromatin during the differentiation process have been well studied. However, parallel studies during the aging process have, to date, been sparse. Analysis of certain core histone subtypes in the *in vitro* aging human diploid fibroblast (HDF) cell system have shown that there is an accumulation of certain variants [153] that were also found in previous studies to increase during differentiation [154]

In this light, the Sekeri-Pataryas and Sourlingas group chose to study the differentiation-associated H1 subtype, H1.0 in the *in vitro* human diploid fibroblast (HDF) aging cell system [reviewed in 110, 112]. Since 1961, when Hayflick and Moorhead showed that HDFs have a limited *in vitro* lifespan [155], the fibroblast system has been widely used as an *in vitro* model cell system with which to study aging and the senescent state. Fibroblast undergo a specific number of cell divisions before exhausting their mitotic potential and then they enter into a viable, but stationary phase known as the post-mitotic (senescent) state [156] which is considered by most researchers to be a G1-like arrested state [157].

The group studied both the H1.0 relative synthesis rate (protein synthesis levels after incorporation of radioactive amino acids) and H1.0 mRNA levels in mitotic fibroblasts of increasing *in vitro* age (cumulative population doublings, CPDs) and in post-mitotic senescent cultures. The specific fibroblast cell system used in these studies (see below) undergoes 50-55 population doublings before entering the post-mitotic senescent state. Analysis of the protein's relative synthesis rate as a function of increasing CPDs in fibroblasts, synchronized by serum starvation in the Go, G1 and S phases of the cell cycle, showed that there were cell cycle phase-specific changes, i.e., maximal synthesis rates were found in the Go phase, while minimal rates were found in the S phase. This was expected and agrees with the previous results from the differentiation studies. However, no changes were found as a function of increasing age of the culture, i.e., when comparing CPD 30 (relatively young cultures), 40 (mid-aged cultures) and 50 (aged cultures, but still mitotically active, albeit with a decreased rate of mitosis) [106]. On the other hand, quite different results were obtained when the H1.0 mRNA levels were analyzed. Again, the analysis was accomplished using cultures with CPDs 30, 40 and 50 and during the Go, G1 and S phases of the cycle. In CPD 30, contrary to synthesis rates, maximal mRNA levels were found during the S phase and not during the Go phase, whereas minimal levels were found in the Go phase and intermediate levels in the G1 phase. The exact same phase-related results were obtained in CPD 40, except that the intensity of the bands for the Go and G1 phases were stronger as compared to those found in CPD 30, but the strongest signal was again obtained in the S phase. Quite different results were obtained in CPD 50 cultures. A strong signal, of more or less equal magnitude, was obtained for all three phases. In others words, at this very late cell culture age, where, however, mitotic activity is still observed, there is a stabilization of H1.0 mRNA levels [107]. These results indicate that there is an uncoupling amongst H1.0 protein synthesis levels and mRNA levels. Lastly, analysis of post-mitotic senescent cultures (after CPD 55) showed that both H1.0 protein and mRNA levels were significantly increased as compared to those of fibroblast cultures in passage 33 [108].

H1.0 mRNA levels seem to be coupled to DNA synthesis as can be seen from cell cycle results in which the levels of H1.0 mRNA are increased during the S phase in actively DNA synthesizing HDF cell populations (CPD 30 and 40). However, when the overall rate of DNA synthesis of the entire population begins to drop, as observed in HDF populations in CPD 50, the H1.0 mRNA levels tend to stabilize during the three phases of the cell cycle analyzed in this late CPD population. Results obtained in this study [107] can be compared to those reported in differentiating systems. More specifically, during induced differentiation of B16 [158] and MEL cells [159], it was found that H1.0 mRNA accumulation is linked to DNA replication, whereas H1.0 protein accumulation was found to occur during the later non-proliferative stages of terminal differentiation. This uncoupling of H1.0 protein accumulation from that of its coding mRNA was also found in differentiated rat cortical neurons [160] and after partial hepatectomy [161].

Thus, the findings of these studies [106, 107, 108, also reviewed in 110, 112] strongly indicate that, during *in vitro* aging, H1.0 behaves in a similar manner as that observed in numerous differentiating cell and/or tissue systems. The reason for this may be that in both systems H1.0 mRNA is accumulated in proliferating cells so as to be available to be used to direct H1.0 protein synthesis when needed by the cells eventually when they become arrested. Moreover, increase in H1.0 mRNA levels in senescent post-mitotic cells is analogous to that found during the last stages of terminal differentiation [160]. Thus it would seem that the timing of expression is of importance, not only in differentiating cell systems, but in aging cell systems as well. An enticing hypothetical explanation may be that, just as in terminal differentiation, during senescence analogous structural chromatin remodeling events take place, thus justifying the necessity of the existence of similar regulatory mechanisms for H1.0 gene expression in these two systems.

In two other studies by Sourlingas, *et al.* [109] and Kypreou *et al.* [111], the authors investigated whether there were quantitative age-related differences in the induction of H1.0 by the histone deacetylase inhibitor, TSA (trichostatin A) in human peripheral blood lymphocytes from donors of different age groups, namely young (25-30 years), mid-aged (40-45 years), senior (60-65 years) and elderly (80-95 years). TSA increases histone acetylation levels in the promoter region of the *H1.0* gene, thus leading to its induction or up-regulation [131]. Lymphocytes are ideal to study the induction or up-regulation of H1.0, since they do not synthesize this H1 subtype in detectable levels [35]. The results showed that there is an observable increase in the H1.0 relative synthesis rate in the elderly age group as compared to the other younger age groups. Interestingly, when comparing the young, mid-aged and senior ages groups, the H1.0 synthesis rates were more or less similar, a result which once more reinforces the idea that timing of the expression of this linker subtype is of major importance (as it appears to be in differentiating and aging fibroblast cell systems) [111]. These results are, yet again, consistent with the supposition that H1.0 has a functional gene regulatory role and is thus necessary in the chromatin remodeling events (i.e., reorganization and/or increase in heterochromatin domains) that must occur in this cell type, as well, during aging and senescence.

Aside from H1.0, the more recent work of Happel *et al.* has also shown that the phosphorylation state of the H1 subtypes also may play a functional role during aging and senescence [162]. These investigators studied the changes that take place in the somatic H1 linker histone subtype profile and their phosphorylation states in human peripheral blood lymphocytes as a function of donor age. The phosphorylation status of the H1 subtypes has been exhaustingly analyzed during growth and development [e.g., 33, 34, 143, 163, 164] and differentiation [reviewed in 161], but no study had previously addressed its relationship with the aging process. Using capillary zone electrophoresis and HPLC, together with mixing experiments with recombinantly expressed H1 subtypes, they found a distinct dephosphorylation of the subtypes, H1.4 and H1.5 as a function of increasing donor age. This correlated with a significant increase in the HP1α protein in the elderly age group. HP1α localizes to heterochromatin regions and can thus be used to grossly assess the heterochromatinization status of the cell. As described in section 6 of this chapter, there is now experimental evidence that phosphorylation and dephosphorylation of H1 decondenses and condenses chromatin respectively, either locally or on a more global scale. Thus the results of Happel *et al.* strongly indicate that dephosphorylation of H1.4 and H1.5 may be associated with heterochromatin formation during the aging process of human peripheral blood lymphocytes [162]. It remains to be seen whether this proves to be the case in other aging cell/tissue systems.

CONCLUSIONS

As can be seen from all of the above cited literature, the H1 chromatin protein class and all of the H1 subtypes have been shown to be associated with, or involved in, or have even been found to have, functionally discrete roles in virtually all biological processes that require chromatin remodeling. The mechanisms by which certain H1 subtypes accomplish their functional roles within the context of certain biological processes have been investigated and, in some cases, have been clarified to a considerable extent (e.g., H1.2 and apoptosis, the oocyte and somatic 5s rRNA genes). In other circumstances, strong indications of the existence of specific functional roles of certain H1 subtypes have been ascertained. In these latter cases, though the precise mechanism(s) may not have been completely elucidated, considerable data have accumulated from a diverse gamma of experiments (such as knock out, knock down, overexpression experiments, chromatin localization studies, analyses of the H1 subtype complement in many cell/tissue systems and other model cell systems (e.g., apoptosis, differentiation, aging, etc.)) that strongly implicate specific H1 histones as having functionally discrete roles in a variety of biological processes. In these cases, more work must be accomplished so as to provide concrete evidence in further support of these postulations. As Woodcock and colleagues point out, it is perhaps a measure of the complexity of the problem that the titles of numerous reviews that describe what is known as to the functional aspects of the H1 subtypes have titles that end with a question mark [58]. The older data obtained for the H1 histones (i.e., *in vitro* structural studies, analyses of the H1 complement during development and differentiation, etc.) gave the foundations with which the more recent work was able to build upon. For example, the results from the investigations of Gurley and colleagues amassed a huge body of data with respect to H1 phosphorylation, the cell cycle and DNA replication, which led to the speculation that H1 phosphorylation was involved chromatin condensation [77, 78, 79, 80]. Newer data showed that the work of the Gurley group can be reappraised and reinterpreted within the context of the more recent results. The point being made is that contemporary research in this area must retrospect, both in order to bring forth new data and also so as to reinterpret the classic H1 data that has accumulated for more than thirty years. It is certain that a reappraisal is required for both the concept of H1 as a general transcriptional repressor and its role as a chromatin architectural protein. In this respect, much more work is necessary in order to unravel the complexities involved in deciphering the functional aspects of the H1 subtypes and their post translational modifications. It is the author's belief that future work will lead to newer data in strong support of the dual aspects of the role of the H1 proteins in chromatin structure and function.

REFERENCES

[1] Luger, K., Mäder, A. W., Richmond, R. K., Sargent, D. F. & Richmond, T. J. (1997). Crystal structure of the nucleosome core particle at 2.8 A resolution. *Nature, 389*, 251–260.

[2] Richmond, T. J. & Davey, C. (2003). The structure of DNA in the nucleosome core. *Nature, 423*, 145–150.

[3] Olins, A. L. & Olins, D. E. (1974). Spheroid chromatin units (v bodies). *Science, 183*, 330–332.

[4] Kornberg, R. & Thomas, J. O. (1974). Chromatin structure. Oligomers and histones. *Science, 184*, 865–868.

[5] Oudet, P., Gross-Bellard M. & Chambon P. (1975). Electron microscopic and biochemical evidence that chromatin structure is a repeating unit. *Cell, 4*, 281–300.

[6] Simpson, R. T. (1978). Structure of the chromatosome, a chromatin particle containing 160 base pairs of DNA and all the histones. *Biochemistry, 17*, 5524–5531.

[7] Wu, C., Bassett, A. & Travers, A. (2007). A variable topology for the 30-nm chromatin fibre. *EMBO Rep, 8*, 1129-1134.

[8] Zhou, Y. B., Gerchman, S. E., Ramakrishnan, V., Travers, A. & Muyldermans, S. (1998). Position and orientation of the globular domain of linker histone H5 on the nucleosome. *Nature,395*, 402–405.

[9] Wolffe, A .P., Khochbin, S. & Dimitrov, S. (1997). What do linker histones do in chromatin *BioEssays, 19*, 249-255.

[10] Allan, J., Hartman, P. C., Crane-Robinson, C. & Aviles, F. X. (1980). The structure of histone H1 and its location in chromatin. *Nature, 288*, 675–679.

[11] Pruss, D., Bartholomew, B., Persinger, J., Hayes, J., Arents, G., Moudrianakis, E. N. & Wolffe, A. P. (1996). An asymmetric model for the nucleosome: a binding site for linker histones inside the DNA gyres. *Science, 274*, 614–617.

[12] Crane-Robinson, C. (1997). Where is the globular domain of linker histone H1 located in the nucleosome? *Trends Biochem Sci, 22*, 75-77.

[13] Bednar, J., Horowitz, R. A., Grigoryev, S. A., Carruthers, L.M., Hansen, J. C., Koster, A. J. & Woodcock, C. L. (1998). Nucleosomes, linker DNA,and linker histone form a unique structural motif that directs the higher-order folding. *Proc. Natl Acad Sci USA, 95*, 14173–14178.

[14] Carruthers, L. M. & Hansen, J. C. (2000). The core histone N termini function independently of linker histones during chromatin condensation. *EMBO Rep, 3*, 37285–37290.

[15] Happel, N. & Doenecke D. (2009). histone H1 and its isoforms: Contribution to chromatin structure and function. *Gene, 431*, 1-12.

[16] Eick, S., Nicolai, M., Mumberg, D. & Doenecke, D. (1989). Human H1 histones: conserved and varied sequence elements in two H1 subtype genes. *Eur J Cell Biol, 49*, 110–115.

[17] Albig, W., Kardalinou, E., Drabent, B., Zimmer, A. & Doenecke, D. (1991). Isolation and characterization of two human H1 histone genes within clusters of core histone genes. *Genomics, 10*, 940–948.

[18] Albig, W., Drabent, B., Kunz, J., Kalff-Suske, M., Grzeschik, K. H. & Doenecke, D. (1993). All known human H1 genes except the H1^0 gene are clustered on chromosome 6. *Genomics, 16*, 649–654.

[19] Albig, W., Meergans, T. & Doenecke, D. (1997). Characterization of the H1.5 gene completes the set of human H1 subtypes genes. *Gene, 184*, 141–148.

[20] Drabent, B., Franke, K., Bode, C., Kosciessa, U., Bouterfa, H., Hameister, H. & Doenecke, D. (1995). Isolation of two murine H1 histone genes and chromosomal mapping of the H1 gene complement. *Mamm Genome, 6*, 505–511.

[21] Gjerset, R., Ibarrando, F., Saragosti, S. & Eisen, H. (1981). Distribution of IP25 in chromatin and its possible involvement in chromatin condensation. *Biochem. Biophys Res Commun, 99*, 349–357.

[22] Gjerset, R., Gorka, C., Hasthorpe, S., Lawrence, J .J. & Eisen, H. (1982). evelopmental and hormonal regulation of protein H1^0 in rodents. *Proc Natl Acad Sci USA, 79*, 2333–2337.

[23] Zlatanova, J. & Doenecke, D. (1994). Histone H1 zero: a major player in cell differentiation? *FASEB J, 8,* 1260–1268.

[24] Yamamoto, T. & Horikoshi, M. (1996). Cloning of the cDNA encoding a novel subtype of histone H1. *Gene, 173*, 281–285.

[25] Happel, N., Schulze, E. & Doenecke, D. (2005). Characterization of human histone H1x. *Biol Chem, 386*, 541–551.

[26] Meistrich, M. L., Bucci, L. R., Trostle-Weige, P. K. & Brock, W. A. (1985). Histone variants in rat spermatogonia and primary spermatocytes. *Dev Biol, 112*, 230–240.

[27] Meistrich, M. L., Trostle-Weige, P .K. & Van Beek, M. E. (1994). Separation of specific stages of spermatids from vitamin A-synchronized rat testes for assessment of nucleoprotein changes during spermiogenesis. *Biol Reprod, 51*, 334–344.

[28] Drabent, B., Bode, C. & Doenecke, D. (1995). Association of histone H4 genes with the mammalian testis-specific H1t histone gene. *DNA Cell Biol, 14*, 591–597.

[29] Martianov, I., Brancorsini, S., Catena, R., Gansmuller, A., Kotaja, N., Parvinen, M., Sassone-Corsi, P. & Davidson, I. (2005). Polar nuclear localization of H1T2, a histone H1 variant, required for spermatid elongation and DNA condensation during spermiogenesis. *Proc Natl Acad Sci USA, 102*, 2808–2813.

[30] Yan, W., Ma, L., Burns, K. H. & Matzuk, M. M. (2003). HILS1 is a spermatid-specific linker histone H1-like protein implicated in chromatin remodeling during mammalian spermiogenesis. *Proc Natl Acad Sci USA, 10*0, 10546–10551.

[31] Tanaka, M., Hennebold, J.D., Macfarlane, J. & Adashi, E. Y. (2001). A mammalian oocyte-specific linker histone gene H1oo: homology with the genes for the oocyte-specific cleavage stage histone (cs-H1) of sea urchin and the B4/H1M histone of the frog. *Development, 28*, 655–664.

[32] Doenecke, D., Albig, W., Bode, C., Drabent, B., Franke, K., Gavenis, K. & Witt, O. (1997). Histones: genetic diversity and tissue-specific gene expression. *Histochem Cell Biol, 107*, 1–10.

[33] Lennox, R. W. & Cohen, L. H. (1988). The production of tissue-specific histone complements during development. *Biochem Cell Biol, 66*, 636–649.

[34] Pina, B., Martinez, R. & Suau, P. (1987). Changes in the H1 complement in differentiating rat-brain cortical neurons. *Eur J Biochem, 164*, 71–76.

[35] ManniRoni, C., Rossi, V., Biondi, A., Ubezio, P., Masera, G., Barbui, T. & D'Incalci, M. (1987). Histone $H1^0$ is synthesized by human lymphocytic leukemic cells but not by normal lymphocytes. *Blood, 70*, 1203–1207.

[36] Jenuwein, T. & Allis, C. D. (2001). Translating the histone code. *Science, 293*, 1074-1080.

[37] Spotswood, H. T. & Turner, B. M. (2002). An increasingly complex code. *J Clin Invest, 110,* 577-582.

[38] Fischle, W., Wang, Y. & Allis, C. D. (2003). Binary switches and modification cassettes in histone biology and beyond. *Nature, 425,* 475-479.

[39] Kouzarides, T. (2007). Chromatin modifications and their function. *Cell 128*, 693–705.

[40] Roth, S.Y. and Allis, C.D. 1992. Chromatin condensation: does histone H1 dephosphorylation play a role? *TIBS, 17*, 93–98.

[41] Riquelme, P. T., Burzio, L. O. & Koide, S. S. (1979). ADP ribosylation of rat liver lysine-rich histone *in vitro*. *J Biol Chem, 254,* 3018-3028.

[42] Aubin, R. J., Dam, V. T., Miclette, J., Brousseau, Y., Huletsky, A. & Poirier, G. G. (1982). Hyper(ADP-ribosyl)ation of histone H1. *Can J Biochem, 60,* 1085-94.

[43] Moore, S. C., Jason, L. & Ausio, J. (2002). The elusive structural role of ubiquitinated histones. *Biochem Cell Biol, 80,* 311–319.

[44] Wisniewski, J. R., Zougman, A., Krüger, S. & Mann, M. (2007). Mass spectrometric mapping of linker histone H1 variants reveals multiple acetylations, methylations, and phosphorylation as well as differences between cell culture and tissue. *Mol Cel. Proteomics, 6*, 72–87.

[45] Ramakrishnan, V., Finch, J. T., Graziano, V., Lee, P. L. & Sweet, R. M. (1993). Crystal structure of globular domain of histone H5 and its implications for nucleosome binding. *Nature, 362*, 219–223.

[46] Graziano, V., Gerchman, S. E., Wonacott, A. J., Sweet, R. M., Wells, J. R., White, S. W. & Ramakrishnan, V. (1990). Crystallization of the globular domain of histone H5. *J Mol Biol, 212,* 253-257.

[47] Brown, D. T. (2003). Histone H1 and the dynamic regulation of chromatin function. *Biochem Cell Biol, 81*, 221-227.

[48] Travers, A. A. (1999). The location of linker histone in the nucleosome. *Trends Biochem Sci, 24,* 4-7.

[49] Duggan, M. M. & Thomas, J. O. (2000). Two DNA-binding sites on the globular domain of histone H5 are required for binding to both bulk and 5 S reconstituted nucleosomes. *J Mol Biol, 304,* 21-33.

[50] Pruss, D. & Wolffe, A. P. (1993). Histone-DNA contacts in a nucleosome core containing a Xenopus 5S rRNA gene. *Biochemistry, 32,* 6810-6814.

[51] Vignali, M. & Workman, J. L. (1998). Location and function of linker histones. *Nature Struct Biol, 5,* 1025-1028.

[52] Allan, J., Mitchell, T., Harborne, N., Bohm, L., & Crance-Robinson, C. (1986). Roles of H1 domains in determining higher order chromatin structure and H1 location. *J Mol Biol, 187,* 591-601. Vignali, M. & Workman, J. L. (1998). Location and function of linker histones. *Nature Struct. Biol. 5,* 1025-1028.

[53] Ponte, I., Vidal-Taboada, J. M. & Suau, P. (1998). Evolution of the vertebrate H1 histone class: evidence for the functional differentiation of the subtypes. *Mol Biol. Evol, 15,* 702-708.

[54] Th'ng, J. P. H., Sung, R., Ye, M. & Hendzel, M. J. (2005). H1 family histone in the nucleus: control of binding and localization by the C-terminal domain. *J Biol Chem 280*, 27809–27814.

[55] Bates, D. L. & Thomas, J. O. (1981). Histones H1 and H5: one molecule or two molecules per nucleosome. *Nucleic Acids Res, 25,* 5883-5894.

[56] Fan, Y., Nikitina, T., Morin-Kensicki E. M., Zhao, J., Magnuson, T. R., Woodcock, C. L. & Skoultchi A. I. (2003). H1 linker histones are essential for mouse development and affect nucleosome spacing in vivo. *Mol Cell Biol, 23*, 4559-4572.

[57] Fan, Y., Nikitina, T., Zhao, J., Fleury, T. J., Bhattacharyya, R., Bouhassira, E. E., Stein, A., Woodcock, C. L. & Skoultchi A. I. (2005). Depletion of histone H1 in mammals alters global chromatin structure but causes specific changes in gene regulation. *Cell, 123,* 1119-1212.

[58] Woodcock, C. L., Skoultchi, A. I. & Fan, Y. (2006). Role of linker histone in chromatin structure and function: H1 stoichiometry and nucleosome repeat length. *Chromosome Res, 14,* 17-25.

[59] Sirotkin, A. M., Edelman, W., Cheng, G., Klein-Szanto, A., Kucherlapati, R. & Skoultchi, A. I. (1995). Mice develop normally without the $H1_0$ linker histone. *Proc Natl Acad Sci, USA 92*, 6434-6438.

[60] Fan, Y., Sirotkin, A., Russell, R. G., Ayala, J. & Skoultchi, A. I. (2001). Individual somatic H1 subtypes are dispensable for mouse development even in mice lacking the H1(0) replacement subtype. *Mol Cell Biol, 2,* 7933-7943.

[61] Misteli, T., Gunjan, A., Hock, R., Bustin, M. & Brown, D. T. (2000). Dynamic binding of histone H1 to chromatin in living cells. *Nature, 408*, 877-881.

[62] Lever, M. A., Th'ng, J. P. H., Sun, X. & Hendzel, M. J. (2000). Rapid exchange of histone H1.1 on chromatin in living cells. *Nature 408*, 873-876.

[63] Widom, J. (1998). Chromatin structure: Linking structure to function with histone H1. *Curr Biol, 8,* R788-R791.

[64] Bustin, M., Catez, F. & Lim, J. H. (2005). The dynamics of histone H1 function in chromatin. *Mol Cell, 17*, 617-620.

[65] De, S., Brown, D. T., Lu, Z. H., Leno, G. H., Wellman, S. E. & Sittman, D. B. (2002). Histone H1 variants differentially inhibit DNA replication through an affinity for chromatin mediated by their carboxyl-terminal domains. *Gene, 292*, 173–181.

[66] Gabrilovich, D. I., Cheng, P., Fan, Y., Yu, B., Nikitina, E., Sirotkin, A., Shurin, M., Oyama, T., Adachi, Y., Nadaf, S., Carbone, D. P. & Skoultchi, A. I. (2002). H1(0) histone and differentiation of dendritic cells. A molecular target for tumor-derived factors. *J Leukocyte Biol, 72***,** 285–296.

[67] Wood, C., Snijders, A., Williamson, J., Reynolds, C. & Baldwin, J. (2009). Post translational modifications of the linker histone variants and their association with cell mechanisms. *FEBS Journal, 276,* 3685-3697.

[68] Alami, R., Fan Y., Pack, S., Sonbuchner, T. M., Besse, A., Lin, Q., Greally, J. M., Skoultchi, A.I. & Bouhassira, EE. (2003). Mammalian linker-histone subtypes differentially affect gene expression *in vivo*. *PNAS, 100*, 5920-5925.

[69] Kuzmichev, A., Jenuwein, T., Tempst, P. & Reinberg, D. (2004). Different EZH2-containing complexes target methylation of histone H1 or nucleosomal histone H3. *Mol Cell 14,* 183–193.

[70] Garcia, B. A., Busby, S. A., Barber, C. M., Shabanowitz, J., Allis, C. D. & Hunt, D. F. (2004). Characterization of phosphorylation sites on histone H1 isoforms by tandem mass spectrometry. *J Proteome Res, 3,* 1219-1227.

[71] Jiang, T., Zhou, X., Taghizadeh, K., Dong, M. & Dedon, P.C. (2007). N-formylation of lysine in histone proteins as a secondary modification arising from oxidative DNA. *Proc Natl Acad Sci USA, 104,* 60-65.

[72] Wisniewski, J. R., Zougman, A. & Mann, M. (2008). Nε-Formylation of lysine is a widespread post-translational modification of nuclear proteins occurring at residues involved in regulation of chromatin function. *Nucleic Acids Res, 36*, 570–577.

[73] Parseghian, M. H. & Hamkalo, B. A. (2001). A compendium of the histone H1 family of somatic subtypes: an elusive cast of characters and their characteristics. *Biochem Cell Biol, 79*, 289–304.

[74] Balhorn, R., Chalkley, R. & Granner, D. (1972). Lysine-rich histone phosphorylation. A positive correlation with cell replication. *Biochemistry, 11*, 1094–1098.

[75] Sarg, B., Helliger, W., Talasz, H., Forg, B. & Lindner, H. H. (2006). Histone H1 phosphorylations occurs stie-specifically during interphase and motosis: identification of a novel phosphorylation site on histone H1. *J Biol Chem, 281,* 6573-6580.

[76] Gurley, L. R., Valdez, J. G. & Buchanan, J. S. (1995). Characterization of the mitotic specific phosphorylation site of histone H1. Absences of a consensus sequence for the p34cdc2/cyclin B kinase. *J Biol Chem. 246,* 27653-27660.

[77] Hohmann, P., Tobey, R. A. & Gurley, L. R. (1976). Phosphorylation of distinct regions of f1 histone. Relationship to the cell cycle. *J Biol Chem, 251,* 3685–3692.

[78] Gurley, L. R., D'Anna, J. A., Barham, S. S., Deaven, L. L. & Tobey, R. A. (1978). Histone phosphorylation and chromatin structure during mitosis in Chinese hamster cells. *Eur J Biochem, 84,* 1–15.

[79] Gurley, L. R., Tobey, R. A., Waters, R. A., Hildebrand, L. E., Hohmann, P. G., D'Anna, J. A., Barham, S. S. & Deaven, L. L. (1978). Cell Cycle Regulation. In Jeter, J. R., Jr., Cameron, I. L., Padilla, G. M. & A. M. Zimmerman, (Eds.) pp. 37–60, New York, Academic Press, Inc.

[80] Gurley, L. R., Walters, R. A. & Tobey, R. A. (1975). Sequential phosphorylation of histone subfractions in the Chinese hamster cell cycle. *J Biol Chem, 250*, 3936–3944.

[81] Yasuda, H., Matsumoto, Y., Mita, S., Marunouchi, T. & Yamada, M. (1981). A mouse temperature-sensitive mutant defective in H1 histone phosphorylation is defective in deoxyribonucleic acid synthesis and chromatin condensation. *Biochemistry, 20,* 4414–4419.

[82] Halmer, L. & Gruss, C. (1996). Effects of cell cycle dependent histone H1 phosphorylation on chromatin structure and chromatin replication. *Nucleic Acids Res, 24,* 1420–1427.

[83] Alexandrow, M. G. & Hamlin, J. L. (2005). Chromatin decondensation in S-phase involves recruitment of Cdk2 by Cdc45 and histone H1 phosphorylation. *J Cell Biol, 168,* 875-886.

[84] Bradbury, E. M., Cary, P. D., Crane-Robinson, C. & Rattle, H. W. (1973). Conformations and interactions of histones and their role in chromosome structure. *Ann NY Acad Sci, 222*, 266–289.

[85] Bradbury, E. M., Inglis, R. J. & Matthews, H. R. (1974). Control of cell division by very lysine rich histone (F1) phosphorylation. *Nature, 247*, 257–261.

[86] Roth, S. Y., Schulman, I. G., Richman, R., Cook, R. G. & Allis, C. D. (1988). Characterization of the phosphorylation sites in histone H1 in the amitotic macronucleus of Tetrahymena during different physiological states. *J Cell Biol, 107,* 2473–2482.

[87] Lin, R., Cook, R. G. & Allis, C. D. (1991). Proteolytic removal of core histone amino termini and dephosphorylation of histone H1 correlate with the formation of condensed chromatin and transcriptional silencing during Tetrahymena macronuclear development. *Genes Dev, 5*, 1601–1610.

[88] Dou, Y. & Gorovsky, M. A. (2000). Phosphorylation of linker histone H1 regulates gene expression *in vivo* by creating a charge patch. *Mol Cell, 6,* 225-231.

[89] Lu, M. J., Mpoke, S. S., Dadd, C.A. & Allis, C. D. (1995). Phosphorylated and dephosphorylated linker histone H1 reside in distinct chromatin domains in Tetrahymena macronuclei. *Mol Biol Cell, 6*, 1077–1087.

[90] Chadee, D. N., Taylor, W. R., Hurta, R. A., Allis, C. D., Wright, J. A. & Davie, J. R. (1995). Increased phosphorylation of histone H1 in mouse fibroblasts transformed with oncogenes or constitutively active mitogen-activated protein kinase kinase. *J Biol Chem, 270,* 20098–20105.

[91] Horn, P. J., Carruthers, L. M., Logie, C., Hill, D. A., Solomon, M. J. & Wade, P. A. (2002). Phosphorylation of linker histones regulates ATP-dependent chromatin remodelling enzymes. *Nat Struct Biol, 9*, 263–267.

[92] Vicent, G. P., Koop, R. & Beato, M. (2003). Complex role of histone H1 in transactivation of MMTV promoter chromatin by progesterone receptor. *J Steroid Biochem Mol Biol, 83*, 15–23.

[93] Dou, Y., Bowen, J., Liu, Y. & Gorovsky, M. A. (2002). Phosphorylation and an ATP-dependent process increase the dynamic exchange of H1 in chromatin. *J Cell Biol, 158,* 1161-1170.

[94] Contreras, A., Hale, T. K., Stenoien, D.L. & Rosen, J. M. (2003). The dynamic mobility of histone H1 is regulated by cyclin/cdk phosphorylation. *Mol Cell Biol, 23*, 8626-8636.

[95] Koop, R., Di Croce, L. & Beato, M. (2003). Histone H1 enhances synergistic activation of the MMTV promoter in chromatin. *EMBO J, 22*, 588–599.

[96] Downs, J. A., Kosmidou, E., Morgan, A. & Jackson, S. P. (2003). Suppression of homologous recombination by the Saccharomyces cerevisiae linker histone. *Mol Cell, 11*, 1685–1692.

[97] Konishi, A., Shimizu, S., Hirota, J., Takao, T., Fan, Y., Matsuoka, Y., Zhang. L., Yoneda, Y., Fujii, Y., Skoultchi, A. I. & Tsujimoto, Y. (2003). Involvement of histone H1.2 in apoptosis induced by DNA double strand breaks. *Cell, 114*, 673-688.

[98] Liu, X., Li, P., Widlak, P., Zou, H., Luo, X., Garrard, W. T. & Wang X. (1998). The 40-kDa subunit of DNA fragmentation factor induces DNA fragmentation and chromatin condensation during apoptosis. *Proc Natl Acad Sci USA, 95,* 8461–8466.

[99] Liu, X., Zou, H., Widlak, P., Garrard, W. T. & Wang, X. (1999). Activation of the apoptotic endonuclease DFF40 (caspase-activated DNase or nuclease). Oligomerization and direct interaction with histone H1. *J Biol Chem, 274,* 13836–13840.

[100] Widlak, P., Peng Li, P., Wang, X. & Garrard, W. T. (2000). Cleavage preferences of the apoptotic endonuclease DFF40 (caspase-activated DNase or nuclease) on naked DNA and chromatin substrates. *J Biol Chem, 275*, 8226–8232.

[101] Widlak, P. (2000). The DFF40/CAD endonuclease and its role in apoptosis. *Acta Biochim Pol, 47,* 1037–1044.

[102] Widlak, P. & Garrard, W. T. (2005). Discovery, regulation, and action of the major apoptotic nucleases DFF40/CAD and endonuclease G. *J Cell Biochem 94*, 1078–1087.

[103] Sourlingas, T. G., Tsapali, D. S., Kaldis, A. D. & Sekeri-Pataryas, K. E. (2001). Histone deacetylase inhibitors induce apoptosis in peripheral blood lymphocytes along with histone H4 acetylation and the expression of the linker histone variant, H1°. *Eur J Cell Biol, 80,* 726-732.

[104] Ninios, Y. P., Sekeri-Pataryas, K .E. & Sourlingas, T. G. (2010). Histone H1 subtype preferences of DFF40 and possible nuclear localization of DFF40/45 in normal and trichostatin A-treated NB4 leukemic cells. *Apoptosis, 15*, 128-138.

[105] Barra, J. L., Rhoumin, L., Rossignol, J. L. & Faugeron, G. (2000). Histone H1 is dispensable for methylation-associated gene silencing in *Ascobolus immersus* and essential for long life span. *Mol Cell Biol, 20*, 61–69.

[106] Tsapali, D. S., Sekeri-Pataryas, K. E. & Sourlingas, T. G. (2000). Study of the H1 linker histone variant, H1°, during the in vitro ageing of human diploid fibroblasts. *Ann NY Acad Sci, 908*, 336–340.

[107] Tsapali, D. S., Sekeri-Pataryas, K. E., Prombona, A. & Sourlingas, T. G. (2000b). mRNA levels of the linker histone variant, H1° in mitotically active human diploid fibroblasts as a function of the phases of the cell cycle and cumulative population doublings. *Mech Ageing Dev, 121*, 101–112.

[108] Tsapali, D. S., Sekeri-Pataryas, K .E. & Sourlingas, T. G. (2001). mRNA levels of the differentiation-associated linker histone variant, H1 zero in mitotically active and postmitotic senescent human diploid fibroblast cell populations. *Exp Gerontol, 36*, 1649–1661.

[109] Sourlingas, T.G., Kypreou, K.P. & Sekeri-Pataryas, K. E. (2002). The effect of the histone deacetylase inhibitor, trichostatin A, on total histone synthesis, H1(0) synthesis and histone H4 acetylation in peripheral blood lymphocytes increases as a function of increasing age: a model study. *Exp Gerontol, 37*, 341–348.

[110] Sourlingas, T. G. & Sekeri-Pataryas, K. E. (2003). Linker histone H1o gene expression during ageing and after the effect of histone deacetylase inhibitors in human diploid fibroblasts and T lymphocytes. In G. Pawelec and M.P. Mattson (Eds.), *Advances in Cell Aging and Gerontology: Basic Biology and Clinical Impact of Immunosenescence* (vol. 13, pp. 231-241). Baltimore, MD, USA: Elsevier.

[111] Kypreou, K. P., Sourlingas, T. G. & Sekeri-Pataryas, K. E. (2004). Age-dependent response of lymphocytes in the induction of the linker histone variant, H1o and histone H4 acetylation after treatment with the histone deacetylase inhibitor, trichostatin A. *Exp Gerontol 39*, 469–479.

[112] Sekeri-Pataryas, K. E. & Sourlingas, T. G. (2007). The differentiation-associated linker histone, H1.0, during the in vitro aging and senescence of human diploid fibroblasts. *Ann NY Acad Sci, 1100*, 361-367.

[113] Shen, X. & Gorovsky, M. A. (1996). Linker histone H1 regulates specific gene expression but not global transcription *in vivo. Cell, 86,* 475-483.

[114] Howe, L., Itoh, T., Katagiri, C. & Ausio, J. (1998). Histone H1 binding doewnot inhibit transcription of nucleosomal *Xenopus laevis* somatic 5S rRNA templates. *Biochemistry, 37*, 7077-7082.

[115] Eisfeld, K., Candau, R., Truss, M. & Beato, M. (1997). Binding of NF1 to the MMTV promoter in nucleosomes: influence of rotational phasing, translational positioning and histone H1. *Nucleic Acids Res, 25,* 3733-3742.

[116] Kermekchiev, M., Workman, J.L. & Pikaard, C. S. (1997). Nucleosome binding by the polymerase I transactivator upstream binding factor displaces linker histone H1. *Mol Cel Biol, 17,* 5833-5842.

[117] Sera, T. and Wolffe, A. P. (1998). Role of H1 as an architectural determinant of chromatin structure and as a specific repressor of transcription on *Xenopus* oocyte 5S rRNA genes. *Mol Cell Biol, 18,* 3668-3680.

[118] Hellauer, K., Sirard, E. & Tucotte, B. (2001). Decreased expression of specific genes in yeast lacking histone H1. *J Biol Chem, 276*, 13587-13592.

[119] Harvey, A. C. and Downs, J. A. (2004). What functions do linker histones provide? *Mol Microbiol, 53*, 771-775.

[120] Brown, D. T., Alexander, B. T. & Sittman, D. B. (1996). Differential effect of H1 variant overexpression on cell cycle progression and gene expression. *Nucleic Acids Res, 24*, 486-493.

[121] Gunjan, A., Alexander, B. T., Sittman, D. B. & Brown, D. T. (1999). Effects of H1 histone variant overexpression on chromatin structure. *J Biol Chem, 274*, 37950-37956.

[122] Sancho, M., Diani, E., Beato, M. & Jordan, A. (2008). Depletion of human histone H1 variants uncovers specific roles in gene expression and cell growth. *PLoS Genetics, 4,* 1-17.

[123] Chubb, J .E. & Rea, S. (2010). Core and linker histone modfications involved in the DNA damage response. *Subcell Biochem, 50*, 17-42.

[124] Rogakou, E. P., Pilch, D. R., Orr, A. H., Ivanova, V. S. & Bonner, W. M. (1998). DNA doublestranded breaks induce histone H2AX phosphorylation on serine 139. *J Biol Chem, 273*, 5858–5868.

[125] Murga, M., Jaco, I., Fan, Y., Soria, R., Martinez-Pastor, B., Cuadrado, M., Yang, S. M., Blasco, M. A., Skoultchi, A. I. & Fernandez-Capetillo, O. (2007). Global chromatin compaction limits the strength of the DNA damage response. *J Cell Biol, 178,* 1101–1108.

[126] Ikura, T., Tashiro, S., Kakino, A., Shima, H., Jacob, N., Amunugama, R., Yoder, K., Izumi, S., Kuraoka, I., Tanaka, K., Kimura, H., Ikura, M., Nishikubo, S., Ito, T., Muto, A., Miyagawa, K., Takeda, S., Fishel, R., Igarashi, K. & Kamiya, K. (2007). DNA damage-dependent acetylation and ubiquitination of H2AX enhances chromatin dynamics. *Mol Cell Biol, 27,* 7028–7040.

[127] van Attikum, H. & Gasser, S. M. (2005). The histone code at DNA breaks: a guide to repair? *Nat Rev Mol Cell Biol 6,* 757–765.

[128] Zhu, P., Zhou, W., Wang, J., Puc, J., Ohgi, K. A., Erdjument-Bromage, H., Tempst, P., Glass, C. K. & Rosenfeld, M. G. (2007). A histone H2A deubiquitinase complex coordinating histone acetylation and H1 dissociation in transcriptional regulation. *Mol Cell, 27*, 609–621.

[129] Bree, R. T., Neary, C., Samali, A. & Lowndes, N. F. (2004). The switch from survival responses to apoptosis after chromosomal breaks. *DNA Repair, 3*, 989-995.

[130] Gillespie, D. A .F. & Vousden, K .H. 2003. The secret life of histones. *Cell, 114*, 655-656.

[131] Girardot, V., Rabilloud, T., Yoshida, M., Beppu, T., Lawrence, J. J. & Khochbin, S. (1994). Relationship between core histone acetylation and histone $H1^0$ accumulation. *Eur J Biochem, 224,* 885- 892.

[132] Oberhammer F., Bursch W., Tiefenbacher R., Froschl G., Pavelka M., Purchio T. & Schulte-Hermann R. (1993). Apoptosis is induced by transforming growth factor-b1 within 5 hours in regressing liver without significant fragmentation of the DNA. *Hepatology, 18,* 1238–1246.

[133] Schulze-Osthoff, K., Walczak, H., Droge, W. & Krammer P. H. (1994). Cell nucleus and DNA fragmentation are not required for apoptosis. *J Cell Biol, 127*, 15–20.

[134] Sahara, S., Aoto, M., Eguchi, Y., Imamoto, N., Yoneda, Y. & Tsujimoto, Y. (1999). Acinus is a caspase-3-activated protein required for apoptotic chromatin condensation. *Nature, 401,* 168–173.

[135] Zhang, J. & Xu, M. (2002). Apoptotic DNA fragmentation and tissue homeostasis. *Trends Cell Biol, 12,* 84–89.

[136] Nagata, S., Nagase, H., Kawane, K., Mukae, N. H. & Fukuyama, N. H. (2003). Degradation of chromosomal DNA during apoptosis. *Cell Death Differ, 10*, 108–116.

[137] Hanus, J., Kalinowska-Herok, M. & Widlok, P. (2008). The major apoptotic endonuclease DFF40/CAD is a deoxyribose-specific and double-strand-specific enzyme. *Apoptosis, 13,* 377-382.

[138] Widlak, P. & Garrard, W. T. (2006). Unique features of the apoptotic endonuclease DFF40/CAD relative to micrococcal nuclease as a structural probe for chromatin. *Biochem Cell Biol, 84,* 405–410.

[139] Wolf, B. B., Schuler, M., Echeverri, F. & Green. D. R. (1999). Caspase-3 is the primary activator of apoptotic DNA fragmentation via DNA fragmentation factor-45/inhibitor of caspase-activated DNase inactivation. *J Biol Chem, 274,* 30651–30656.

[140] Sakahira, H., Enari, M.& Nagata, S. (1998). Cleavage of CAD inhibitor in CAD activation and DNA degradation during apoptosis. *Nature, 391,* 6–99.

[141] Panyim, S. & Chalkley, R. G., (1969). A new histone found only in mammalian tissues with little cell division. *Biochem Biophys Res Commun, 37*, 1042–1049.

[142] Lea, M. A. (1987). Relationship of H1o histone to differentiation and cancer. *Cancer Biochem Biophys, 9,* 199–209.

[143] Lennox, R. W. & Cohen, L. H. (1983). The histone H1 complements of dividing and nondividing cells of the mouse. *J Biol Chem 258*, 262–268.

[144] Wang, Z. F., Sirotkin, A. M. & Buchold, G. M. (1997). The mouse histone H1 genes: gene organization and differential regulation. *J Mol Biol, 271,* 124-138.

[145] Macieira-Coelho, A. (1991). Chromatin reorganization during senescence of proliferating cells. *Mutat Res, 256,* 81-104.

[146] Howard, B. H. (1996). Replicative senescence: considerations relating to the stability of heterochromatin domains. *Exp Gerontol, 31,* 281-293.

[147] Chaly, N., Munro, S. B. & Swallow, M. A. (1996). Remodelling of nuclear periphery during muscle cell differentiation. *J Cell Biochem, 62*, 76-89.

[148] Qumsiyeh, M. B. (1999). Structure and function of the nucleus: anatomy and physiology of chromatin. *Cell Mol Life Sci, 55*, 1129-1140.

[149] Lezhava, T. (2001). Chromosome and ageing: conception of ageing. *Biogenrontology, 2,* 253-260.

[150] Narita, M., Nunez, S., Heard, E., Narita, M., Lin, A.W., Hearn, S. A., Spector, D. L., Hannon, G. J. & Lowe, S. W. (2003). RB-mediated heterochromatin formation and silencing of E2F target genes during cellular senescence. *Cell, 113*, 703–716.

[151] Adams, P. D. (2007). Remodeling chromatin for senescence. *Aging Cell, 6*, 425-427.

[152] Grigoryev, S. A., Bulyniko, Y. A. & Popova, E. Y. (2006). The end adjusts the means: heterochromatin remodelling during terminal cell differentiation. *Chromosome Res, 14,* 53-60.

[153] Rogakou, E. P. & Sekeri-Pataryas, K. E. (1999). Histone variants of H2A and H3 families are regulated during *in vitro* aging in the same manner as during differentiation. *Exp Gerontol, 34,* 741-754.

[154] Grove, G. W. & Zweidler, A. (1984). Regulation of nucleosmal core histone variant levels in differentiating muring erythroleukemia cells. *Biochemistry, 23,* 4436-4443.

[155] Hayflick, L. & Moorhead, P. S. (1961). The serial cultivation of human diploid strains. *Exp Cell Res, 25,* 585-621.

[156] Cristofalo, V. J. & Pignolo, R. J. (1996). Molecular markers of senescence in fibroblast-like cultures. *Exp Gerontol 31,* 111-123.

[157] Pignolo, R. J., Martin, J. H., Horton, J. H., Kalbach, A. N. & Cristofalo, V.J. (1998). The pathway of senescence: WI-38 cells arrest in late G1 are unable to traverse the cell cycle from a true Go state. *Exp Gerontol 33,* 67-80.

[158] Rousseau, D., Khochbin, S., Gorka, C. & Lawrence, J. J. (1992). Induction of H1o-gene expression in B16 murine melanoma cells. *Eur J Biochem 208,* 775-779.

[159] Grunwald, D. Khochbin, S. & Lawrence, J .J. (1991). Cell cycle related accumulation of H1o mRNA: induction in murine erythroleukkemia cells. *Exp Cell Res, 194,* 174-179.

[160] Scaturro, M., Cestelli, A., Castiglia, D., Nastasi, T. & Di Liegro, I. (1995). Posttranscriptional regulation of H1 zero and H3.3B histones in differentiating rat cortical neurons. *Neurochem Res, 20,* 969-976.

[161] Khochbin, S. (2001). Histone H1 diversity: bridging regulatory signals to linker histone function. *Gene, 271*, 1–12.

[162] Happel, N., Doenecke, D., Sekeri-Pataryas, K. E. & Sourlingas, T. G. (2008). H1 histone subtype constitution and phosphorylation state of the ageing cell system of human peripheral blood lymphocytes. *Exp Gerontol, 43,* 184-199.

[163] Lennox, R. W. & Cohen, L. H. (1984). The alterations in H1 histone complement during mouse spermatogenesis and their significance for H1 subtype function. *Dev Biol, 103*, 80–84.

[164] Pina, B., Martinez, R., Simon, L. & Suau, P. (1984). Differential kinetics of histone H1o accumulation in neuronal and glial cells from rat cerebral cortex during postnatal development. *Biochem. Biophys. Res. Commun. 123,* 697–702.

In: Histones: Class, Structure and Function
Editor: Chang-Hui Shen
ISBN: 978-1-62100-274-1

Chapter 2

LINKER HISTONES AND CHROMATIN STRUCTURE

Chang-Hui Shen*

Department of Biology, College of Staten Island, City University of New York, Staten Island, NY, US

Institute for Macromolecular Assemblies, City University of New York, Staten Island, NY, US

PhD Program in Biology/Biochemistry, The Graduate Center, City University of New York, NY, US

ABSTRACT

The manner in which linker histone globular domains associate with the nucleosome is a key issue in understanding the location of the linker histone tails, and thus their functional roles in regulating gene expression. Understanding of the location of the linker histone globular domain within the nucleosome can provide insight into the nucleosome architecture in the context of the location relative to the dyad on a statistically significant nucleosome. It can also indicate any possible association between linker histones and different nucleosomes in the higher order chromatin structure. Therefore, it would contribute to the implication of gene regulation and higher-order structure models. However, studies on the binding of the linker histone globular domain to the nucleosome have not yielded a unified picture. In this chapter, I will discuss the current findings in this field including the linker histone variants, the structure of the linker histone globular domain, the location of the linker histone within a chromatosome, the functional domains of the linker histone tail, the contribution of linker histones to higher order chromatin structure and its connection to gene expression.

* E-mail: ChangHui.Shen@ csi.cuny.edu.

1. INTRODUCTION

In the eukaryotic genome, DNA is packaged inside nuclei by association with histone proteins to form a nucleoprotein complex known as chromatin. This complex consists of roughly a 2:1 mass ratio of protein to DNA and a 1:1 mass ratio of histones to DNA. The fundamental subunit of chromatin is the nucleosome, which is comprised of two molecules of each of the core histones H2A, H2B, H3 and H4 (the histone octamer) around which is wrapped by 1.75 turns of DNA. In higher eukaryotic cells, a linker histone binds to the nucleosome core particle to form the chromatosome and promote the organization of nucleosomes into the 30nm filament [1]. The chromatosome is a particle which is transiently produced during nuclease digestion of chromatin and its appearance is diagnostic for the presence of linker histone. This packaging of the DNA into chromatin, which occurs at a series of different levels, provides the compaction and organization of the DNA required to accommodate processes such as transcription, recombination, replication, and mitosis.

Linker histones constitute a family of structurally related proteins that act to stabilize the nucleosome, have an essential role in organizing nucleosomes into higher order structures, and may have a role as repressors of transcription [2]. Nevertheless, a recent study has demonstrated that linker histones did not preclude ATP-dependent chromatin remodeling activity in reconstituted minichromosomes, suggesting that the role of linker histones may be that of a chromatin organizer rather than a repressor [3].

Although in most eukaryotes, linker histones present a characteristic three-domain organization, there are at least two organisms, *Tetrahymena* and yeast, in which linker histones do not have this classical structure [4-5]. All the metazoans have linker histones with 3 structural domains: a short, basic unstructured N- terminal domain (~35amino acids), a long C-terminal domain (~100 amino acids) and a folded central globular domain of ~80 amino acids [6-8]. The different parts of the linker histones may have different functions in the binding of the linker histone to chromatin. The N-terminal tail may be involved in the positioning of the globular domain on the nucleosome whereas the C-terminal tail is involved in chromatin condensation [9-10]. The central globular domain is the most conserved region of this molecule compared to the tails, and is positioned close to the nucleosome dyad axis, but its exact position is still debated. Reconstitution experiments using either intact H1 or isolated fragments thereof have suggested that the globular domain of the linker histone is necessary and sufficient for the protection of the 168 bp chromatosomal DNA [11]. The globular domain of the linker histone is also responsible for positioning the basic N- and C-terminal tails of the H1 molecule so that they can effectively induce higher order structure [9]. However, the ways in which the tails affect the tight packaging of the nucleosome and linker histone to self-associate via its globular domain is still not known. The structure of the linker histone globular domain and its location within a chromatosome will be discussed later in this chapter.

2. Linker Histone Variants

The number of linker histone variants, as well as their degree of divergence from one another, is much greater than that of the core histones [12]. The most common class of linker histone is H1. There are many H1 variants found within any one species or cell type and they differ in molecular mass, amino acid composition, sequence, and physico-chemical and immunochemical properties [13]. Eleven different H1 subtypes have been identified to date in mammals [14-15]: Seven somatic variants (H1.1-H1.5, H1.0 and H1.X) [16-19], three testis-specific variants (H1t, H1T2 and HILS1) [20-23] and an oocyte-specific variant (H1oo) [24; Table 1]. The content of each variant differs in the timing of expression [25], extent of phosphorylation [26], turnover rate [27-29], chromatin binding affinity [30-31], evolutionary stability [32], and DNA condensing capacity [30, 33-34]. For example, the expression of H1.1 is restricted to thymus, testis, spleen, lymphocytic and neuronal cells [35-36]. Histone H1.0 is found in mammalian tissues showing little cell division with a highly differentiated state. Histone H1.0 has, therefore, been implicated in the establishment and maintenance of the terminally differentiated state [2, 37]. One of the most intensively studied linker histones, histone H5, is structurally related to H1.0. Histone H5 is the avian-specific histone and is expressed only in erythrocytes [38]. It accumulates during the process of terminal differentiation in nucleated erythrocytes [25]. Histone H5 is believed to be specifically involved in chromatin compaction and transcriptional inactivation. Histone H1t is a tissue-specific variant, which only appears in a stage-specific manner during the development of the spermatocytes in vertebrates and invertebrates testis [16, 20]. Its expression was shown to take place essentially in pachytene cells [16].

In addition to cell-type classification, linker histones can also be classified as being either DNA replication-dependent or independent in their expression patterns [14, 25]. Five somatic variants are DNA replication-dependent, including H1.1-H1.5. They are expressed only in S phase. Two other somatic variants, H1.0 and H1.X, and those testis- and oocyte-specific variants are DNA replication-independent. Both H1.0 and H1.X are expressed primarily during G_0 and G_1 phase, respectively. These types of linker histone have also been referred to as replacement histones since they serve to replace the replication-dependent ones during various times in the cell cycle [14, 39]. Since histone replacement is most evident during the process of development, these DNA replication-dependent linker histones are all development-dependent histones. This is because they are expressed at defined periods of embryonic development and cell differentiation. However, it has been shown that both H1.2 and H1.4 mRNA continue to be expressed in non-dividing cells. Therefore, both H1.2 and H1.4 may also be development-independent [40].

3. The Structure of the Linker Histone Globular Domain and Its Location within the Chromatosome

The interaction of the linker histone with the nucleosome core particle is an ongoing debatable issue. In the absence of the linker histone, micrococcal nuclease digestion of chromatin results in 146 bp nucleosome core particle. In the presence of the linker histone an additional 20 bp of DNA is protected from digestion, forming a chromatosome [1, 41]. It has been shown that the central globular domain of linker histone H5 (GH5) is essential for protection of the additional 20 bp of DNA [11]. This is the primary element for recognition of the nucleosome. Furthermore, linker histone H5 may be involved in chromatin condensation [3] and chromatin compaction [42]. Since it is important in transcriptional regulation, many studies have been working to determine the structure of GH5 and the exact location of its binding site on the nucleosome.

Clore *et al.* were the first to use distance geometry and restrained molecular dynamics calculations to determine the NMR structure of GH5 [43]. They found a structural core made up of residues 3-18, 23-34, 37-60, and 71-79 (numbered with respect to the N-terminus of the trypsin-produced globular domain), and two loops composed of residues 19-22 and 61-70. The structure of this core is dominated by three helices (helix I, II, and III). The structure appears to be stabilized by hydrophobic interactions involving Tyr32, Tyr37 and Phe72 as well as long-chain aliphatic residues. This structure was observed to have a similar fold to the C-terminal DNA-binding domain of the cAMP receptor protein. Because of the number of lysine residues protected from selective radio-labeling during reductive methylation of histone H5 bound to the nucleosome [44], Clore *et al.* predicted that the polar residues at the N-terminal end of helix III and helix II may be involved in protein-DNA contacts. Overall, this structure was in agreement with a model in which GH5 sits within a cage of three double helical DNA strands at the entry and exit points of the chromatosome. In this model, the globular domain is proposed to interact with the central DNA strand and the inner surface of the two outer strands.

Another GH5 structural study used multi-wavelength anomalous diffraction of crystals of the selenomethionylated protein [45]. The three-helix topology and the assignment of the residues that make up the helices from this crystal study are in agreement with the NMR study. On the basis of the homology between GH5 and catabolite gene activator protein (CAP), the highly conserved residues Lys69 and Arg73 are positioned to interact with one strand of DNA, and Lys85 is positioned to interact with the other strand and these interactions form the primary binding site [45]. Lys40, Arg42, Lys52, and Arg94 make up a putative secondary binding site, and His25 and His62 are likely to be other subsidiary DNA binding sites. Another NMR structure of the globular domain of linker histone H1 (GH1) also revealed that it is structurally homologous to CAP [46]. By using site-directed mutagenesis, Buckle *et al.* proved a critical role for Lys85 in determining the effective interaction between GH5 and the nucleosome [47]. Lys52 and Lys69 were also thought to play a major part in this interaction as these residues are protected from chemical modification [44]. Biochemical analysis has confirmed that at least two of the DNA binding sites on the GH5 molecule are required to position the globular domain correctly on the nucleosome and to provide chromatosomal stabilization [48-49]. Recently, a computational docking tool was used to investigate the interaction of GH5 and linear B-DNA [42]. The result also agreed with the NMR structure, crystal structure and biochemical analysis in that there are three GH5 DNA-binding sites.

On the basis of nuclease digestion experiments, the globular domain of the linker histone alone was found to protect linker DNA immediately adjacent to the core structure. This protection was thought to be symmetrical. Allan *et al.* proposed that GH1 binds to the nucleosome at the point where the DNA enters and exits the nucleosome, and that the location of GH1 is orientated away from the nucleosomal surface at the dyad axis [11]. As shown in Figure 1A, the globular domain of the linker histone binds to the outside of the superhelix of

DNA at the dyad of the nucleosome and makes three, symmetrical interactions with the nucleosomal DNA which protects 10 bp of linker DNA on either side of the nucleosome core. Many experiments have demonstrated support for this proposal. For example, DNase I footprinting of linker histone-containing dinucleosome complexes showed that the globular domain of the linker histone binds directly to the dyad axis of nucleosome [50]. Furthermore, a site-directed mutagenesis study supports this symmetrical model [49]. Both NMR structure calculations and crystal structures also suggest that the globular domain of the linker histone binds symmetrically to the dyad axis of the nucleosome, interacting with DNA at the entry and exit points of the nucleosome [43, 45].

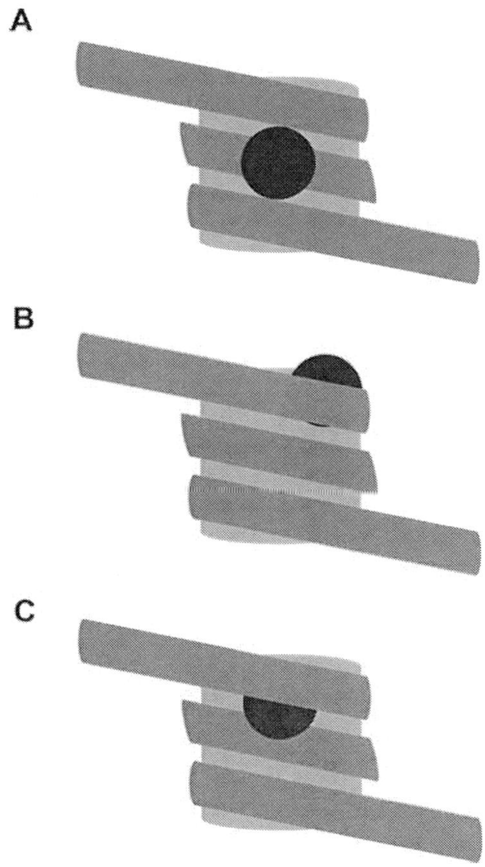

Figure 1. Models for the locations of the linker histone globular domain on the nucleosome. A. Symmetrical model from Allan *et al.* [11]; B. Asymmetrical model from Hayes & Wolffe [51] and Pruss *et al.* [52]; C. Boundary asymmetrical model from Travers & Muyldermans [55]. The filled black circle is the linker histone.

An alternative hypothesis based on studies of nucleosomes incorporating specific DNA sequences suggests that a single molecule of linker histone interacts asymmetrically with DNA in the nucleosome [51]. From both nuclease protection and histone-DNA crosslinking methodologies, Hayes *et al.* provided evidence for an asymmetric interaction of the linker histone with *Xenopus borealis* 5S DNA in a positioned nucleosome. Their results indicated an asymmetry of protection of DNA in the 5S nucleosome with 5 bp and 15 bp being protected at either side of the nucleosome core by the linker histone globular domain [Figure 1B]. By measuring the crosslinking of the linker histone to nucleosomal DNA, Pruss *et al.* proposed that GH5 binds to the major groove of the DNA at an open site ~65 bp from the dyad [52]. At this site, GH5 faces in towards the core histones and is in contact with histones H2A and H3. A site-directed chemical mapping method has been employed which suggests that the location of the linker histone binding site is located on the inside of the superhelical gyre of DNA, near the periphery of the nucleosome complex. This means that a single globular domain interacts asymmetrically with DNA in the nucleosome [53]. Furthermore, Eisfeld *et al.* found that the binding of NF1 to linker DNA in the MMTV promoter was unaffected by incorporation of histone H1 into the MMTV nucleosome particle, thus supporting this proposal [54].

In addition to the above model, Travers & Muyldermans have proposed a third possible interaction between the linker histone globular domain and nucleosomal DNA [55]. Their model is based on studies of the sequences of chromatosome DNAs isolated from chicken erythrocytes. They noted that the elements AGGA and AAA/TTT were frequently located at one of the chromatosomal DNA termini and midpoint, respectively. On the assumption that this specificity of sequence location was brought about by linker histones they concluded that the linker histone globular domain contacts chromatosomal DNA termini asymmetrically [Figure 1C]. Neutron scattering data also suggested that GH5 might bind asymmetrically, forming a bridge between one terminus of chromatosomal DNA and a site close to the dyad [56]. In a protein-DNA crosslinking and DNA cleavage approach, Ser71 was mapped to one terminus of the chromatosomal DNA and Ser41 to a site close to the midpoint [57], which also supports this model.

On the basis of structural studies and biochemical studies, much uncertainty remains as to the location and binding of the linker histone globular domain within the chromatosome. However, a recent computational study has been used to examine the interaction of GH5 with the nucleosome [42]. The predictions of the interactions were based on the sum of van der Waals and electrostatic intermolecular energies. It showed that three binding models could be observed experimentally, depending on certain conditions and experimental methods. Nevertheless, a more defined experimental approach such as the monomer extension method should be useful in resolving such conflicts. This is because the monomer extension method can accurately determine nucleosome positions without ambiguity [58, 59], and thus it can reveal the location and binding of the linker histone globular domain within the chromatosome [Figure 2]. Briefly, isolated core DNA and chromatosomal DNA can be extracted and purified from a gel, and end-labeled with T4 polynucleotide kinase. Labeled core particle DNA and chromatosomal DNA can be used as a primer for extension by the Klenow fragment with single-stranded plasmid DNA as template. The replicated DNA is then digested with different restriction enzymes to resolve different regions of the target DNA sequence. This produces a population of labeled DNA fragments which all share a common 3′ boundary (restriction site) but are distinct in having unique 5′ ends which reflect the location of specific core particle or chromatosome boundaries. The lengths of the resulting DNA fragments can then be determined accurately in sequencing gels: each band defines the distance from the border of a nucleosome or a chromatosome to the chosen restriction site. The borders of each nucleosome or chromatosome can now be located precisely, with one using the positive strand as template and the other using the negative strand. The data from both template strands define all of the "upstream" and "downstream" nucleosome borders

Table 1. A list of known linker histone variants

Organism	*H. sapiens*		*X. laevis*	*G. gallus*	Replication dependency	Distinctive characteristics	Chromatin compacting capacity
Cell Type	Numeric	Alphabetic					
	H1.1	H1a			yes	Found in thymus, testis, spleen, lymphocytic and neuronal cells	weak compacting
	H1.2	H1c		H1.11L	yes		weak compacting
	H1.3	H1d	H1d	H1.01	yes		intermediate compacting
Somatic	H1.4	H1e		11RH1; H1.02*	yes		highly compacting
	H1.5	H1b			yes		highly compacting
	H1.0	$H1^0$	H1.0-A; H1.0-B	H5	no	Restricted to terminally differentiated cells	highly compacting
	H1.X	H1x	H1x	H1x*	no		highly compacting
Testis-specific		H1t	H1t				
		H1t2					
		HILS1					
Oocyte-specific		H1oo(H1Foo)	B4	H1Foo*			

* Predicted protein.

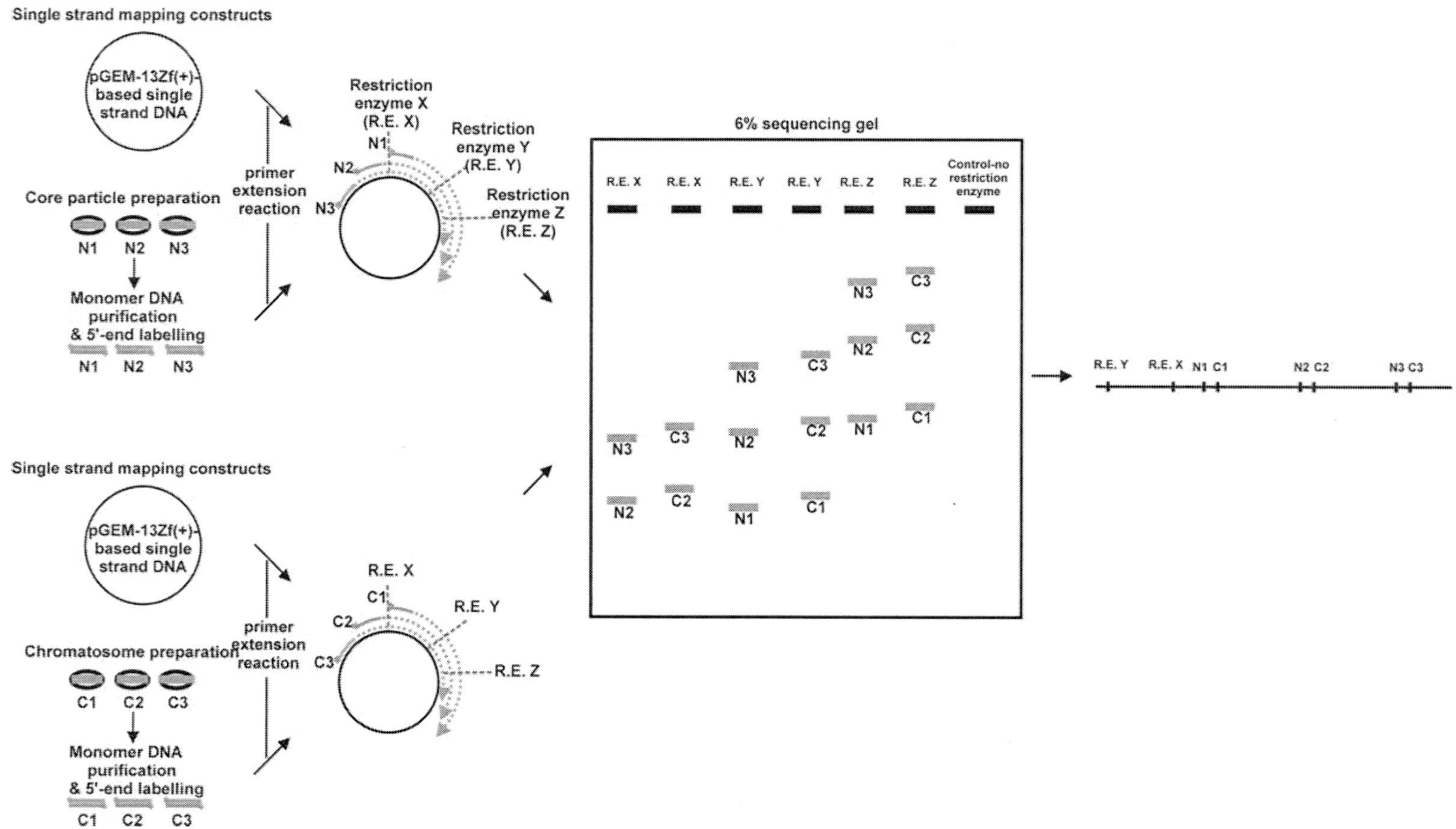

Figure 2. Schematic outline of monomer extension analysis. In the 6% sequencing gel, each of the band reflects the distance between farthest boundary of individual nucleosome and the specific restriction site. The length of the fragments can be referred back to its positioning sites on the DNA fragment. N1: core particle 1; C1: chromatosome 1; N2: core particle 2; C2: chromatosome 2; N3: core particle 3; C3: chromatosome 3.

unambiguously. In this way, the positions at which the histone octamers or chromatosomes are positioned on the target DNA can be determined at high resolution. The distance of the positions between the chromatosome and histone octamer from the same nucleosome will determine how the linker histone globular domain protects the extra 20 bp DNA. Therefore, this technique can help to clarify the conflict in this regard.

4. THE FUNCTIONAL SUBDOMAINS WITHIN THE LINKER HISTONE TAILS

Linker histone tails are generally thought to bind to linker DNA, stabilize chromatin structure, and are important for chromatin condensation [9, 60]. They are the least conserved parts of the molecule and might be largely unstructured and highly extended in solution, probably due to the electrostatic repulsion between the closely-spaced, positively-charged residues [61]. The linker histone tails might also behave as an unstructured polycationic domain that binds linker DNA and neutralizes the negative charge. This thereby permits the close approach of neighboring nucleosomes and the stabilization of condensed chromatin structure. However, trypsin digestion of sea urchin sperm H1 in chromatin has suggested that a proline-free segment of the C-terminal tail immediately adjacent to the globular domain is in an α-helical conformation and this may be important for stabilization of the tail interaction with linker DNA [13]. Other recent studies have also shown that the peptides derived from the C-terminal tail adopt a detectable α-helical structure when interacting with DNA [62-63]. Therefore, the C-terminal tail has some structural features and thus is not completely disordered.

Allan *et al.* proposed that the ability to stabilize extensively folded secondary chromatin structures lies exclusively in the C-terminal tail [9]. By measuring the fluorescence recovery after photobleaching profiles, it has been shown that the deletion of the N-terminal tail only modestly reduced binding affinity *in vivo*. Deletion of the C-terminal tail almost completely eliminates histone H1.1 binding [64]. Therefore, the C-terminal tail of histone H1 is responsible for the high affinity binding of histone H1 to chromatin.

Based on the structural analysis of the N-terminal tail of sea urchin sperm histone H1, Suzuki identified that repeats of Ser-Pro-Lys(Arg)-Lys(Arg) (SPKK) residues are responsible for the binding of the tail to the minor groove of linker DNA. The SPKK motif was thought to be able to undergo a dynamic conversion between a β-turn structure and a ρ-turn structure [65-66]. This suggests that the β-turn structure could be stabilized by hydrogen bonding between the hydroxyl group and carboxyl group of Ser and the amino group of Lys. These peptide repeats are also found in the C-terminal tails of H1. It has been shown that the octapeptide KSPKKAKK (8-mer) is present in many histone H1 subtypes, and the imperfect repeat ATPKKSTKKTPKKAKK (16-mer TPKK) occurs in the C-terminus of rat histone H1d [67]. These studies also suggest that the β-turn structure of the S/TPKK motif is important for binding to the DNA and that the pairing of S/TPKK motifs is a required feature for bringing about DNA and chromatin condensation.

The binding of Hoechst 33258 to DNA is inhibited by SPKK repeat motifs, suggesting that the binding mode of the SPKK motifs to AT rich DNA may resemble the minor groove binding characterized for Hoechst 33258 [68]. Using conditions of high co-operativity, histone H1 binds preferentially to scaffold-associated regions (SAR)-containing DNA fragments *in vitro* [69-70]. Furthermore, Khadake & Rao have shown that the motif ATPKKSTKKTPKKAKK (16-mer TPKK), which occurs in the C-terminus of rat histone H1d, can condense SAR DNA fragments effectively [67]. Therefore, within a polynucleosomal fiber, the SAR sequences present in the internucleosomal linker DNA provide high affinity binding sites for histone H1, which can bring about compaction presumably through utilizing the SPKK repeat motifs present in the C-terminus of histone H1. However, it has also been shown that histone H1e from murine erythroleukemic cells and an H1e peptide (25-mer) containing the highly conserved sequence S/TPKKAKKP, binds preferentially to GC-rich regions of DNA [71]. In light of this contradictory evidence, the binding preference of S/TPKK motifs to DNA still remains to be clarified.

The sequence motif SPKK is generally found in the C-terminus of H1, while the sequence motif SPXX, where X can be Ser, Thr, or Ala but rarely a basic residue, is found in gene regulatory proteins [72]. Examples include Gal 4 [73] and the Trp repressor [74]. To regulate gene expression, the DNA-binding constant of the SPKK motif must be much higher than that of the SPXX motif, which lacks the basic residues that would increase its DNA-binding ability. Therefore, Suzuki *et al.* proposed that while linker DNA is protected by SPKK-rich H1 tails, transcription factors can only slide along DNA rather than bind the same DNA sequences with their SPXX motifs [75]. However, if the DNA binding ability of H1 SPKK motifs is weakened by phosphorylation of Ser residues, SPXX motifs of transcription factors could bind to linker DNA and thereby enter the nucleosome core to find the specific DNA sequence that will satisfy its specific DNA-binding region. This hypothesis has been indirectly supported by studies on the differential DNA-binding affinity of phosphorylated and non-phosphorylated sea urchin testis-specific H1 and H2B [75-76].

The nature of the evenly distributed, high density of positively charged amino acids within the C-terminal tail has led to the agreement that chromatin condensation is mediated through charge-neutralization of the negatively charged linker DNA. However, a recent study has shown that the ability of histone H1.0 to stabilize chromatin folding is not evenly distributed; rather it is localized to two specific subdomains in the C-terminal tail [77]. These two subdomains are localized into two discontinuous 24-/25-residue C-terminal tail regions (H1.0 residues 98-122 and 147-170) which contain S/TPKK motifs. In this intrinsic disorder model, the C-terminal tail binds to linker DNA through a charge neutralization-based mechanism, but once bound, the C-terminal tail become structured and those two subdomains subsequently mediate alteration of linker DNA structure and chromatin condensation [77]. It also has been shown that S/TPKK motif in each subdomain is needed to stabilize extensively folded secondary chromatin structures. Furthermore, using fluorescence recovery after photobleaching (FRAP), it has been confirmed that these two S/TPKK motifs are important in chromatin condensation [64].

5. Linker Histones and the 30 nm Chromatin Structure

Studies of the arrangement of nucleosomes in the higher order chromatin fiber have been facilitated by virtue of the reversible salt-induced transition between the unfolded beaded chain of nucleosomes and the condensed 30 nm fiber [78-79]. Electron micrographs of chromatin undergoing this transition often show a fairly regular zigzag pattern of chromatosome - linker DNA - chromatosome [79-80] but as compaction proceeds, the individual nucleosomes are no longer resolved and their arrangement in the 30 nm fiber becomes less accessible to direct observation. Therefore, although the structure of the nucleosome core particle has been determined to high resolution [81], the next level, the 30 nm chromatin structure still lacks direct structural information and remains controversial. The linker histone, on binding to the nucleosome core particle, not only constrains an additional 20 bp of DNA, but also determines the trajectory of the DNA entering and exiting the nucleosome. In turn, the linker histone will direct the relative positioning of successive nucleosomes. Therefore, the linker histones have a pivotal role in forming higher order chromatin structure. There have been many different models of higher order chromatin structure proposed. Most models share a common feature in that an open zigzag string of nucleosomes self-assembles into a helical structure with a 30 nm diameter. Two major types of 30 nm chromatin structure are widely accepted and have been discussed recently, including the one-start helix solenoid and two-start cross-linker models [82].

Based on the assumption that nucleosomes are 11 nm diameter spherical particles, Finch & Klug postulated that a nucleosome chain can fold into a solenoid-like shape which is a single-start contact helix with a pitch of 11 nm and a diameter of about 30 nm, containing from 4 to 10 nucleosomes per turn, with a mode of about 6 or 7 [83]. However, the observation of a zigzag fiber at low ionic strength makes simple helical coiling unlikely, so Thoma *et al.* proposed an alternative hypothesis concerning the condensation of chromatin fibers which interpreted the zigzag fiber as a contact helix with two nucleosomes per turn stabilized by H1-H1 interactions in the center [79]. Condensation of the zigzag fiber into a compact fiber was postulated to occur by twisting this helix around its axis, increasing the number of nucleosomes per turn to 6 to 8, decreasing the angle between nucleosome disks while keeping the pitch constant at about 11 nm and maintaining the radial orientation of the nucleosomes. In the formation of higher order structures, H1-binding regions on neighboring nucleosomes come closer together so that an H1 polymer may be formed in the center of the superhelical structure. In a neutron scattering study, the result showed that the H1 is located in the interior of the filament [84]. Furthermore, solenoidal models can accommodate variable lengths of linker DNA, which may be internally located provided the linker is allowed to form reversed loops, kinks or sharp bends [85]. Recent structural analysis using an *in vitro* reconstituted model nucleosome arrays based on the strong 601 nucleosome positioning DNA sequence have led to the proposal of the one-start helix model [86-87]. This model was derived from tight constraints obtained from measurements of the physical dimensions of long nucleosome arrays visualized by an electron microscopical study [86]. In this model, consecutive nucleosomes follow each other, and the linker DNA between the nucleosomes is oriented towards the center of the solenoid fiber. Different nucleosome lengths permit an unambiguous definition of fiber orientation and dimensions. This is a left-handed one-start helix with 5.4 nucleosomes per helical turn, in which nucleosomes from successive gyres which interdigitate to give rise to the high packing ratio. The interdigitation of nucleosomes leads to regular nucleosome face-to-face contacts [86, 88]. The linker DNA is bent and accommodated in the interior of the fiber. The packing of nucleosomes in this model is consistent with the X-ray diffraction pattern of aligned native 30 nm fibers [89] and with nucleosome repeat length analysis [90].

The feature of the cross-linker models is a non-sequential arrangement of nucleosome cores connected by linker DNA, which transverses the fiber axis. Furthermore, these models can be characterized by the number of helical ramps in each helical repeat. Based upon the tendency of nucleases to produce dinucleosomes rather than mononucleosomes, Staynov proposed two cross-linker models that form single start left-handed helices [91]. Makarov *et al.* proposed a triple helix according to their flow dichroism measurements [92]. Their model suggested that nucleosomal disc faces are tilted relative to the fiber axis, the orientation of nucleosomes does not change upon folding and unfolding of chromatin, and that the orientation of nucleosomes is maintained by the globular domain of histone H1. Williams *et al.* collected electron microscopy images and x-ray scattering data from nuclei and isolated chromatin fibers of seven different tissues, and suggested a linker length dependence of fiber diameter and mass per unit length [93]. A structure with left-handed helical symmetry, a pitch of about 26 nm and a pitch angle of about 32° was calculated from three optical diffraction patterns of electron micrographs of negatively stained fibers. They proposed that the zigzag fibers form a two-start contact helix with the zig nucleosomes starting on one side and forming a ramp, and the zag nucleosomes starting on the opposite side to form the second ramp. The linker DNA would cross the center of the fiber. Recently, the crystal structure of an oligonucleosome (a compact tetranucleosome) was solved at 9 Å resolution [94]. It showed that linker DNA zigzags back and forth between two stacks of nucleosome cores, which form a truncated two-start helix, and does not follow a path compatible with a one-start solenoidal helix. The length of linker DNA is most probably buffered by stretching of the DNA contained in the nucleosome cores. In this zigzag arrangement of nucleosomes, the linker DNA connects back and forth two rows of nucleosomes. A crosslinking analysis has shown that neither nucleosome repeat lengths nor the presence or absence of histone H1 would influence the development of two stacks of nucleosomes which is in support of the two-start zigzag model [81].

The major difference between these two 30 nm chromatin fiber models is that the one-start model was derived from nucleosome arrays saturated with linker histones, whereas the two-start model derived from a tetra-nucleosome array in the absence of a linker histone. Using AFM, it has been shown that linker histone H1 is required for stabilizing higher order chromatin folding, but the extent to which it does so depends on the linker histone subtype [3; Table 1]. Therefore, the controversy may be resolved by considering different levels of chromatin compaction, such as a very compact structure that is only formed in the presence of the linker histone versus a less condensed higher order chromatin structure that may occur in the absence of histone H1 [15, 82].

The linker histone binds to the nucleosome and is essential for the organization of nucleosomes into the 30 nm filament of chromatin, but how the linker histone influences the folded state by virtue of their location in the higher order chromatin structure is still not defined. From solid-phase immunoassay and inhibition experiments, results suggest that the GH5 is internally located in the 30 nm chromatin fiber [95]. Similarly, investigations by radioimmunoassay and electron microscopy indicated an inaccessible location for GH5 in the higher order chromatin fiber [96]. However, other results suggest that the antigenic determinants of GH5 are accessible to the antibody both in folded and unfolded chromatin, while those of the same region of H1 are masked in both states [97]. The use of proteolytic enzymes immobilized on membranes revealed that both GH1 and GH5 were inaccessible in folded chromatin fibers [98]. By altering the H_2O/D_2O ratio and deuterating H1, neutron scattering analysis was used to examine the H1 radial location in the filament [84]. The result showed that chromatin 30 nm filaments have ~6 nucleosomes per 11 nm, with histone H1 at about the same radial location as the inner face of the nucleosome. This result together with the fact that H1 interacts with the nucleosome at the entry and exit points of nucleosomal DNA [11], is consistent with the one-start helix model that H1 binds to the face of the nucleosome that is presented to the interior of the filament. Besides, the capacity of the GH5 to self-associate in solution may also imply that linker histone globular domain occupies an axial position within the higher order chromatin fiber. The spatial juxtaposition of the GH5 domains at this location would be expected to promote their association and exert a stabilizing effect upon higher order chromatin structure [99].

6. Linker Histones and Gene Expression

It has been shown that most, if not all eukaryotic promoters and regulatory elements are organized into precise architectures within extensive nucleoprotein complexes. To activate transcription, the transcription factors have to compete with histones for DNA sites, which may be blocked or obscured by nucleosomes on the promoter [100-104]. The wrapping of nucleosomal DNA in 1.75 superhelical turns over the surface of the histone octamer may be sterically incompatible with the assembly of a stable transcription initiation complex. Therefore, the issue about how nucleosome structure is destabilized in order to facilitate access to sequence-specific transcription factors and the general transcriptional machinery has been a focus of vigorous research in many years.

Nucleosomes can be directed to precise positions on DNA by signals in the sequence [105-106], a phenomenon termed nucleosome positioning. Positioned nucleosomes can regulate the exposure of *cis*-acting DNA elements, preventing their interaction with *trans*-acting transcription factors and thereby inhibiting transcription. Nucleosome positioning signals may also play an important role in gene regulation by virtue of their proposed ability to direct the folding of a gene into an inactive higher order chromatin structure [107]. It follows that factors, which modulate nucleosome positioning or the fundamental properties of chromatin will also have the potential to regulate gene activity. Similarly, the binding of linker histones to nucleosomes promoting the formation of higher order chromatin structure will have the potential to influence transcriptional activation. However, how linker histones modulate gene activity in the context of chromatin structure is still a debatable issue. This is at least partly due to the lack of the information concerning the location and binding mode of linker histones in chromatin.

A role for linker histones as repressive components of positioned nucleosomes has been demonstrated by a series of biochemical experiments using the mouse mammary tumor virus (MMTV) long terminal repeat (LTR) and the *Xenopus* 5S rRNA genes [108-112]. These investigations indicated that removal of histone H1 might be sufficient to allow the transcriptional machinery to function effectively in chromatin. It has been shown that the mobility of histone octamers positioned on constructs of 5S rRNA genes is suppressed by the binding of histone H1/H5 [113]. Mobile histone octamers on the 5S rRNA genes are accessible to the pol III transcriptional machinery, but in the presence of linker histone H1 this accessibility is severely restricted and transcription is repressed [114-115]. The removal of linker histone H1 from the chromatin of the MMTV LTR, when the glucocorticoid receptor initiates the process of chromatin rearrangement, might increase the local mobility of histone octamers with respect to DNA sequences facilitating transcription factor access to regulatory elements [116]. Therefore, the reversible restraint on chromatin dynamics is important in local regulation of processes that require access to the DNA, and such reversible restraint depends on the presence of linker histones [113].

It has been shown that the structures of GH5 and hepatocyte nuclear factor (HNF)-3γ, which is a mammalian tissue-specific transcription factor, are similar [45, 48]. This might suggest that the linker histone might compete with the transcription factor for the access to the *cis*-acting sequence. Furthermore, both linker histone H1 and poly (ADP-ribose) polymerase I bind to the nucleosome at overlapping sites in a competitive and mutually exclusive manner [117-118]. They both are nucleosome binding proteins, which could modulate chromatin architecture and transcription [119-120]. A recent study showed that linker histone H1 and poly (ADP-ribose) polymerase I exhibit a reciprocal pattern of chromatin binding at many RNA polymerase II-transcribed promoters [121]. Therefore, the repressive role of linker histones could result from the structural similarity between linker histones and transcription factors or a similar binding preference between linker histones and nucleosome-binding proteins.

Since the linker histone binds to DNA much less tightly than the core histones, it is the easiest component of chromatin to be removed. The removal of histone H1 from chromatin might involve active displacement by molecular machines such as RNA polymerase [122-123] or the SWI/SNF complex [122, 124]. Alternatively, transcription factors might simply compete for their recognition sites in the nucleosome and displace H1 by virtue of their greater binding constant to DNA. Indeed, the binding of histone H1 to the nucleosome is very dynamic and can be redistributed [125-126]. It has been shown that H1 is a highly mobile chromatin component in comparison to the core histones and the interaction of an H1 molecule with a nucleosome is transient through FRAP experiments [127-130]. Therefore, the displacement of histone H1 by transcription factors is possible. In addition, this competition might also be influenced by the post-translational modification of histone H1 or by the presence of molecular chaperones. Many studies have shown that phosphorylation of linker histones weakens the interaction of linker histones with DNA *in vitro* [131] and with chromatin *in vivo* [132], and that phosphorylated linker histones are preferentially associated with transcriptionally competent chromatin [133]. As for the molecular chaperone, nucleoplasmin could direct the selective dissociation of linker histones from the somatic chromatin [134] and therefore activate transcription [135].

Conclusions

The manner in which the linker histone globular domain associate with the nucleosome is a key issue in understanding the location of the linker histone tails, and thus their functional roles in regulating gene expression [136]. However, studies on the binding of the linker histone globular domain to the nucleosome have not yielded a unified picture [11, 51, 55]. The main difference between these studies concerns the precise location of the globular domain with respect to the nucleosome dyad axis. Therefore, a more defined experimental approach as I mentioned earlier in this chapter can help in resolving such a conflict.

Recent studies have challenged the idea that H1 is essential for higher order structure of eukaryotic chromatin. The protozoan *Tetrahymena thermophila* contains two types of nuclei, a large transcriptionally active macronucleus and a small transcriptionally inert micronucleus. When the gene encoding either the macronuclear-specific or micronuclear-specific H1 variant is deleted, no effect is observed on viability or growth rate, and the transcription of most genes does not change [137-138]. Although the nucleus lacking H1 does increase in size, it appears that H1 plays a non-essential role in chromatin compaction and only controls the expression of a subset of genes [137]. Furthermore, in cell-free extracts of *Xenopus* eggs, linker histones were found to be nonessential for the assembly of nuclei, which were deemed normal by cytological criteria and their ability to replicate DNA and to import proteins into the nucleus [139]. It remains to be determined whether loss of linker histones has more subtle rather than global effects on transcriptional regulation. One solution to these conflicts is to examine the location of the linker histone globular domain within the nucleosome, which can provide further insight into the nucleosome architecture both in the context of the location relative to the dyad on a statistically significant nucleosome and possible association with different nucleosomes. Therefore, this would contribute greatly to the implication of gene regulation and higher-order structure models.

Acknowledgments

I thank Mohammed Hassan Bhuiyan and Michelle Esposito for critical reading of this chapter. This work was supported by an NSF Grant (MCB-0919218) and PSC-CUNY awards (62180-0040 and 64243-0042).

References

[1] Simpson, R.T. (1978). Structure of the chromatosome, a chromatin particle containing 160 base pairs of DNA and all the histones. *Biochemistry, 17*, 5524-5531.

[2] van Holde, K.E. (1989). *Chromatin*. New York, New York: Springer-Verlag.

[3] Clausell, J., Happel, N., Hale, T.K., Doenecke, D. & Beato, M. (2009). *PLoS One*, *4*, e0007243.

[4] Wu, M.., Allis, C.D., Richman, R., Cook, R.G. & Gorovsky, M.A. (1986). An intervening sequence in an unusual histone H1 gene of *Tetrahymena thermophila*. *Proc. Natl. Acad. Sci. USA*, *83*, 8674-8678.

[5] Landsman, D. (1996). Histone H1 in *Saccharomyces cerevisiae*: a double mystery solved? *Trends Biochem. Sci*, *21*, 287-288.

[6] Aviles, J., Chapman, G.E., Kneale, G.G., Crane-Robinson, C. & Bradbury, E.M. (1978). The conformation of histone H5. *Eur J Biochem, 88*, 363-371.

[7] Hartman, P.G., Chapman, G.E., Moss, T. & Bradbury, E.M. (1977). Studies of the role and mode of operation of the very lysine rich histone H1 in eukaryote chromatin. *Eur J Biochem, 77*, 45-51.

[8] Ramakrishnan, V. (1997). Histone structure and the organization of the nucleosome. *Annu. Rev. Biophys. Biomol. Struct*, *26*, 83–112

[9] Allan, J., Mitchell, T., Harborne, N., Bohm, L. & Crane-Robinson, C. (1986). Roles of H1 domains in determining higher-order chromatin structure and H1 location. *J. Mol. Biol, 187*, 591-601.

[10] Belikov, S. & Karpov, V. (1998). Linker histones: paradigm lost but questions remain. *FEBS Lett*, *441*, 161–164.

[11] Allan, J., Hartman, P.G., Crane-Robinson, C. & Aviles, F.X. (1980). The structure of histone H1 and its location in chromatin. *Nature, 288*, 675-679.

[12] Brown, D.T. (2001). Histone variants: are they functionally heterogeneous? *Genome Biol, 2*, 0006.1–0006.6.

[13] Cole, R.D. (1984). A minireview of microheterogeneity in H1 histone and its possible significance. *Anal Biochem, 136*, 24-30.

[14] Ausio, J. (2006). Histone variants–the structure behind the function. *Brief. Funct. Genomic Proteomic*, *5*, 228–243.

[15] Happel, N. & Doenecke, D. (2009). Histone H1 and its isoforms: contribution to chromatin structure and function. *Gene*, *431*, 1–12.

[16] Doenecke, D., Albig, W., Bode, C., Drabent, B., Franke, K., Gavenis, K. & Witt, O. (1997). Histones: genetic diversity and tissue-specific gene expression. *Histochem. Cell Biol*, *107*, 1–10.

[17] Happel, N., Schulze, E. & Doenecke, D. (2005). Characterisation of human histone H1x. *Biol. Chem*, *386*, 541–551.

[18] Albig, W., Meergans, T. & Doenecke, D. (1997). Characterization of the H1.5 gene completes the set of human H1 subtype genes. *Gene*, *184*, 141–148.

[19] Parseghian, M.H., Henschen, A.H., Krieglstein, K.G. & Hamkalo, B. A. (1994). A proposal for a coherent mammalian histone H1 nomenclature correlated with amino acid sequences. *Protein Sci*, *3*, 575–587.

[20] Seyedin, S.M. & Kistler, W.S. (1980). Isolation and characterization of rat testis H1t. An H1 histone variant associated with spermatogenesis. *J. Biol. Chem*, *255*, 5949–5954.

[21] Martianov, I., Brancorsini, S., Catena, R., Gansmuller, A., Kotaja, N., Parvinen, M., Sassone-Corsi, P. & Davidson, I. (2005). Polar nuclear localization of H1T2, a histone H1 variant, required for spermatid elongation and DNA condensation during spermiogenesis. *Proc. Natl. Acad. Sci. USA*, *102*, 2808–2813.

[22] Yan, W., Ma, L., Burns, K.H. & Matzuk, M.M. (2003). HILS1 is a spermatidspecific linker histone H1-like protein implicated in chromatin remodeling during mammalian spermiogenesis. *Proc. Natl. Acad. Sci. USA*, *100*, 10546–10551.

[23] Iguchi, N., Tanaka, H., Yomogida, K. & Nishimune, Y. (2003). Isolation and characterization of a novel cDNA encoding a DNA binding protein (Hils1) specifically expressed in testicular haploid germ cells. *Int. J. Androl*, *26*, 354–365.

[24] Tanaka, M., Kihara, M., Meczekalski, B., King, G.J. & Adashi, E.Y. (2003). H1oo: a preembryonic H1 linker histone in search of a function. *Mol. Cell Endocrinol*, *202*, 5–9.
[25] Khochbin, S. (2001). Histone H1 diversity: bridging regulatory signals to linker histone function. *Gene*, *271*, 1–12.
[26] Lennox, R.W., Oshima. R.G. & Cohen, L.H. (1982). The H1 histones and their interphase phosphorylated states in differentiated and undifferentiated cell lines derived from murine teratocarcinomas. *J. Biol. Chem*, *257*, 5183–5189.
[27] Hall, J.M. & Cole, R.D. (1985). Modulation in proportions of H1 histone subfractions by differential changes in synthesis and turnover during butyrate treatment of neuroblastoma cells. *Biochemistry*, *24*, 7765–7771.
[28] Dominguez, V., Pina, B. & Suau, P. (1992). Histone H1 subtype synthesis in neurons and neuroblasts. *Development*, *115*, 181–185.
[29] Th'ng, J.P., Sung, R., Ye, M. & Hendzel, M.J. (2005). H1 family histones in the nucleus. Control of binding and localization by the C-terminal domain. *J. Biol. Chem*, *280*, 27809–27814.
[30] Talasz, H., Sapojnikova, N., Helliger, W., Lindner, H. & Puschendorf, B. (1998). In vitro binding of H1 histone subtypes to nucleosomal organized mouse mammary tumor virus long terminal repeat promotor. *J. Biol. Chem*, *273*, 32236–32243.
[31] Orrego, M., Ponte, I., Roque, A., Buschati, N., Mora, X. & Suau, P. (2007). Differential affinity of mammalian histone H1 somatic subtypes for DNA and chromatin. *BMC Biol*, *5*, 22.
[32] Ponte, I., Vidal-Taboada, J.M. & Suau, P. (1998). Evolution of the vertebrate H1 histone class: evidence for the functional differentiation of the subtypes. *Mol. Biol. Evol. 15*, 702–708.
[33] Liao, L.W. & Cole, R.D. (1981). Condensation of dinucleosomes by individual subfractions of H1 histone. *J. Biol. Chem*, *256*, 10124–10128.
[34] Khadake, J.R. & Rao, M.R. (1995). DNA- and chromatin-condensing properties of rat testes H1a and H1t compared to those of rat liver H1bdec; H1t is a poor condenser of chromatin. *Biochemistry*, *34*, 15792–15801.
[35] Franke, K., Drabent, B. & Doenecke, D. (1998). Expression of murine H1 histone genes during postnatal development. *Biochim. Biophys. Acta*, *1398*, 232–242.
[36] Lennox, R.W. & Cohen, L.H. (1983). The histone H1 complements of dividing and nondividing cells of the mouse. *J. Biol. Chem*, *258*, 262–268.
[37] Zlatanova, J. & Doenecke, D. (1994). Histone H1 degrees - a major player in cell-differentiation. *Faseb J, 8*, 1260-1268.
[38] Khochbin, S. & Wolffe, A.P. (1994). Developmentally regulated expression of linker-histone variants in vertebrates. *Eur. J. Biochem*, *225*, 501-510.
[39] Godde, J.S. & Ura, K. (2009). Dynamic alterations of linker histone variants during development. *Int. J. Dev. Biol*, *53*, 215-224.
[40] Wang, Z.F., Sirotkin, A.M., Buchold, G.M., Skoultchi, A.I. & Marzluff, W.F. (1997). The mouse histone H1 genes: gene organization and differential regulation. *J. Mol. Biol*, *271*, 124-138.
[41] Noll, M. & Kornberg, R.D. (1977). Action of *Micrococcal nuclease* on chromatin and the location of histone H1. *J Mol Biol,* 109, 393-404.
[42] Fan, L. & Roberts, V.A. (2006). Complex of linker histone H5 with the nucleosome and its implications for chromatin packing. *Proc. Natl. Acad. Sci. USA*, *103*, 8384-8389.

[43] Clore, G.M., Gronenborn, A.M., Nilges, M., Sukumaran, D.K. & Zarbock, J. (1987). The polypeptide fold of the globular domain of histone H5 in solution - a study using nuclear magnetic resonance, distance geometry and restrained molecular dynamics. *EMBO J, 6*, 1833-1842.

[44] Thomas, J.O. & Wilson, C.M. (1986). Selective radiolabelling and identification of a strong nucleosome binding site on the globular domain of histone H5. *EMBO J, 5*, 3531-3537.

[45] Ramakrishnan, V., Finch, J.T., Graziano, V., Lee, P.L. & Sweet, R.M. (1993). Crystal-structure of globular domain of histone H5 and its implications for nucleosome binding. *Nature, 362*, 219-223.

[46] Cerf, C., Lippens, G., Ramakrishnan, V., Muyldermans, S., Segers, A., Wyns, L., Wodak, S.J. & Hallenga, K. (1994). Homonuclear and heteronuclear 2-dimensional NMR studies of the globular domain of histone H1 - full assignment, tertiary structure, and comparison with the globular domain of histone H5. *Biochemistry, 33*, 11079-11086.

[47] Buckle, R.S., Maman, J.D. & Allan, J. (1992). Site-directed mutagenesis studies on the binding of the globular domain of linker histone H5 to the nucleosome. *J. Mol. Biol, 223*, 651-659.

[48] Goytisolo, F.A., Gerchman, S.E., Yu, X., Rees, C., Graziano, V., Ramakrishnan, V. & Thomas, J.O. (1996). Identification of 2 DNA-binding sites on the globular domain of histone H5. *EMBO J, 15*, 3421-3429.

[49] Duggan, M.M. & Thomas, J.O. (2000). Two DNA-binding sites on the globular domain of histone H5 are required for binding to both bulk and 5 S reconstituted nucleosomes. *J. Mol. Biol, 304*, 21-33.

[50] Staynov, D.Z. & Crane-Robinson, C. (1988). Footprinting of linker histones H5 and histones H1 on the nucleosome. *EMBO J, 7*, 3685-3691.

[51] Hayes, J.J. & Wolffe, A.P. (1993). Preferential and asymmetric interaction of linker histones with 5S DNA in the nucleosome. *Proc. Natl. Acad. Sci. USA, 90*, 6415-6419.

[52] Pruss, D., Bartholomew, B., Persinger, J., Hayes, J., Arents, G., Moudrianakis, E.N. & Wolffe, A.P. (1996). An asymmetric model for the nucleosome - a binding site for linker histones inside the DNA gyres. *Science, 274*, 614-617.

[53] Hayes, J.J. (1996). Site-directed cleavage of DNA by a linker histone-Fe(II) EDTA conjugate - localization of a globular domain binding site within a nucleosome. *Biochemistry, 35*, 11931-11937.

[54] Eisfeld, K., Candau, R., Truss, M. & Beato, M. (1997). Binding of NF1 to the MMTV promoter in nucleosomes: influence of rotational phasing, translational positioning and histone H1. *Nucl. Acids. Res, 25*, 3733-3742.

[55] Travers, A.A. & Muyldermans, S.V. (1996). A DNA-sequence for positioning chromatosomes. *J. Mol. Biol, 257*, 486-491.

[56] Lambert, S., Muyldermans, S., Baldwin, J., Kilner, J., Ibel, K. & Wijns, L. (1991). Neutron scattering studies of chromatosomes. *Biochem. Biophys. Res. Commun, 179*, 810-816.

[57] Zhou, Y.B., Gerchman, S.E., Ramakrishnan, V., Travers, A. & Muyldermans, S. (1998). Position and orientation of the globular domain of linker histone H5 on the nucleosome. *Nature, 395*, 402-405.

[58] Shen, C.-H., Leblanc, B., Alfieri, J. & Clark, D. (2001). Remodeling of the yeast *CUP 1* chromatin involves activator-dependent re-positioning of nucleosomes over the entire gene and flanking sequences. *Mol. Cell. Biol, 21*, 534-547.

[59] Shen, C.-H. & Clark, D. (2001). DNA sequence plays a major role in determining nucleosome positions in yeast chromatin. *J. Biol. Chem, 276*, 35209-35216.

[60] Thoma, F., Losa, R. & Koller, T. (1983). Involvement of the domains of histones H1 and H5 in the structural organization of soluble chromatin. *J. Mol. Biol, 167*, 619-640.

[61] Lewis, E.A. & Reams, R.R. (1983). Histone H1 – ultra centrifugation studies of the effects of ionic strength and denaturants on the solution conformation. *Arch. Biochem. Biophys, 223*, 185-192.

[62] Vila, R., Ponte, I., Jiménez, M.A., Rico, M. & Suau, P. (2000). A helix-turn motif in the C-terminal domain of histone H1. *Protein Sci, 9*, 627-636.

[63] Vila, R., Ponte, I., Collado, M., Arrondo, J.L. & Suau, P. (2001). Induction of secondary structure in a COOH-terminal peptide of histone H1 by interaction with the DNA: an infrared spectroscopy study. *J. Biol. Chem, 276*, 30898-30903.

[64] Hendzel, M.J., Lever, M.A., Crawford, E. & Th'ng, J.P. (2004). The C-terminal domain is the primary determinant of histone H1 binding to chromatin *in vivo. J. Biol. Chem, 279*, 20028-20034.

[65] Suzuki, M. (1988). The DNA-binding unit in sea-urchin spermatogenous histones .2. the position of the DNA-binding unit in the nucleosome structure. *Proc. Japan Acad. Series B-Physical and Biological Sciences, 64*, 33-36.

[66] Suzuki, M., Gerstein, M. & Johnson, T. (1993). An NMR-study on the DNA-binding SPKK motif and a model for its interaction with DNA. *Protein Eng, 6*, 565-574.

[67] Khadake, J.R. & Rao, M.R.S. (1997). Condensation of DNA and chromatin by an SPKK-containing octapeptide repeat motif present in the C-terminus of histone H1. *Biochemistry, 36*, 1041-1051.

[68] Churchill, M.E.A. & Suzuki, M. (1989). 'SPKK' motifs prefer to bind to DNA at A/T-rich sites. *EMBO J, 8*, 4189-4195.

[69] Kas, E., Izaurralde, E. & Laemmli, U.K. (1989). Specific inhibition of DNA binding to nuclear scaffolds and histone H1 by distamycin - the role of oligo(dA)•oligo(dT) tracts. *J. Mol. Biol, 210*, 587-599.

[70] Izaurralde, E., Kas, E. & Laemmli, U.K. (1989). Highly preferential nucleation of histone H1 assembly on scaffold associated regions. *J. Mol. Biol, 210*, 573-585.

[71] Wellman, S.E., Sittman, D.B. & Chaires, J.B. (1994). Preferential binding of H1e histone to GC-rich DNA. *Biochemistry, 33*, 384-388.

[72] Suzuki, M. (1989). SPKK, a new nucleic acid binding unit of protein found in histone. *EMBO J, 8*, 797-804.

[73] Johnston, M. & Dover, J. (1987). Mutations that inactivate a yeast transcriptional regulatory protein cluster in an evolutionarily conserved DNA-binding domain. *Proc. Natl. Acad. Sci. USA, 84*, 2401-2405.

[74] Otwinowski, Z., Schevitz, R.W., Zhang, R.G., Lawson, C.L., Joachimiak, A., Marmorstein, R.Q., Luisi, B.F. & Sigler, P.B. (1988). Crystal-structure of Trp repressor operator complex at atomic resolution. *Nature, 335*, 321-329.

[75] Suzuki, M., Sohma, H., Yazawa, M., Yagi, K. & Ebashi, S. (1990). Histone H1 kinase specific to the SPKK motif. *J Biochem,* 108, 356-364.

[76] Green, G.R., Lee, H.J. & Poccia, D.L. (1993). Phosphorylation weakens DNA-binding by peptides containing multiple SPKK sequences. *J. Biol. Chem, 268*, 11247-11255.

[77] Lu, X. & Hansen, J.C. (2004). Identification of specific functional subdomains within the linker histone $H1^0$ C-terminal domain. *J. Biol. Chem, 279*, 8701-8707.

[78] Thoma, F. & Koller, T.H. (1977). Influence of histone H1 on chromatin structure. *Cell, 12*, 101-107.

[79] Thoma, F., Koller, T.H. & Klug, A. (1979). Involvement of histone H1 in the organization of the nucleosome and of the salt-dependent superstructures of chromatin. *J. Cell Biol, 83*, 403-427.

[80] Worcel, A., Strogatz, S. & Riley, D. (1981). Structure of chromatin and the linking number of DNA. *Proc. Natl. Acad. Sci. USA-Biological Sciences, 78*, 1461-1465.

[81] Dorigo, B., Schalch, T., Kulangara, A., Duda, S., Schroeder, R.R. & Richmond, T.J. (2004). Nucleosome arrays reveal the two-start organization of the chromatin fiber. *Science, 306*, 1571-1573.

[82] Robinson, P.J. & Rhodes, D. (2006). Structure of the '30 nm' chromatin fibre: a key role for the linker histone. *Curr. Opin. Struct. Biol, 16*, 336-343.

[83] Finch, J.T. & Klug, A. (1976). Solenoidal model for superstructure in chromatin. *Proc. Natl. Acad. Sci. USA, 73*, 1897-1901.

[84] Graziano, V., Gerchman, S.E., Schneider, D.K. & Ramakrishnan, V. (1994). Histone H1 is located in the interior of the chromatin 30-nm filament. *Nature, 368*, 351-354.

[85] Butler, P.J.G. (1984). A defined structure of the 30-nm chromatin fiber which accommodates different nucleosomal repeat lengths. *EMBO J, 3*, 2599-2604.

[86] Robinson, P.J., Fairall, L., Huynh, V.A. & Rhodes, D. (2006). EM measurements define the dimensions of the ‘‘30-nm’’ chromatin fibre: evidence for a compact, interdigitated structure. *Proc. Natl. Acad. Sc.i USA, 103*, 6506-6511.

[87] Huynh, V.A., Robinson, P.J. & Rhodes, D. (2005). A method for the in vitro reconstitution of a defined ‘‘30 nm’’ chromatin fibre containing stoichiometric amounts of the linker histone. *J. Mol. Biol, 345*, 957-968.

[88] Daban, J.R. & Bermudez, A. (1998). Interdigitated solenoid model for compact chromatin fibers. *Biochemistry, 37*, 4299-4304.

[89] Widom, J. & Klug, A. (1985). Structure of the 300 Å chromatin filament: X-ray diffraction from oriented samples. *Cell, 43*, 207-213.

[90] Routh, A., Sandin, S. & Rhodes, D. (2008). Nucleosome repeat length and linker histone stoichiometry determine chromatin fiber structure. *Proc. Natl. Acad. Sci. USA, 105*, 8872-8877.

[91] Staynov, D.Z. (1983). Possible nucleosome arrangements in the higher order structure of chromatin. *Intl. J. Biol. Macromolec, 5*, 3-9.

[92] Makarov, V., Dimitrov, S., Smirnov, V. & Pashev, I. (1985). A triple helix model for the structure of chromatin fiber. *FEBS Lett, 181*, 357-361.

[93] Williams, S.P., Athey, B.D., Muglia, L.J., Schappe, R.S., Gough, A.H. & Langmore, J.P. (1986). Chromatin fibers are left-handed double helices with diameter and mass per unit length that depend on linker length. *Biophys. J, 49*, 233-248.

[94] Schalch, T., Duda, S., Sargent, D.F. & Richmond, T.J. (2005). X-ray structure of a tetranucleosome and its implications for the chromatin fibre. *Nature, 436*, 138-141.

[95] Dimitrov, S.I., Russanova, V.R. & Pashev, I.G. (1987). The globular domain of histone H5 is internally located in the 30-nm chromatin fiber - an immunochemical study. *EMBO J, 6*, 2387-2392.

[96] Cattini, P.A., Harborne, N. & Allan, J. (1988). The linker histone stoichiometry of chicken erythrocyte chromatin *in situ*. *J. Electron. Microscopy, 37*, 31-37.

[97] Russanova, V.R., Dimitrov, S.I., Makarov, V.L. & Pashev, I.G. (1987). Accessibility of the globular domain of histone H1 and histone H5 to antibodies upon folding of chromatin. *Eur. J. Biochem, 167*, 321-326.

[98] Leuba, S.H., Zlatanova, J. & van Holde, K. (1993). On the location of histone H1 and histone H5 in the chromatin fiber - studies with immobilized trypsin and chymotrypsin. *J. Mol. Biol, 229*, 917-929.

[99] Maman, J.D., Yager, T.D. & Allan, J. (1994). Self-association of the globular domain of histone H5. *Biochemistry, 33*, 1300-1310.

[100] Knezetic, J.A. & Luse, D.S. (1986). The presence of nucleosomes on a DNA template prevents initiation by RNA polymerase II *in vitro*. *Cell, 45*, 95-104.

[101] Lorch, Y., Lapointe, J.W. & Kornberg, R.D. (1987). Nucleosomes inhibit the initiation of transcription but allow chain elongation with the displacement of histones. *Cell, 49*, 203-210.

[102] Losa, R. & Brown, D.D. (1987). A bacteriophage RNA polymerase transcribes *in vitro* through a nucleosome core without displacing it. *Cell, 50*, 801-808.

[103] Wu, T.C. & Lichten, M. (1994). Meiosis induced double-strand break sites determined by yeast chromatin structure. *Science, 263*, 515-518.

[104] Grosschedl, R., Giese, K. & Pagel, J. (1994). HMG domain proteins - architectural elements in the assembly of nucleoprotein structures. *Trends Genet, 10*, 94-100.

[105] Simpson, R.T. (1991). Nucleosome positioning - occurrence, mechanisms, and functional consequences. *Prog. Nucl. Acid. Res. Mol. Biol, 40*, 143-184.

[106] Thoma, F. (1992). Nucleosome positioning. *Biochim. Biophys Acta, 1130*, 1-19.

[107] Davey, C., Pennings, S., Meersseman, G., Wess, T.J. & Allan, J. (1995). Periodicity of strong nucleosome positioning sites around the chicken adult β-globin gene may encode regularly spaced chromatin. *Proc. Natl. Acad. Sci. USA, 92*, 11210-11214.

[108] Bouvet, P., Dimitrov, S. & Wolffe, A.P. (1994). Sp.ecific regulation of *Xenopus* chromosomal 5S rRNA gene transcription *in vivo* by histone H1. *Genes Dev, 8*, 1147-1159.

[109] Kandolf, H. (1994). The H1a histone variant is an *in vivo* repressor of oocyte type 5S gene transcription in *Xenopus laevis* embryos. *Proc. Natl. Acad. Sci. USA, 91*, 7257-7261.

[110] Perlmann, T. & Wrange, O. (1988). Specific glucocorticoid receptor binding to DNA reconstituted in a nucleosome. *EMBO J, 7*, 3073-3079.

[111] Pina, B., Bruggemeier, U. & Beato, M. (1990). Nucleosome positioning modulates accessibility of regulatory proteins to the mouse mammary tumor virus promoter. *Cell, 60*, 719-731.

[112] Richard-Foy, H. & Hager, G.L. (1987). Sequence-specific positioning of nucleosomes over the steroid-inducible MMTV promoter. *EMBO J, 6*, 2321-2328.

[113] Pennings, S., Meersseman, G. & Bradbury, E.M. (1994). Linker histones H1 and H5 prevent the mobility of positioned nucleosomes. *Proc. Natl. Acad. Sci. USA, 91*, 10275-10279.

[114] Ura, K., Hayes, J.J. & Wolffe, A.P. (1995). A positive role for nucleosome mobility in the transcriptional activity of chromatin templates - restriction by linker histones. *EMBO J, 14*, 3752-3765.

[115] Ura, K. & Wolffe, A.P. (1996). Reconstruction of transcriptionally active and silent chromatin. *Methods Enzymol, 274*, 257-271.

[116] Bresnick, E.H., Bustin, M., Marsaud, V., Richardfoy, H. & Hager, G.L. (1992). The transcriptionally active MMTV promoter is depleted of histone H1. *Nucl. Acids Res, 20*, 273-278.

[117] Vignali, M. & Workman, J.L. (1998). Location and function of linker histones. *Nat. Struct. Biol, 5*, 1025-1028.

[118] Kim, M. Y., Mauro, S., Gevry, N., Lis, J. T. & Kraus, W.L. (2004). NAD+-dependent modulation of chromatin structure and transcription by nucleosome binding properties of PARP-1. *Cell, 119*, 803-814.

[119] Woodcock, C. L., Skoultchi, A. I. & Fan, Y. (2006). Role of linker histone in chromatin structure and function: H1 stoichiometry and nucleosome repeat length. *Chromosome Res, 14*, 17-25.

[120] Kim, M.Y., Zhang, T. & Kraus, W.L. (2005). Poly(ADP-ribosyl)ation by PARP-1: "PAR-laying" NAD+ into a nuclear signal. *Genes Dev, 19*, 1951-1967.

[121] Krishnakumar, R., Gamble, M.J., Frizzell, K.M., Berrocal, J.G., Kininis, M. & Kraus, W.L. (2008). Reciprocal binding of PARP-1 and histone H1 at promoters specifies transcriptional outcomes. *Science, 319*, 819-821.

[122] Cote, J., Quinn, J., Workman, J.L. & Peterson, C.L. (1994). Stimulation of Gal4 derivative binding to nucleosomal DNA by the yeast SWI/SNF complex. *Science, 265*, 53-60.

[123] Wilson, C.J., Chao, D.M., Imbalzano, A.N., Schnitzler, G.R., Kingston, R.E. & Young, R.A. (1996). RNA polymerase II holoenzyme contains SWI/SNF regulators involved in chromatin remodelling. *Cell, 84*, 235-244.

[124] Imbalzano, A.N., Kwon, H., Green, M.R. & Kingston, R.E. (1994). Facilitated binding of TATA-binding protein to nucleosomal DNA. *Nature, 370*, 481-485.

[125] Caron, F. & Thomas, J.O. (1981). Exchange of histone H1 between segments of chromatin. *J. Mol. Biol, 146*, 513-537.

[126] Thomas, J.O. & Rees, C. (1983). Exchange of histone H1 and histone H5 between chromatin fragments - a preference of H5 for higher order structures. *Eur. J. Biochem, 134*, 109-115.

[127] Misteli, T., Gunjan, A., Hock, R., Bustin, M. & Brown, D.T. (2000). Dynamic binding of histone H1 to chromatin in living cells. *Nature, 408*, 877–881.

[128] Lever, M.A., Th'ng, J.P., Sun, X. & Hendzel, M.J. (2000). Rapid exchange of histone H1.1 on chromatin in living human cells. *Nature, 408*, 873–876.

[129] Phair, R.D., Scaffidi, P., Elbi, C., Vecerova, J., Dey, A., Ozato, K., Brown, D.T., Hager, G., Bustin, M. Misteli, T. (2004). Global nature of dynamic protein–chromatin interactions in vivo: three-dimensional genome scanning and dynamic interaction networks of chromatin proteins. *Mol. Cell. Biol, 24*, 6393–6402.

[130] Raghuram, N., Carrero, G., Th'ng, J. & Hendzel, M. (2009). Molecular dynamics of histone H1[1]. *Biochem. Cell Biol, 87*, 189-206.

[131] Hill, C.S., Rimmer, J.M., Green, B.N., Finch, J.T. & Thomas, J.O. (1991). Histone - DNA interactions and their modulation by phosphorylation of -ser-pro-x-lys/a. *EMBO J, 10*, 1939-1948.

[132] Aubert, D., Garcia, M., Benchaibi, M., Poncet, D., Chebloune, Y., Verdier, G., Nigon, V., Samarut, J. & Mura, C.V. (1991). Inhibition of proliferation of primary avian fibroblasts through expression of histone H5 depends on the degree of phosphorylation of the protein. *J. Cell Biol, 113*, 497-506.

[133] Roth, S.Y., Schulman, I.G., Richman, R., Cook, R.G. & Allis, C.D. (1988). Characterization of phosphorylation sites in histone H1 in the amitotic macronucleus of tetrahymena during different physiological states. *J. Cell Biol, 107*, 2473-2482.

[134] Dimitrov, S. & Wolffe, A.P. (1996). Remodeling somatic nuclei in *Xenopus laevis* egg extracts -molecular mechanisms for the selective release of histones H1 and H1-degrees from chromatin and the acquisition of transcriptional competence. *EMBO J, 15*, 5897-5906.

[135] Wolffe, A.P., Khochbin, S. & Dimitrov, S. (1997). What do linker histones do in chromatin? *Bio. Essays, 19*, 249-255.

[136] Crane-Robinson, C. (1997). Where is the globular domain of linker histone located on the nucleosome? *Trends Biochem. Sci, 22*, 75-77.

[137] Shen, X.T., Yu, L.L., Weir, J.W. & Gorovsky, M.A. (1995). Linker histones are not essential and affect chromatin condensation *in vivo*. *Cell, 82*, 47-56.

[138] Shen, X.T. & Gorovsky, M.A. (1996). Linker histone H1 regulates specific gene-expression but not global transcription *in vivo*. *Cell, 86*, 475-483.

[139] Dasso, M., Dimitrov, S. & Wolffe, A.P. (1994). Nuclear assembly is independent of linker histones. *Proc. Natl. Acad. Sci. USA, 91*, 12477-12481.

In: Histones: Class, Structure and Function
Editor: Chang-Hui Shen
ISBN: 978-1-62100-274-1

Chapter 3

THE LINKER HISTONE AND CHROMATIN OF YEAST *SACCHAROMYCES CEREVISIAE*

***George Miloshev** and *Milena Georgieva*†**
Laboratory of Yeast Molecular Genetics, Institute of Molecular Biology
Bulgarian Academy of Sciences, Sofia, Bulgaria

ABSTRACT

The advantages of using *Saccharomyces cerevisiae* as a model for chromatin studies imply its relatively small number of genes, about 6000, which enables uncomplicated genome-wide study. On the other hand, its short cell cycle - around two hours for genome duplication is another benefit for research. It has been shown that *S. cerevisiae* cells possess about 28% homology of genes with human. Interestingly 20% of genes known to take part in the development and progression of certain human diseases have their homologs in yeast cells. Furthermore, yeast genome possesses a full complement of histones and other chromatin proteins which express vast homology with the respective proteins in human cells. These characteristics of *S. cerevisiae* allow obtained results on them to be easily approximated for human cells. Chromatin has always been a matter of extensive scientific research. A good part of our knowledge on its structure and function has been gathered by experimenting on *S. cerevisiae*. Strangely enough, among the plethora of proteins which build up yeast chromatin the linker histone catches a specific attention with its odd features. *S. cerevisiae* linker histone, Hho1p, is the only histone known to possess two globular domains instead of one. Previous hypotheses suggested that it is responsible for the organization of yeast chromatin mostly in euchromatin. Nonetheless, recent data, including ours, have shown that Hho1p functions in yeast chromatin are far more diverse. It has been proved to be involved in the building and maintenance of the higher-order chromatin organization in yeast, from the 30 nm fiber, chromatin loop organization up to further levels of chromatin compaction. Altogether these data brings this histone protein closer to linker histones of higher eukaryotic cells and points out that linker histones and higher-order chromatin structures are more evolutionary conserved than it is generally believed.

* Corresponding author: miloshev@bio21.bas.bg.

1. Introduction

It is hard to overestimate the valuable role of *Saccharomyces cerevisiae* as an experimental model for chromatin research. Yeast cells' simplicity, fast growing and well-studied genetics allowed a good part of our knowledge of chromatin to be obtained by experimenting on them. This was possible because most of the genes of multicellular organisms have their homologous counterparts in yeast and therefore the information gathered in yeast cells can be easily approximated to human cells. Generally, the chromatin of *S. cerevisiae* possesses all of the features of human cell chromatin, though some differences do exist. The intention of this chapter is not to present a complete summary of all existing data about the yeast chromatin, which would be a huge task, but to briefly underline some of its odd features. One such feature is the existence of less abundant and specific linker histone in *S. cerevisiae* and its characteristic functions. In addition, we also tried to review recently accumulated data on histones post-translational modifications, specific for yeast cells.

2. The Linker Histones

Linker histones are the fifth class of histone proteins in eukaryotic cells and play important roles in building up the higher-order chromatin structure. Several subtypes of linker histones exist in higher eukaryotes, making studying of their functions in chromatin really hard, especially at the higher levels of chromatin compaction. The exact way by which histone H1 binds to the linker DNA of the nucleosomal fiber is yet unclear, though there are several competing models for this interaction [1, 2].

3. The Linker Histone of Saccharomyces Cerevisiae

The discovery of *Saccharomyces cerevisiae* linker histone was an interesting tale with misleading experimental results and controversial conclusions. As a popular model for genetic research *S. cerevisiae* was one of the first organisms in which histone proteins were isolated and characterized. However, the initial isolations failed to find any trace of a linker histone in it [3, 4, 5]. Therefore, on the basis of these early studies many authors have denied the existence of linker histone in yeast. Others have found some electrophoretic fractions in yeast with resemblance to histone H1 of multicellular organisms [6, 7, 8]. Unfortunately, in latter research some of these fractions have been questioned and assigned as members of the High Mobility Group of proteins [9]. In one particular case it was shown that the suspective for the linker histone fraction was a mitochondrial protein [10]. However, it should be emphasized that for more than 30 years the opinion that the yeast *Saccharomyces cerevisiae* is the only known eukaryote lacking linker histone prevailed in general. Moreover, on the basis of this assumption the existence of more euchromatin and the shorter nucleosomal repeat in *S. cerevisiae* were explained [11]. The results showing presence of a protein fraction in budding yeast with biochemical and immunochemical characteristics of mammalian H1 [12], as well as the existence of a fragment possessing more than 80% homology with the canonical H1 histone sequence in the budding yeast genome did little to persuade that H1 histone does exist in the yeast [13]. It was not before 1996 when the sequencing of the whole *S. cerevisiae* genome has led to the identification of the putative linker histone H1 gene [14]. From this first unequivocal evidence for the existence of the yeast linker histone some odd features of this histone started to emerge. Unlike most of the other histone proteins this linker histone, Hho1p, was shown to be coded by a single-copy gene - *HHO1*, located on the XVI chromosome of *S. cerevisiae* [15]. The most surprising, however, was the predicted tertiary structure of Hho1p. In general, H1 histone proteins have a three domain structure: a short amino-terminal lysine-rich peptide, a central globular domain and a carboxy-terminal lysine/alanine/proline-rich tail domain. The *Saccharomyces cerevisiae* homologue of the linker histone, Hho1p, has shorter basic N-terminal tail than usual. It possesses two globular domains that are similar in sequence to the globular domain of H1 and are connected by a lysine-rich segment, called L [16]. The sequence between the two globular domains (L) is much shorter than other H1 carboxy-terminal tail sequences, but probably could carry out its functions because it contains 12 lysines, 10 alanines and four prolines out of 42 residues, which is similar to carboxy-terminal regions of other H1s. It is believed that the globular domain is the most important for the functions of linker histones [17]. The globular domains of Hho1p, GI and GII under ionic conditions in which they are both folded are very similar, reflecting their high level of sequence identity. *HHO1* gene is highly expressed in the polyA+ RNA fraction [15] and is co-expressed with the core histone genes in S-phase [18].

4. Structure of Saccharomyces Cerevisiae Linker Histone

As was already mentioned, the most puzzling characteristic of Hho1p is that it possesses two well-defined globular domains instead of one (Figure 1). This feature makes Hho1p unique as until now there are no other reports of a linker histone with two globular domains. Bearing in mind the important role of linker histones in higher-order chromatin structure [19, 20] it is reasonable to suppose that Hho1p with its unusual configuration would have a different interaction with DNA and with the nucleosome probably leading to differences in chromatin compaction. Indeed, Landsman in its initial *in silico* analysis of *S. cerevisiae* linker histone gene and the protein that it codes proposed a model in which the two globular domains of Hho1p bind two neighbor nucleosomes [14]. How this binding would influence the building of the higher-order chromatin structures remained to be elucidated. In order to make predictions about the type of these interactions a better understanding of the tertiary structure of Hho1p is needed. Fortunately, it was done in three consequent works [21, 22, 23]. By using NMR spectroscopy and circular dichroism these authors obtained valuable information about the *in vitro* behavior of major parts of Hho1p molecule. In the first globular domain of Hho1p, Ono and co-workers have found a winged helix-turn-helix motif, an amino acid sequence that is characteristic for the DNA binding and recognition domains of several proteins and particularly of the globular domain of histone H5 [24]. These results and the amino acid sequence homology led to the conclusion that the first globular domain of Hho1p (designated by the authors as HD1) resembled more the globular domain of histone H5 (GH5) than that of the other linker histones. The second globular domain - HD2 had a flexible and disordered structure [21]. In a more detailed study by Thomas's lab it was shown that GII (the second globular domain) is likely to be significantly unfolded at "physiological" ionic strength (10 mM sodium phosphate) and can fold into a globular structure only at 500 mM sodium phosphate [22, 23]. However, it was proven that the second globular GII has a potential to adopt a structure similar to that of GI when Hho1p is bound to chromatin *in vivo*, suggesting that both globular domains might be functional. These authors also found that GI and GII differ in their relation to DNA and to reconstituted chromatin fragments. Full-length Hho1p or recombinant GI domain confers chromatosome protection during micrococcal nuclease digestion and, in addition, can form complexes with four-way DNA junctions. The other domain GII, in contrast, does not perform chromatosome protection however, it binds with higher affinity to four-way junctions than GI [23].

One of the important features of Hho1p is the absence of a long basic C-terminal tail in it. It should be noted, however, that the region between GI and GII in Hho1p is similar in amino acid composition and sequence to that of the canonical H1 C-terminal domain and therefore might perform a similar role [23]. The shorter length of linker DNA in yeast [25, 26] could be a result of the unusual structure of its linker histone Hho1p. Whether Hho1p performs a structural role by bridging two neighbor nucleosomes or it binds to only one nucleosome sealing its structure, remains to be determined.

From the results discussed above it is becoming clear that Hho1p is the yeast homologue of the linker histones of higher eukaryotes. It is relatively abundant in yeast [27, 28] and it stabilizes yeast nucleosomes.

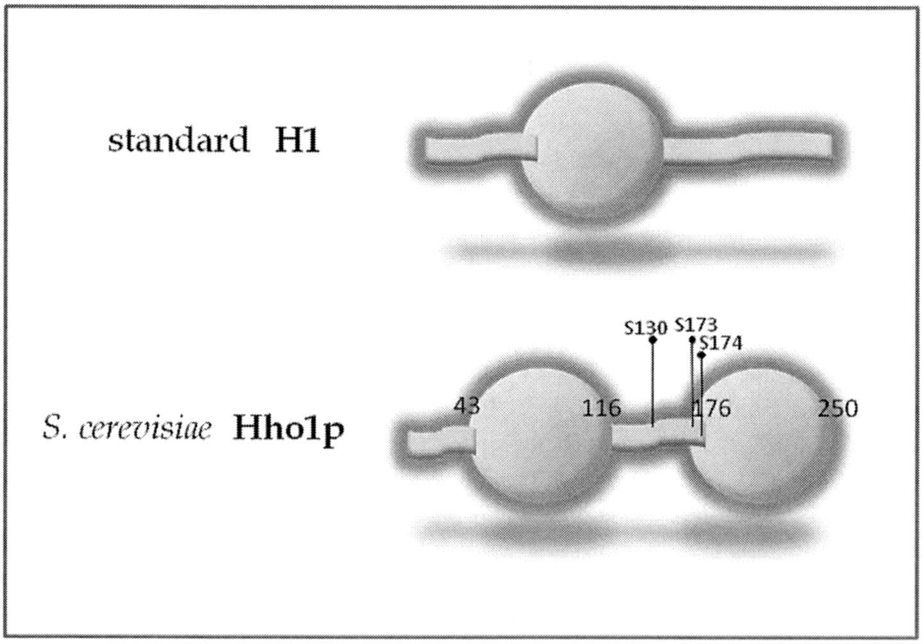

Figure 1. Schematic representation of mammalian and yeast linker histones. On the molecule of Hho1p with numbers are marked the beginning and the end of the globular domains [22, 23]. The proposed sites for phosphorylation are presented with vertical bars. Note that all proposed sites for modification in Hho1p are in the L-region, i.e. between the globular domains.

5. Hho1p Roles in Yeast Chromatin Organization

The next question is whether Hho1p plays roles in any higher-order chromatin structures in yeast? Little is known about how chromatin folds in its native state in the nucleus. As a whole the information about the higher-order chromatin structures in *S. cerevisiae* is little, moreover information about the involvement, if any, of the linker histone Hho1p in the structures above the chromatosome is really scarce.

Initially, due to high transcriptional activity (around 40% of the genome is transcribed [29]) even the existence of higher-order chromatin structures in *S. cerevisiae* has been questioned. Latter on it was shown that part of yeast chromatin can be found in more compacted structures. The electron microscopy observations of spreads from budding yeast chromatin described loops and filaments [30]. The study showed that this higher-order chromatin structures in yeast resemble very much that of the multicellular organisms. In a detailed research involving light scattering and x-ray diffraction of purified chromatin fragments, Lowary and Widom proved the ability of the yeast chromatin to form "30 nm" fibre [31]. These structures formed by yeast chromatin have been compared with native and linker histone depleted chicken erythrocyte chromatin under different ionic conditions. The results evidenced that the higher-order structures of *Saccharomyces cerevisiae* chromatin were assembled in a very similar way to that of the chicken erythrocytes [31].

Quite an interesting question starts to rise from the above mentioned data and it concerns the role of Hho1p in the building of the high-order chromatin structures of yeast?

6. Hho1p and the "30 nm" Fibre in *Saccharomyces Cerevisiae*

Information gathered by reconstitution of chromatin fragments *in vitro* show that linker histones *per se* play an important role in building and maintaining chromatin structures like "30 nm" fibre in metazoan organisms [32]. As a proof of this are the results obtained in our lab by Atomic Force Microscopy (AFM) on purified chromatin fragments from yeast Hho1p knocked-out mutants (*Δhho1p*). The data demonstrate a complete loss of the "30 nm" fibre in *hho1Δ* mutants *in vivo*. It should be emphasized that, in fact, our results confirm that Hho1p is indeed the linker histone of *Saccharomyces cerevisiae* and that it is intrinsically involved in the building and/or maintainting of the "30 nm" fibre in this organism.Chromatin condensation above the "30 nm" fibre in structures such as loops and chromosome territories is poorly understood in all organisms including yeast. It is widely believed that the chromatin fibre is organized into loops during interphase. However, the estimated loop sizes vary widely. Loop organization of nuclear chromatin has been assumed to depend on different attachment elements such as SARs (Scaffold Attached Regions), MARs (Matrix Attached Regions) [33] or BUR (Base-Unpairing Regions) [34] in DNA. There is no consensus element in these sequences and probably because of this the loops they anchor are found to be extremely different in size. The experimentally measured chromatin loop sizes in the eukaryotic nuclei range from 5 kb to 200 kb [35], or are in the range of 70 kb [36, 37]. The most frequently observed chromatin loop sizes are ~100 kb [reviewed in 38] even reaching up to 3 Mb [39]. Obviously, nuclei contain a range of loops with different sizes. The diversity in loop sizes can reflect structural requirements, or their existence could be implicated as an important unit of nuclear function or both. On the other hand, however, the major discrepancy in chromatin loop size could come from the particular method employed in measuring [35]. The most frequently used methods in loop sizing are the fluorescence *in situ* hybridization (FISH), chromosome conformation capture (3C) assay [40] and chromatin immunoprecipitation (ChIP)-loop assay [41].

To our knowledge chromatin loop sizes have never been measured in budding yeast. Recently we have developed a new technique specifically designed for chromatin loop estimation. The method is a modification of Comet assay, which we called Chromatin Comet Assay (ChCA) [42]. Shortly, the cells to be studied are included in low-melt agarose and spread as microgels on microscopic slides. After solidifying of the gels the nuclear chromatin is digested mildly with nucleases in order to make rare cuts in DNA. In a short electrophoresis the chromatin loops leave the nucleus and form a comet tail. By the liner size of this tail the size of chromatin loops can be calculated. Using this method we succeeded in calculating the size of loops in human erythroleukemia cells (K562) and compared them with the size of chromatin loops in the yeast. The estimated median sizes of chromatin loops were around 0.4 Mb for K562 cells and 0.3 Mb for the yeast cells, respectively. Interestingly, in yeast *hho1Δ* mutant the chromatin loops increased in size up to 0.4 Mb, arguing that this histone is involved in chromatin organization in loops in budding yeast.

7. Functions of Hho1p

Though biochemical analyses confirm that the functions of Hho1p globular domains are similar to those of canonical linker histones, sparse remains the information about its exact *in vivo* functions. Hho1p linker histone does not position at every two nucleosomes [27, 28] and predominantly localizes in the ribosomal locus being crucial for the proper compaction of DNA in it and for the accurate functioning of DNA-dependent RNA polymerase I [43].

Surprisingly, in sharp contrast to the higher eukaryotes, where the reduction of H1 content in mice with 50% lead to severe embryonic defects and finally to death [44], the budding yeast linker histone appears to be non-essential for yeast cells. Something more, it is not even a general repressor of transcription. On the contrary, the deletion of its gene leads to a slight decrease in the transcriptional levels of 27 of all 6216 yeast genes [45]. These 27 genes with changed expression profile participate in various functions: in the process of translation (*EFT1, EFT2, SUI3*, etc), in cellular metabolism (*PEK1, PEK2, GPH1, GAD1*), even in cellular response to high temperature stress (*HSP82, SSE1*). Therefore, the authors quite logically propose that Hho1p is a fine regulator of gene transcription [45].

Noteworthy the interest in the yeast linker histone has increased in the recent eight years. It has been shown that Hho1p has an inhibitory effect on the processes of homologous recombination and on the double-strand DNA breaks repair [28]. Quite unexpectedly it has been proven that Hho1p promotes the activity of several chromatin barrier elements and inhibits transcriptional silencing in *S. cerevisiae* [46]. Later on, Hho1p has been shown to interact with core histone H4 and to negatively regulate *de novo* establishment of heterochromatin [47]. These results unequivocally validate the role of Hho1p as a fine regulator of chromatin structure and of gene expression.

Very intriguing emerges the recently discovered function of Hho1p in the processes of aging [48]. The authors have shown that Hho1p plays an important role in cells which are about to enter a stationary phase of the cell growth. During this phase its binding to chromatin is increasing, thus influencing the transcription of genes and cell fate in general.

8. Post-translational Modifications of the Yeast Histones – Chromatin Epigenetic Information

Literally the term epigenetics means above and beyond genetics and depicts heritable phenotypic alterations that are not determined by changes in DNA sequence [49]. It is a key element in organism adaptation to environmental changes. The major epigenetic phenomena involve DNA methylation [50, 51], histone post-translational modifications [52], histone variants [53], non-coding RNAs [54] and chromatin remodeling [55]. Normally every epigenetic change is associated with changes in chromatin structure and in gene expression.

Histone proteins act as structural elements of chromatin but also play a fine tuning role in the regulation of all processes that happen in the nucleus. A diverse array of post-translational modifications occurs on tail domains including phosphorylation, acetylation, methylation, ADP-ribosylation, ubiquitinylation, sumoylation, proline isomerization, citrullination, butyrylation and propionylation (the last two being abbreviated KButy and KProp). All histone modifications are carried out by histone-modifying enzyme complexes in a dynamic manner, thus re-organizing chromatin structure and influencing cellular processes [52, 56, 57].

It has been shown that *Saccharomyces cerevisiae* possesses simpler histone code with more histone marks associated with transcriptional activation in comparison with mammals [58, 59]. Notably, these “activation” marks in the budding yeast epigenome are relatively conserved among species and the low abundance of “repressive” marks might be due to the well known more active yeast genome [60] and of course to the simplicity of this unicellular eukaryote.

9. Budding Yeast Core Histone Post-translational Modifications

Among the most studied histone modifications are acetylation, methylation and phosphorylation.

Acetylation is always referred to transcriptional activation. The sites of acetylation in *Saccharomyces cerevisiae* include four lysines in histone H4 (K5, K8, K12 and K16), six in histone H3 (K9, K14, K18, K23, K27 and K56), five in histone H2B (K6, K11, K16, K17, and K22), and two in histone H2A (K4, K7). Despite the fact that all these amino-acid moieties in yeast core histones can be potentially acetylated, obviously some preferential ones exist. For example, H4 can be acetylated at K5, K8, K12 and K16, but 12% of total H4 is not acetylated at all, 36% of total H4 is acetylated at one lysine, 28% at two, 13% at three and 12% at four lysine residues [61]. K16 is the most favoured site which is acetylated in 80% of all H4 molecules [62]. Recent results have demonstrated that H4K20ac and H3K36ac are novel yeast histone modifications [58]. Both of these modifications are of interest because they entail lysine residues on which generally methylation occurs, thus giving intriguing speculations about the complex interplay among different modifications.

Histone methylation in yeast extensively occurs on H3 at K4, K36 and K79, the first one being part of the globular domain of H3 molecule [52]. All these three lysine residues on H3, when methylated, are associated with active transcription or permissive chromatin structure. It has long been considered that in *Saccharomyces cerevisiae* methylation of H3K9 and H3K27 does not exist. In 2006, Garcia and co-workers have shown that though in low amounts H3K9me and H3K27me3 can also be found in budding yeast epigenome [58]. In mammalian cells these two epigenetic marks are characteristic for the transcriptionally silent chromatin. The low abundance of methylation at H3K9 and H3K27 in the yeast, is probably due to the the nature of transcriptionally active yeast genome. What is important here is that the presence of these epigenetic marks in the budding yeast proves that histone post-translational modifications are evolutionary conserved among organisms. Therefore, simpler organisms could be used successfully as models for studies in the field of epigenetics.

Histone methylation is a complex post-translational modification. Results from previous studies present evidence that the site specificity of lysine methylation, as well as the number of methyl groups attached to a particular lysine, have distinct consequences for transcription. In the case of H3K4me, a gradient in the degree of methylation along the coding region of a transcribing gene has been detected - from trimethylation at the 5′-end to monomethylation at the 3′-end [63]. In contrast to H3K4me, the H3K36me3 methylation mark was detected all along the coding region peaking at the 3’ end of the transcription unit [64]. The methylation of these particular lysine residues on H3 in yeast appears quite often in the yeast epigenome, usually represented by the above-mentioned specific distribution of mono-, di- and tri-methylated K4 and K36 along the transcription unit. Monomethylation of K79 is estimated to occur in 90% of all yeast H3 histones and is associated with telomeric silencing [65, 66] while H3K79me3 is linked to transcriptional activation and is distributed along the transcribed region of a gene.

Compared to the wealth of information about acetylation and methylation, less is known about the function of phosphorylation. It has been proved that this post-translational modification is very specific and is involved in transcriptional regulation, DNA repair, apoptosis, and chromosome condensation. H2A phosphorylation at residues S122, T126, and S129 in budding yeast is a specific post-translational modification that is in response to different damage conditions [67, 68]. H3S10ph is involved in chromosome condensation. Interestingly, it has been shown that mutation in S10 in histone H3 in budding yeast does not change the chromosome condensation [69]. The authors suggest that H2B phosphorylation could compensate H3S10ph. Recently, it has been proved that phosphorylation of H4S1 is a specific epigenetic mark in budding yeast, which has a role in DNA repair and can regulate sporulation of yeast cells through chromatin compaction [70, 71]. Remarkably in the two processes this particular modification of H4 behaves quite differently. During repair of DNA double strand breaks it usually follows a decrease in the H4 acetylation levels [70], while during sporulation it is found over the transcription start site regions of the genes correlating with acetylation of the same histones [72]. These kinds of studies are important for the elucidation of the complex “cross-talk” among different histone modifications. It will greatly help for the better understanding of the so-called bivalent epigenetic marks that combine active and repressive histone modifications at regulator genes in mammalian embryonic stem cells [73, 74].

Work on yeast post-translational modifications is accelerating and growing amount of new data is being collected. The main objective now is the old and the new post-translational marks to be compared among organisms in order to better understand the presumptive "histone code" [75]. In this light Berger and co-workers [59] have detected specific sites for sumolyation in budding yeast core histones - H2A (K126), H2B (K6/7) and H4 (all five lysines in the N-terminal). Moreover they have linked these sites with transcriptional repression, suggesting a dynamic interplay among this modification witheither acetylation or ubiquitylation.

Prior to this study it has been long considered that the budding yeast does not contain any negative histone marks. The discovery of sumolyation [76] as well as the discovery of H3K9me and H3K27me3 [58] in the budding yeast has proved the fact that histone marks are more conserved than it has been thought before. Furthermore as in the budding yeast, biochemical approaches are easily combined with genetic analyses, it is helping a lot in the study of histone modifying complexes, their *in vivo* substrate specificity, their target genetic loci, and after all the functional consequences for the organism. In 2009, the group of Zhao [77] has discovered the existence of some novel budding yeast modifications - lysine propionylation and butyrylation in histones H2B, H3, and H4. The sites for lysine propionylation were shown to be H3K23 (H3K23prop) and H3K56 (H3K56prop), while for butyrylation – it includes H3K14 (H3K14buty), H3K27/36 (H3K27/36buty), H4K8 (H4K8buty) and H2BK21 (H2BK21buty). These results were obtained by a systematic study of post-translational modifications of budding yeast histones. Despite the fact that a more detailed study of the functional consequences of these modifications is still needed, these results are adding valuable information to the histone code in different organisms.

10. Specificity of Histone Post-translational Modifications in *Saccharomyces Cerevisiae*

New studies have shown that although conserved among organisms some histone modification sites differ in multicellular and unicellular eukaryotes, thus drawing rather unique epigenetic picture. The situation gets even more complicated by the fact that each core histone has multiple sites at which modifications are found whereas most post-translational marks interact with each other, providing a complex "cross-talk" [78, 79]. Significantly, the existence of species-specific histone marks has been recently reported [58, 80, 81].

A genomewide comparative study of histone post-translational modifications in *Saccharomyces cerevisiae*, human, mouse and *Tetrahymena* has been recently published [58]. Interestingly, the authors have identified the following species-specific histone epigenetic marks: H3K14me2 (yeast), H3K56me3 (human), H3K64me1 (human), H3K64ac (yeast), H3K79me3 (yeast), H4K20ac (yeast), H3K79ac (human), and H3K122me1 (human). They have also found some novel modifications: H3K4ac (human, and to a lesser extent, mice, and *Tetrahymena*), H3K56me3 (human), H3K64ac (yeast), H4K20ac (yeast), and H3K79ac (human). The comparison between yeast and human potentially species-specific marks is given in Table 1. Collectively, these data point out the need for further comparative analyses in the field of post-translational modifications in different organisms.

Table 1. Potential human and yeast - specific posttranslational modifications of core histones [58]

Potential specific PTMs*	human	budding yeast
H3K4ac	+	-
H3K79ac	+	-
H3K23me1	+	+
H3K14me2	-	+ (in low abundance)
H3K64ac	-	+
H3K64me1	+	-
H3K56me3	+ (in low abundance)	-
H3K56me1	+	-
novel marks		
H4K20me2	+	+ (in low abundance)
H4K20ac	+	+
H4K20me	+	-

* PTMs: posttranslational modifications.

11. Yeast Linker Histone Modifications

Phosphorylation is probably the most characteristic post-translational modification of the linker histones. This modification is linked mainly to the cell cycle. In human cells, H1 is phosphorylated at up to five sites by the cyclin-dependent kinase 1 (CDK1) [82]. Up to four of these sites are located in the C-terminal tails of the H1, and one is located in the N-terminal portion of the histone. In addition to these sites, phosphorylation of non-CDK1 sites within the N-terminal tails was also reported. This modification is very specific with great importance in cellular processes [83]. Novel sites for acetylation, methylation and ubiquitinilation have been discovered lately in linker histones. They are specific for different linker histones sub-variants [83, 84].

In contrast to our knowledge about higher eukaryotic linker histones and their epigenetic modifications, almost nothing is known about *S. cerevisiae* linker histone – Hho1p. In the past eight years a lot of data start gathering about the roles of this protein in chromatin structure and functions. All of these data outline the importance of Hho1p in the fine regulation of gene expression, DNA repair, transcriptional silencing, and chromatin compaction in budding yeasts. In contrast, Hho1p post-translational modifications are scarcely studied. Recently a phosphorylation at S173 has been reported (Figure 1) [85, 86].

This phosphorylation site is situated between the first and the second globular domain of Hho1p in the so-called L-part of the molecule which exhibits strong preferential binding to linker DNA [22, 23]. Obviously, a post-translational modification at this particular site could induce changes in the binding of Hho1p to the linker DNA region. Additionally, we can discuss the Mass-Spec derived results from a project aimed to support system biology signaling research. It is about protein phosphorylation in 4 different organisms. The programme is PhospoPep version 2.0 and by now contains data from the fly (*Drosophila melanogaster*), human (*Homo sapiens*), worm (*Caenorhabditis elegans*), and yeast (*Saccharomyces cerevisiae*). The proposed phosphorylation sites for the budding yeast linker histone, Hho1p, are S130, S173, S174, referred as certain and one ambiguous site at S177 [87]. These potential sites for phosphorylation in Hho1p seem quite logic in relevance to the known phosphorylation sites in higher eukaryotic cells, which as has been already mentioned appear at C- and N- tails of H1. Evidence exists showing that H1 variants bind chromatin dynamically *in vivo* and that both the globular domain and C-terminal domain contribute to chromatin binding [88]. The roles of the two globular domains of Hho1p are similar to those of canonical linker histones. Moreover, it is well believed that the L-part of Hho1p molecule is the one that binds in the linker region, assuring protection against micrococcal nuclease [22, 23]. Therefore modifications at this particular L-part, like in the case with S130, means changes in Hho1p binding to chromatin and further influencing of chromatin compaction. The other three sites – S173, 174 and 177 are in the second globular domain and could also play a role in Hho1p targeting and binding to chromatin.

Further research is needed in order to be elucidated the roles of Hho1p in all the processes listed above. Future studies on Hho1p post-translational modifications will be of great help for better understanding of these functions. What is promising, though, is the fact that yeast *Saccharomyces cerevisiae* pertains all characteristics of all eukaryotes thus highlighting its advantages as a model in any kind of research.

CONCLUSIONS

There is no doubt that the information about the involvement of chromatin structures in the regulation of the genome activity will continue to grow. In this light the experiments with *Saccharomyces cerevisiae* will play tremendous role. Therefore, it is very important to learn as much as possible and in more details about the peculiarities of yeast chromatin. The unusual yeast linker histone Hho1p is one of these odd features. With the existence in yeast of this atypical member of the family of the linker histone proteins several questions emerge. Some of them are: How important is the abundance of Hho1p for the yeast genome? Could it be replaced by another one linker histone, as it is the situation with the linker histone variants in multicellular organisms? Would post-translational modifications of the linker histone Hho1p interplay with the core histones modifications? The answers of these questions will be searched in the near future. It has to be boldly stated thatin order to have the whole picture of the genome regulation through chromatin structures it is very important to consider the fine role of linker histones in these processes.

ACKNOWLEDGMENTS

G.M. is supported by the Bulgarian Science Fund, project number: DID 02-35. M.G. is supported by the Bulgarian Science Fund, project number: DMU 02/8 and by the World Federation of Scientists, National Fellowship Program.

REFERENCES

[1] Pruss, D., Bartholomew, B., Persinger, J., Hayes, J., Arents, G., Moudrianakis, E.N. & Wolffe, A.P. (1996). An asymmetric model for the nucleosome: a binding site for linker histones inside the DNA gyres. *Science, 274*, 614–617.

[2] Stasevich, T., Mueller, F., Brown, D. & McNally, J. (2010). Dissecting the binding mechanism of the linker histone in live cells: an integrated FRAP analysis. *The EMBO J, 29*, 1225–1234.

[3] Tonino, G.J.M. & Rozijn, T.H. (1966). On the occurrence of histones in yeast. *Biophys Biochem Acta, 124*, 427-429.

[4] Franco, L., Johns, E.W. & Navlet, S.M. (1974). Histones from baker's yeast. Isolation and characterization. *Eur J Biochem, 45*, 83-89.

[5] Brandt, W.F. & von Holt, C. (1976). The occurrence of histone H3 and H4 in yeast. *FEBS Lett*, *65*, 386-390.

[6] Suchiliene, S.P. & Gineitis, A.A. (1978). Histones from *Saccharomyces cerevisiae*. *Exp Cell Res, 114 (2)*, 454-458.

[7] Sommer, A. (1978). Yeast chromatin: Search for H1. *Mol & General Genet, 161*, 323-331.

[8] Pastink, A., Berkhout, T.A., Mager, W.H. & Planta, R.G. (1979) Analysis of histones from *Saccharomyces carlsbergensis*. *Biochem J, 177*, 917-923.

[9] Weber, S. & Isenberg, I. (1980). High Mobility Group proteins of *Saccharomyces cerevisiae*. *Biochemistry, 19*, 2236-2240.

[10] Certa, U., Colavito-Shepansky, M. & Grunstein, M. (1984). Yeast may not contain histone H1: The only known "histone H1-like" protein in *Saccharomyces cerevisiae* is a mitochondrial protein. *NAR, 12*, 7975-7985.

[11] van Holde, K.E. (1989). *Chromatin*, New York, Springer-Verlag.

[12] Srebreva, L., Zlatanova, J., Miloshev, G. & Tsanev, R. (1987). Immunological evidence for the existence of H1-like histone in yeast. *Eur J Biochem, 156*, 449-454.

[13] Miloshev, G. & Zlatanova, J. (1995). Existence of yeast genomic fragments hybridizing with the sea urchin histone H1 gene. *J Gen Appl Microbiol, 40*, 359-363.

[14] Landsman, D. (1996). Histone H1 in Saccharomyces cerevisiae: a double mystery solved? *Trends Biochem Sci, 21*, 287-288.

[15] Ushinsky, S.C., Bussey, H., Ahmed, A.A., Wang, Y., Friensen, J., Williams, B.A., Storms, R.K. (1997). Histone H1 in *Saccharomyces cerevisiae*. *Yeast, 13*, 151-161.

[16] Baxevanis, A.D. & Landsman, D. (1998). Homology model building of Hho1p supports its role as a yeast histone H1 protein. *In Silico Biol*, *1*, 5-11.

[17] Allan, J., Mitchell, T., Harborne, N., Bohm, L. & Craine-Robinson, C. (1986). Roles of H1 domains in determining higher order chromatin structure and H1 location. *J Mol Biol, 187*, 591-601.

[18] Spellman, P.T., Sherlock, G., Zhang, M.Q., Iyer, V.R., Anders, K., Eisen, M.B. *et al.* (1998). Comprehensive identification of cell cycle-regulated genes of the yeast *Saccharomyces cerevisiae* by microarray hybridization. *Mol Biol Cell, 9*, 3273–3297.

[19] Thoma, F., Koller, T. & Klug, A. (1979). Involvement of histone H1 in the organization of the nucleosome and of the salt-dependent superstructures of chromatin. *J Cell Biol, 83*, 403–427.

[20] Huynh, V.A.T., Robinson, P.J.J. & Rhodes, D. (2004). A method for the in vitro reconstitution of a defined "30 nm" chromatin fibre containing stoichiometric amounts of the linker histone. *J Mol Biol, 345*, 957-968.

[21] Ono, K., Kusano, O., Shimotakahara, S., Shimizu, M., Yamazaki, T. & Shindo, H. (2003). The linker histone homolog Hho1p from *Saccharomyces cerevisiae* represents a winged helix- turn-helix fold as determined by NMR spectroscopy. *NAR, 31*, 7199-7207.

[22] Ali, T. & Thomas, J.O. (2004). Distinct properties of the two putative "globular domains" of the yeast linker histone, Hho1p. *J Mol Biol, 337*, 1123-1135.

[23] Ali, T., Coles, P., Stevens, T.J., Stott, K. & Thomas, J.O. (2004). Two homologous domains of similar structure but different stability in the yeast linker histone, Hho1p. *J Mol Biol, 338*, 139-148.

[24] Ramakrishnan, V., Finch, J.T., Graziano, V., Lee, P.L. & Sweet, R.M. (1993). Crystal structure of globular domain of linker histone H5 and its implications for nucleosome binding. *Nature, 362*, 219-223.

[25] Lohr, D., Tatchell, K. & Van Holde, K.E. (1977). On the occurrence of nucleosome phasing in chromatin. *Cell, 12 (3)*, 829-836.

[26] Thomas, J.O., Furber, V. (1976). Yeast chromatin structure. *FEBS Lett, 66 (2)*, 274-280.

[27] Freidkin, I. & Katcoff, D. (2001). Specific distribution of the *Saccharomyces cerevisiae* linker histone homologue Hho1p in the chromatin. *NAR, 29*, 4043-4051.

[28] Downs, J.A., Kosmidou, E., Morgan, A. & Jackson, S.P. (2003). Suppression of homologous recombination by the *Saccharomyces cerevisiae* linker histone. *Mol Cell, 11*, 1685-1692.

[29] Hereford, L. & Rosbash, M. (1977). Regulation of a set of abundant mRNA sequences. *Cell, 10*, 453-462.

[30] Rattner, J.B., Saunders, C., Davie, J.R. & Hamkalo, B.A. (1982). Ultrastructural organization of yeast chromatin. *J Cell Biol, 92*, 217-222.

[31] Lowary, P.T. & Widom, J. (1989). Higher-order structure of *Saccharomyces cerevisiae* chromatin. *Proc Natl Acad Sci USA, 86,* 8266-8270.

[32] Hizume, K., Yoshimura, S.H. & Takeyasu, K. (2005). Linker histone H1 per se can induce three- dimensional folding of chromatin fibre. *Biochemistry, 44*, 12978-12989.

[33] Bode, J., Benham, C., Knopp, A. & Mielke, C. (2000). Transcriptional augmentation: modulation of gene expression by scaffold/matrix-attached regions (S/MAR elements). *Crit Rev Eukaryot Gene Expr 10(1)*, 73-90.

[34] Kohwi-Shigematsu, T. & Kohwi, Y. (1990). Torsional stress stabilizes extended base unpairing in suppressor sites flanking immunoglobulin heavy chain enhancer. *Biochemistry, 29*, 9551–9560.

[35] Jackson, D.A., Dickinson, P. & Cook, P.R. (1990). The size of chromatin loops in HeLa cells. *EMBO J, 9*, 567-571.

[36] Gasser, S.M. & Laemmli, U.K. (1987). A glimpse at chromosomal order. *Trends in Genetics, 3*, 16-22.

[37] Vogelstein, B., Pardoll, D.M. & Coffey, D.S. (1980). Supercoiled loops and eucaryotic DNA replication. *Cell, 22,* 79-81.

[38] Tsanev, R., Russev, G., Pashev, I. & Zlatanova, J. (1993). *Replication and transcription of chromatin*. Boca Raton, FL, CRC Press.

[39] Sachs, R.K., Van den Engh, G., Trask, B., Yokota, H. & Hearst, J.E. (1995). A random-walk/giant-loop model for interphase chromosomes. *Proc Natl Acad Sci USA, 92*, 2710-2714.

[40] Dekker, J., Rippe, K., Dekker, M. & Kleckner, N. (2002). Capturing chromosome conformation. *Science, 295*, 1306–1311.

[41] Horike, S., Cai, S., Miyano, M., Cheng, J.F. & Kohwi-Shigematsu, T. (2005). Loss of silent chromatin looping and impaired imprinting of DLX5 in Rett syndrome. *Nat Genet, 37*, 31–40.

[42] Georgieva, M., Harata, M. & Miloshev, G. (2008). The nuclear actin-related protein Act3p/Arp4 influences yeast cell .shape and bulk chromatin organization. *J Cellular Biochem, 104*, 59-67.

[43] Levy, A., Eyal, M., Hershkovits, G., Salmon-Divon, M., Klutstein, M. & Katcoff, D.J. (2008). Yeast linker histone Hho1p is required for efficient RNA polymerase I processivity and transcriptional silencing at the ribosomal DNA. *Proc Natl Acad Sci USA, 105(33),* 11703-11708.

[44] Fan, Y. & Skoultchi, A.I. (2004). Genetic analysis of H1 linker histone subtypes and their functions in mice. *Methods Enzimol, 377*, 85-107.

[45] Hellauer, K., Sirard, E. & Turcotte, B. (2001). Decreased expression of specific genes in yeast cells lacking histone H1. *J Biol Chem, 276*, 13587-13592.

[46] Veron, M., Zou, Y., Yu, Q., Bi, X., Selmi, A., Gilson, E. & Defossez, P.A. (2006). Histone H1 of *Saccharomyces cerevisiae* inhibits transcriptional silencing. *Genetics, 173*, 579- 87.

[47] Yu, Q., Kuzmiak, H., Zou, Y., Olsen, L., Defossez, P.A. & Bi, X. (2009). *Saccharomyces cerevisiae* linker histone Hho1p functionally interacts with core histone H4 and negatively regulates the establishment of transcriptionally silent chromatin. *JBC, 284 (2),* 740-750.

[48] Schäfer, G., McEvoy, C.R. & Patterton, H.G. (2008). The *Saccharomyces cerevisiae* linker histone Hho1p is essential for chromatin compaction in stationary phase and is displaced by transcription. *Proc Natl Acad Sci U S A, 105(39)*, 14838-14843.

[49] Jaenisch, R. & Bird A. (2003). Epigenetic regulation of gene expression: how the genome integrates intrinsic and environmental signals. *Nat Genet, 33(Suppl),* 245–254.

[50] Bird, A. (2002). DNA methylation patterns and epigenetic memory. *Genes Dev, 16*, 6–21.

[51] Bernstein, B., Meissner, A. & Lander, E. (2007). The mammalian epigenome. *Cell, 128*, 669– 681.

[52] Kouzarides T. (2007). Chromatin modifications and their function. *Cell, 128*, 693–705.

[53] Ausió, J. (2003). Histone variants - The structure behind the function. *Briefings in Functional Genomics and Proteomics, 5 (3)*, 228-243.

[54] Cathew, R. (2006). Gene regulation by microRNAs. *Current Opinion in Genetics and Development, 16*, 203–208.

[55] Becker, P.B. & Hörz, W. (2002). Atp-dependent nucleosome remodeling. *Annual Review of Biochemistry, 71*, 247-273.

[56] Khorasanizadeh S. (2004). The nucleosome: from genomic organization to genomic regulation. *Cell, 116*, 259–272.

[57] Chen, Y., Sprung, R., Tang, Y., Ball, H., Sangras, B., Kim, S.C., Falck, J.R., Peng, J., Gu, W. & Zhao, Y. (2007). Lysine propionylation and butyrylation are novel post-translational modifications in histones. *Mol Cell Proteomics, 6(5)*, 812–819.

[58] Garcia, B.A., Hake, S., Diaz, R., Kauer, M., Morris, S., Recht, J., Shabanowitz, J., Mishra, N., Strahl, B., Allis, D. & Hunt, D. (2006). Organismal differences in post-translational modifications in histones H3 and H4. *The Journal of Biological Chemistry, 282(10)*, 7641–7655.

[59] Nathan, D., Ingvarsdottir, K., Sterner, D.E., Bylebyl, G.R., Dokmanovic, M., Dorsey, J.A., Whelan, K.A., Krsmanovic, M., Lane, W.S., Meluh, P.B., Johnson, E.S. & Berger, S.L. (2006). Histone sumoylation is a negative regulator in *Saccharomyces cerevisiae* and shows dynamic interplay with positive-acting histone modifications. *Genes Dev, 20(8)*, 966-976.

[60] Hereford, L.M. & Rosbash, M. (1977). Number and distribution of polyadenylated RNA sequences in yeast. *Cell 10(3)*, 453-62.

[61] Smith, C.M., Gafken, P.R., Zhang, Z., Gottschling, D.E., Smith, J.B. & Smith, D.L. (2003). Mass spectrometric quantification of acetylation at specific lysines within the aminoterminal tail of histone H4. *Anal Biochem, 316*, 23–33.

[62] Clarke, D.J., O'Neill, L.P. & Turner, B.M. (1993). Selective use of H4 acetylation sites in the yeast *Saccharomyces cerevisiae*. *Biochem J, 294*, 557–561.

[63] Millar, C. & Grunstein, M. (2006). Genome-wide patterns of histone modifications in yeast. *Nat Rev Mol and Cell Biol, 7*, 657-666.

[64] Strahl, B.D., Grant, P.A., Briggs, S.D., Sun, Z.W., Bone, J.R., Caldwell, J.A., Mollah, S., Cook, R.G., Shabanowitz, J., Hunt, D.F. & Allis, C.D. (2002). Set2 is a nucleosomal histone H3-selective methyltransferase that mediates transcriptional repression. *Mol Cell Biol, 22*, 1298–1306.

[65] van Leeuwen, F., Gafken, P.R. & Gottschling, D.E. (2002). Dot1p modulates silencing in yeast by methylation of the nucleosome core. *Cell, 109*, 745–756.

[66] van Welsem, T., Frederiks, F., Verzijlbergen, K., Faber, A., Nelson, Z., Egan, D., Gottschling, D. & van Leeuwen, F. (2008). Synthetic lethal screens identify gene silencing processes in yeast and implicate the acetylated amino terminus of Sir3 in recognition of the nucleosome core. *Molecular and Cellular Biology, 28 (11)*, 3861–3872.

[67] Downs, J.A., Lowndes, N.F. & Jackson, S.P. (2000). A role for *Saccharomyces cerevisiae* histone H2A in DNA repair. *Nature, 408*: 1001–1004.

[68] Moore, J.D., Yazgan, O, Ataian, Y. & Krebs, JE. (2007). Diverse roles for histone H2A modifications in DNA damage response pathways in yeast. *Genetics, 176,* 15–25.

[69] Hsu, Y., Sun, Z.W., Li, X., Reuben, M., Ttchell, K. & Bishop, D.K. (2000). Mitotic phosphorylation of histone H3 is governed by Ipl1/aurora kinase and Glc7/PP1 phosphatase in budding yeast and nematodes. *Cell, 102,* 279-291.

[70] Utley, R.T., Lacoste, N., Jobin-Robitaille, O., Allard, S. & Cote, J. (2005). Regulation of NuA4 histone acetyltransferase activity in transcription and DNA repair by phosphorylation of histone H4. *Mol Cell Biol, 25*, 8179–8190.

[71] Krishnamoorthy, T., Chen, X., Govin, J., Cheung, W.L., Dorsey, J., Schindler, K., Winter, E., Allis, C.D., Guacci, V., Khochbin, S. *et al.* (2006). Phosphorylation of histone H4 Ser1 regulates sporulation in yeast and is conserved in fly and mouse spermatogenesis. *Genes Dev, 20*, 2580–2592.

[72] Govin, J., Schug, J., Krishnamoorthy, T., Dorsey, J., Khochbin, S. & Berger, S. (2010). Genome-wide mapping of histone H4 serine-1 phosphorylation during sporulation in *Saccharomyces cerevisiae. NAR, 38 (14)*, 4599–4606.

[73] Bernstein, B.E., Mikkelsen, T.S., Xie, X., Kamal, M., Huebert, D.J., Cuff, J., Fry, B., Meissner, A., Wernig, M., Plath, K. *et al.* (2006). A bivalent chromatin structure marks key developmental genes in embryonic stem cells. *Cell, 125*, 315–326.

[74] Georgieva, M. & Miloshev, G. (2007). Chromatin dictates embryonic stem cells fate, *Embryology 2 (2)*, 11-15.

[75] Strahl, B.D. & Allis, C.D. (2000). The language of covalent histone modifications *Nature, 403*, 41-45.

[76] Nathan, D., Ingvarsdottir, K., Sterner, D., Bylebyl, G., Dokmanovic, M., Dorsey, J., Whelan, K., Krsmanovic, M., Lane, W., Meluh, P., Johnson, E. & Berger, S. (2006). Histone sumoylation is a negative regulator in *Saccharomyces cerevisiae* and shows dynamic interplay with positive-acting histone modifications. *Genes and Development, 20,* 966– 976.

[77] Zhang, K., Chen, Y., Zhang, Z. & Zhao, Y. (2009). Identification and verification of lysine propionylation and butyrylation in yeast core histones using PTMap software. *Proteome Res, 8 (2)*, 900–906.

[78] Boyne, M.T., Pesavento, J.J., Mizzen, C.A. & Kelleher, N.L. (2006). Precise characterization of human histones in the H2A gene family by top down mass spectrometry. *J Proteome Res, 5*, 248–253.

[79] Li, S. & Shogren-Knaak, M.A. (2008). Cross-talk between histone H3 tails produces cooperative nucleosome acetylation. *Proc Natl Acad Sci USA, 105 (47),* 18243-18248.

[80] Martin, C. & Zhang, Y. (2005). The diverse functions of histone lysine methylation. *Nature Rev Mol Cell Biol, 6*, 838–849.

[81] Pokholok, D., Harbison, C., Levine, S., Cole, M., Hannett, N., Lee, T.L., Bell, W, G., Walker, K., Rolfe, A., Herbolsheimer, E., Zeitlinger, J., Lewitter, F., Gifford, D. & Young, R. (2005). Genome-wide map of nucleosome acetylation and methylation in yeast. *Cell, 122*, 517–527.

[82] Sarg, B., Helliger, W., Talasz, H., Forg, B. & Lindner, H.H. (2006). Histone H1 phosphorylation occurs site-specifically during interphase and mitosis. Identification of a novel phosphorylation site on histone H1. *J Biol Chem, 281*, 6573–6580.

[83] Wisniewski, J.R., Zougman, A., Kruger, S. & Mann, M. (2007). Mass spectrometric mapping of linker histone H1 variants reveals multiple acetylations, methylations, and phosphorylation as well as differences between cell culture and tissue. *Mol Cell Proteomics, 6*, 72-87.

[84] Weiss, T., Hergeth, S., Zeissler, U., Izzo, A., Tropberger, F., Zee, B., Dundr, M., Garcia, B., Daujat, S. & Schneider, R. (2010). Histone H1 variant-specific lysine methylation by 9a/KMT1C and Glp1/KMT1D. *Epigenetics & Chromatin,* 3-7.

[85] Smolka, M.B., Albuquerque, C.P., Chen, S.H. & Zhou, H. (2007). Proteome-wide identification of *in vivo* targets of DNA damage checkpoint kinases. *Proc Natl Acad Sci U.S.A, 104*, 10364-10369.

[86] Li, X., Gerber, S.A., Rudner, A.D., Beausoleil, S.A., Haas, W., Villen, J., Elias, J.E. & Gygi, S.P. (2007). Large-scale phosphorylation analysis of alpha-factor-arrested *Saccharomyces cerevisiae. J Proteome Res, 6*, 1190-1197.

[87] Bodenmiller, B., Campbell, D., Gerrits, B., Lam, H., Jovanovic, M., Picotti, P., Schlapbach, R. & Aebersold, R. (2008). PhosphoPep—a database of protein phosphorylation sites in model organisms. *Nature Biotechnology, 26*, 1339 – 1340.

[88] Misteli, T., Gunjan, A., Hock, R., Bustin, M. & Brown, D.T. (2000). Dynamic binding of histone H1 to chromatin in living cells. *Nature, 408*, 877–881.

In: Histones: Class, Structure and Function
Editor: Chang-Hui Shen

ISBN: 978-1-62100-274-1

Chapter 4

HISTONES IN METAPHASE CHROMOSOMES

Susumu Uchiyama and Kiichi Fukui*

Department of Biotechnology, Graduate School of Engineering, Osaka University, Osaka, Japan

ABSTRACT

Recent quantitative proteome analyses of human metaphase chromosomes have revealed various types of histones and their amounts involved in metaphase chromosomes. Core histones, and their variants such as H2A.Z and MacroH2A account for 70 % of the total amount of chromosome proteins. Significant amounts of ubiquinized H2A are also found in chromosomes. In addition, different types of linker histone H1 and its variants, including H1.X, are found in total 6 % of chromosome proteins. Unlike authentic linker histones that localize over the entire chromatin region, histone H1.X accumulates in the nucleolus during interphase and is at the chromosome periphery during mitosis, suggesting a distinct structural role for H1.X. Knockdown analysis of H1.X results in the failure of alignment and/or segregation of mitotic chromosomes. HP1-BP74 which has an H1 globular domain at the central region has been identified from proteome analysis. The HP1-BP74 globular domain binds to the linker DNA between nucleosomes, facilitating the formation of the higher-order structure, chromatosome. Proteome analyses of metaphase chromosomes shed light on the importance of histones in the formation of higher-order chromosome structure.

1. INTRODUCTION

The major components of metaphase chromosome are DNA and proteins. The protein constituents of metaphase chromosomes are divided into histones and non-histone proteins (NHPs). Histones are bound to DNA to form, nucleosome, a complex which is repeated to form a chromatin fiber. The metaphase chromosome is a condensed chromatin fiber constructed during only cell division, possibly with a help of NHPs. Topoisomerase IIα and

* Corresponding author e-mail: kfukui@bio.eng.osaka-u.ac.jp.

condensin complexes are well known NHPs that have roles in the formation or maintenance of chromosome higher-order structure.

The electrostatic properties of chromatin are similar to those of DNA being highly dependent on ionic conditions, since DNA's negative charge is not neutralized by histones. The negatively charged chromatin fiber is self-repulsive in low ionic conditions, the repulsion is reduced in the presence of cations, with polyvalent cations having a greater effect. Indeed, addition of Mg^{2+} results in the aggregation of chromatin fibers [1]. Therefore, in addition to NHPs, divalent cations should be considered as important factors in the chromosome condensation.

Extraction of histones from isolated metaphase chromosomes can be carried out using the acetic acid method [2, 3]. The classical approach of histone isolation and fractionation can also be utilized [4]. In both techniques, histones purified in acidic conditions are obtained as powders that are freely soluble in distilled water. Histones extracted using the acetic acid method can be solubilized into physiological buffers such as phosphate buffered saline, and therefore become highly effective antigens for antibody production against histones [5]. The dissociation of core and linker histones from metaphase chromosomes as a function of salt concentration is very similar to that from bulk chromatin [6]; histone H1 is removed first, followed by H2A and H2B at about 1 M, with H3 and H4 being completely removed above 1.5 M.

In addition to previous studies on individual histones, recent proteome analyses of metaphase chromosomes have presented a comprehensive catalogue of histones involved in the proteome of human metaphase chromosomes [3, 7, 8, 9] and chicken metaphase chromosomes [10]. It should be noted that, although a recent proteome analysis identified more than 4,000 proteins from isolated chromosomes, only 20 different types of histones were identified [10]. The general properties of histones have yet described in detail previously [11] and thus we have focus on histones in metaphase chromosomes.

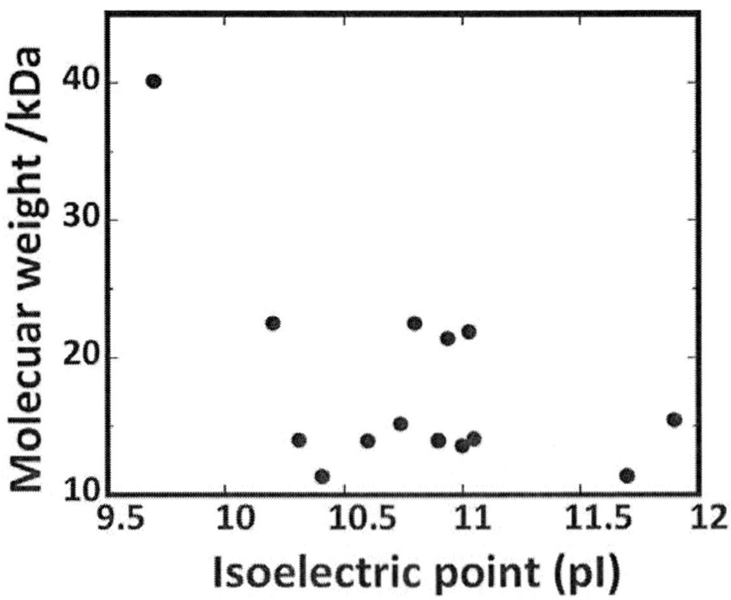

Figure 1. Isoelectric point (pI) and molecular mass of histones in metaphase chromosomes.

Table 1. Histones in metaphase chromosomes

Protein	pI	Mw	Mol % of metaphase chromosomes	Mol % (total hisotnes =100)	Weight %
MacroH2A	9.7	40058	1.98	0.44	1.21
Histone H1.b (H1.4)	11.03	21852	35.34 (H1.b and H.d)	7.83	11.75
Ubiquinized H2A.5	10.2	22475	5.58	1.24	1.91
H1 histone family, member X	10.8	22474	0.82	0.18	0.28
Histone H1.d (H1.2)	10.94	21352	35.34 (H1.b and H.d)		
H3 histone, family 3A	11.9	15394	95.96	21.27	22.48
Histone H2A.X	10.74	15135	4.53	1.00	1.04
Histone H2A.5	11.05	14047	99.11 (H2A total)	21.97	21.18
H2B histone family, member E	10.31	13981	98.89 (H2B total)	21.92	21.04
H2A histone family, member Q	10.9	13980	99.11 (H2A total)		
H2A histone family, member E	10.9	13928	99.11 (H2A total)		
H2B histone family, member N	10.6	13899	98.89 (H2B total)		
H2A histone family, member Z	11	13545	8.87	1.97	1.83
H4 histone family, member A	11.7	11360	100.00	22.17	17.29
Histone H2B.1	10.41	11324	98.89 (H2B total)		

These information is derived from proteome analyses of human metaphase PA chromosomes (Uchiyama et al. 2005).

The other histones identified in the proteome of metaphase chromosomes are H1.03, H1.11R, H1.01, H1.11L, and H1.10 (Ohta et al., 2010).

Table 1 shows the histones involved in human metaphase chromosomes. As shown in Figure 1, histones are highly basic, with a pI ranging from 10 and 12, and have small molecular mass, typically less than 25 kDa. These properties have made biochemical analysis difficult. We have successfully separated histones extracted from isolated metaphase chromosomes by using RFHR (radical free highly reducing) two-dimensional electrophoresis (2-DE). RFHR 2-DE has been employed in the analysis of ribosomal proteins and has great advantages in separating small basic proteins like histones.

In this chapter, based on the previous and recent findings, properties of histones and histone variants in metaphase chromosomes are reviewed.

2. Canonical Histones (Core Histones)

Further properties of histones in metaphase chromosomes are indicated in Table 1 and Figure 2. Four canonical histones (core histones), H2A, H2B, H3, and H4, are most abundant (87.3% mol percent of all histones) and are essential for the construction of nucleosome. Here the quantitative information of histones in a human cell is demonstrated (Figure 3). The mass of DNA per human cell before cell division in the mitotic phase can be estimated as 13.2 pg based on the following calculation; 3 Gbp×2×2×M_{bpDNA} / N_A, where M_{bpDNA} is the molecular mass of DNA base pair (660), and N_A is the Avogadro constant. According to the previous report, the mass of histones per unit mass of DNA is 1.08 for HeLa cells [12]; thus the total mass of histones per cell can be calculated as 14.2 pg. This amounts to 71 % of the total mass of chromosome proteins and 37 % of the total chromosome mass. Quantitative human chromosome proteome analysis revealed the mass percentage of histone H4 out of total histones to be 17.3% [3], from which the mass of histone H4 in one cell is estimated to be 2.5 pg. This value corresponds to 1.3×10^8 histone H4 molecules per cell, leading to the estimation that two molecules of H4 every 184 bp. In the case of human chromosome 1, a replicated chromosome contains DNA with a length of 0.247 Gbp being equal to 0.54 pg plus histones for a total mass of 0.59 pg.

Canonical histones (H2A, H2B, H3, and H4) and their variants are 92% (mol percent) or 88% (mass percent) of total chromosome proteins. During mitosis, although a small fraction of total core histones are localized at the cytoplasmic region, most of them are incorporated into chromatin fiber [13]. A precise localization analysis can be carried out using monoclonal antibodies that specifically recognize histones [5]. Canonical histones have been studied extensively in terms of function, structure, and stability.

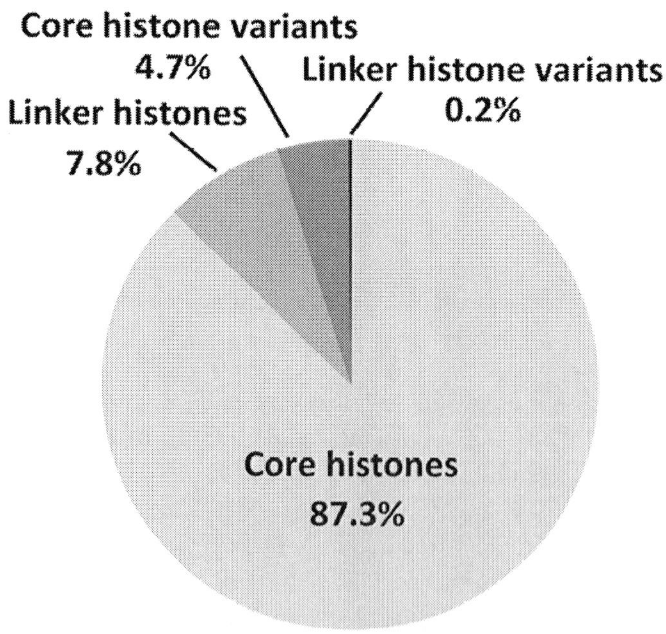

Figure 2. Properties (mol %) of histones in human metaphase chromosomes.

A FRAP experiment using GFP-fused histones as monitoring probes showed that histones are exchanged in mitosis between chromatin fiber and pooled histones located outside the chromatin region, as they do in interphase [14], and with a higher exchange rate at anaphase and telophase than at prophase and metaphase [15].

3. Core Histone Variants

Several types of core histone variants have been identified. It has been shown that H2A and H2B show greater variability than H3 and H4 [11]. Consistently, variants of H2A are frequently identified in proteome analysis of human metaphase chromosomes [3,7]. Core histone variants amounts to 4.7 mol percent of total histones. Proteomic analysis of metaphase chromosomes revealed H2A.Z is the most abundant (2 mol percent) of the histone variants in metaphase chromosomes. H2A.Z has been shown to stabilize the histone octamer in nucleosome structure [16], while H2A.Z-containing nucleosomes exhibit altered linker histone binding [17]. MacroH2A shows unique localization at specific regions of the inactive X chromosome (Xi), including at the X inactivation center, resulting in banding pattern of the metaphase chromosomes [18].

Histone variant H3.3 undergoes phosphorylation at serine 31 in mitotic mammalian cells, predominantly in regions bordering centromeres during mitosis. In a biochemical study of reconstituted nucleosomes using H3.3, subtle differences in the compaction and stability of the particles were observed [17].

H2A.X, a variant of histone H2A, is stably incorporated into chromatin in interphase similarly to canonical H2A [13] and phosphorylated by a kinase ATM in human M-phase cells independently of DSB and DNA damage signaling. A functional contribution to genomic integrity during mitosis is therefore suggested for H2A.X [19].

An important variant of histone H3, CENP-A, which is specifically prerequisite for the construction of centromere, was not identified in the proteome analysis of human metaphase chromosomes due to extremely low abundance compared to other histones.

4. Linker Histones

More than 10 human linker histones have been reported, among which 6 histone H1 proteins (H1.1 (H1a), H1.2 (H1c), H1.3(H1d), H1.4(H1e), H1.5(H1b), and H1.X) are ubiquitously expressed. The role of linker histones in the formation of chromatin higher order structure is currently unclear. Ohsumi *et al.* showed that chromatin condensation occurs without significant histone H1 *in vitro* [20]. Sirotkin *et al.* also indicated that linker histones are not essential for cell viability and deletion of linker histones has no global effect on chromatin condensation [21]. However, more recently it was demonstrated that H1 histones play an important role in the structure and function of chromosomes during mitosis [22]. Fan *et al.* also showed that triple knockout (H1c, H1d, and H1e) mouse embryonic stem cells caused a reduction of linker histone to nucleosome ratio and changed global chromatin structure [23]. These studies imply a structural role for linker histones in chromatin higher order structure.

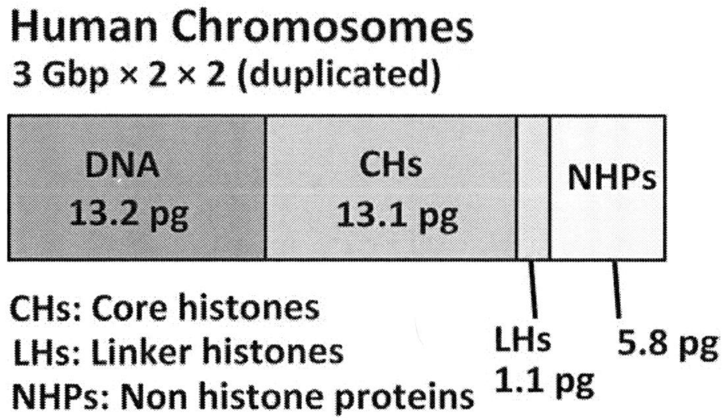

Figure 3. Mass of each component for the case of duplicated human chromosomes.

Linker histones identified in metaphase chromosomes are summarized in Table 1. They account for 7.8 mol percent (12 % by weight) of total histones, which corresponds to 5.8 mol percent of total chromosome proteins and is close to 1/3 of core histones in mol terms. This value means that linker histones are not positioned at every linker DNA segment connecting nucleosomes.

Linker histones have a core region composed solely of α-helices as determined by X-ray crystal analysis [24]. The core region is bound to the naked DNA between nucleosomes, leading to the formation of higher order chromatin structure, the chromatosome. An electron microscopic study showed that histone H1 plays a role in promoting chromatin higher order structure [25]. The N- and C- terminal regions of linker histones interact with the linker DNA and further stabilize chromatosome structure. The characteristic properties of H1-containing chromatin is that entering and exiting DNA segments are brought together, possibly promoting the formation of the zig-zag arrangement.

Histone H1d and H1b are two major linker histones of metaphase chromosomes. Localization analyses of H1d indicated its localization over the entire chromosome region with a slight tendency for localization at chromosome peripheral regions [3].

5. Linker Histone Variants

A variant of histone H1, H1.X, has been investigated in terms of both localization and function in mitosis [26]. Unlike other histones, H1.X localizes at the nucleolus in interphase and covers metaphase chromosomes at chromosome peripheral regions as do chromosome peripheral proteins. Depletion of H1.X by RNAi results in mis-steps during mitosis, such as chromosome misalignment and chromosome non-alignment. Because chromosome structure seems to be identical to that under normal conditions expressing endogenous H1.X, it suggests that the function of H1.X is the regulation of chromosome segregation rather than chromosome condensation.

A protein newly identified from the proteome analysis of human metaphase chromosomes, HP1-BP74 (heterochromatin protein 1-binding protein 74), is unique. Like linker histone H1, HP1-BP74 dissociates from metaphase in the presence of 0.4 M NaCl, at which metaphase chromosomes are converted to a rather swollen morphology [6]. It has 554 amino acid residues including a domain (H1 domain) composed of approximately 100 residues with high similarity to histone H1. NMR analysis of the HP1-BP74 H1 domain showed a high three dimensional structural similarity to H1 [6]. Chromatosome formation is induced by the binding of HP1-BP74 H1 domain to reconstituted nucleosome [6]. These results suggest that HP1-BP74 could be classified as a variant of histone H1. Additionally, HP1-BP74 has a HP1 binding ability via an HP1 binding motif located near the H1 domain. It is possible that HP1-BP74 organizes chromatin in concert with HP1 proteins.

6. Post-Translational Modification (PTM) of Histones

Most prominent post-translational modification (PTM) occurring in core histones during mitosis is the phosphorylation of serine 10 in histone H3 (H3S10) by mitotic kinases, including Aurora B [27] and Vaccinia-Related Kinase 1 [28]. Phosphorylation of histone H3 is depicted in Figure 4. Phosphorylation of histone H3 seems to be important for the proper condensation and segregation of chromosomes [29, 30]. In mammalian cells, phosphorylation of H3S10 starts to increase in late G2 in pericentromeric regions and then the phosphorylation becomes more prominent at the tips of chromosome arms as mitosis proceeds [31]. The temporal and spatial relationship between chromosome condensation and phosphorylation of H3S10 is clearly correlated [32]. Dephosphorylation of H3S10 begins in late anaphase and is completed in early telophase at which chromosomes still retain condensed state [33].

However, another study showed that phosphorylation of H3S10 is required for the initiation of chromosome condensation rather than the maintenance of condensed states during mitosis [34]. Furthermore, de la Barre *et al.* showed that the tail of H3 is not important for chromosome condensation based on experimental results that deletion of the histone H3 tail region has no effect on the chromosome condensation in the *Xenopus* reconstitution system [35]. A yeast strain with a single H3S10 mutation showed cell cycle progression identical to wild-type strain [36]. Thus, the exact role of histone H3 phosphorylation in chromosome condensation during mitosis is still enigmatic.

Mass spectrometry analyses of human histone H3 during mitosis revealed that phosphorylation and methylation can occur on adjacent residues of the same H3, such as Lys9/Ser10 and Lys27/Ser28 [37]. These results support the "methylation/phosphorylation" binary switch hypothesis. Methylation of Lys 9 and phosphorylation of Ser 10 on histone H3, could modulate the binding of effector molecules, such as HP1, to histones. As cells enter mitosis, phosphorylation of H3S10 promotes the dissociation of HP1β that is bound to di- or tri- methylated histone H3 at Lys9, without change in the level of H3 tri-methylation. Methylation of Lys 9 in histone H3 is known to recruit HP1 and mediate heterochromatin formation [38].

The recent discovery of H2A, H2B, and H4 post-translational modification by O-GlcNAc (β-N-acetylglucosamine) sheds light on the significance of this novel modification in chromatin higher order structure, as they observed change of chromatin structure by O-GlcNAcylation during mitosis [39] and failure in faithful chromosome segregation under the presence of over-expressed N-acetylglucosamine (O-GlcNAc) transferase [40].

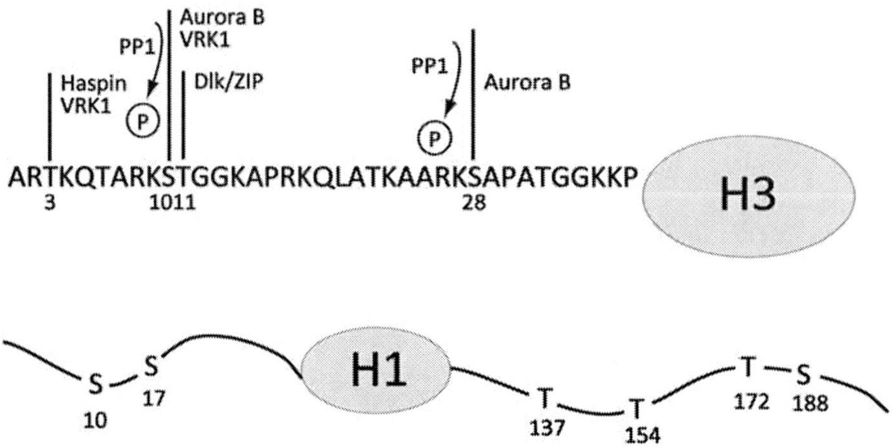

Figure 4. Phosphorylation sites of histone H3 and H1 (H1.5) during mitosis. Phosphorylation of H1.3 is indicated based on the report by Sarg *et al.* [41].

Phosphorylation is the main PTM of linker histones. Six different sites have been reported as amino acid residue to be modified in histone H1.5 [41]. The role of histone H1 phosphorylation remains unclear. It might loosen the association of histone H1 with DNA and thereby promote the interaction of other chromatin condensation factors with chromatin [42].

Conclusions

Proteome analysis of metaphase chromosomes indicates that histones are the main protein constituents of metaphase chromosomes. Recent studies suggest that histones might be the only proteins that are essential for the formation of metaphase chromosomes. Some results have indicated the PTM of histones plays a role for chromosome condensation. Weak forces derived from PTM may be accumulated, resulting in the global structural conversion of chromosomes. Alternatively, the change in chromosome environmental conditions might facilitate chromosome condensation. For example, shielding of repulsive forces between chromatins by increasing cation concentration might be important for chromosome condensation. Further investigations of chromatin fibers in cells are required for complete understanding the structural details of metaphase chromosomes.

Acknowledgments

This study was supported in part by Special Coordination Funds of the Ministry of Education, Culture, Sports, Science and Technology, Japan, granted to SU and KF. SU is also grateful to Senri Life Science Foundation for the research funding.

REFERENCES

[1] Ausio, J. (2000) Analytical ultracentrifugation and the characterization of chromatin structure. *Biophys. Chem. 86*, 141-153.

[2] Wada, A. (1986) Analysis of Escherichia coli ribosomal proteins by an improved two dimensional gel electrophoresis. I. Detection of four new proteins. *J. Biochem. 100*, 1583-1594.

[3] Uchiyama, S., Kobayashi, S., Takata, H., Ishihara, T., Hori, N., Higashi, T., Hayashihara, K., Sone, T., Higo, D., Nirasawa, T., Takao, T., Matsunaga, S., & Fukui, K. (2005) Proteome analysis of human metaphase chromosomes. *J. Biol. Chem. 280*, 16994-17004.

[4] Johns, E. W. (1977) The isolation and purification of histones. *Methods Cell Biol., 16*, 183-203.

[5] Higashi, T., Miyakawa, S., Uchiyama, S., Matsunaga, M., Takata, H., Fujimoto, S., Noda, N., Terauchi, A., Shimizu, S., Oda, O., Azuma, T., & Fukui, K. (2005) Generation of monoclonal antibodies against chromosomal antigens that have a high sequence similarity between humans and mice. *J. Biotechnology 120*, 262-72.

[6] Hayashihara, K., Uchiyama, S., Shimamoto, S., Kobayashi, S., Tomschik, M., Wakamatsu, H., No, D., Suagahara, H., Hori, N., Noda, M., Ohkubo, T., Zlatanova, J., Matsunaga, S., & Fukui, K. (2010) The middle region of a HP1-binding protein, HP1-BP74, associates with linker DNA at the entry/exit site of nucleosomal DNA. *J. Biol. Chem. 285*, 6498-6507.

[7] Takata, H., Uchiyama, S., Nakamura, N., Nakashima, S., Kobayashi, S., Sone, T., Kimura, S., Lahmers, S., Granzier, H., Labeit, S., Matsunaga, S., & Fukui, K. (2007) A comparative proteome analysis of human metaphase chromosomes isolated from two different cell lines reveals a set of conserved chromosome-associated proteins. *Genes to Cells 12*, 269-284.

[8] Fukui, K. & Uchiyama, S. (2007) Chromosome protein framework from proteome analysis of isolated human metaphase chromosomes. *Chemical Rec. 7*, 230-7.

[9] Uchiyama, S. (2009) The proteome of human metaphase chromosomes. *Chromosome Science 12*, 13-25.

[10] Ohta, S., Bukowski-Wills, J. C., Sanchez-Pulido, L., Alves, Fde. L., Wood, L., Chen, Z. A., Platani, M., Fischer, L., Hudson, D. F., Ponting, C. P., Fukagawa, T., Earnshaw, W. C., & Rappsilber, J. (2010) The protein composition of mitotic chromosomes determined using multiclassifier combinatorial proteomics. *Cell 142*, 810-821.

[11] van Holde, K. E. (1989). Chromatin. New York; Springer-Verlag,

[12] Bhorjee, J. S. & Pederson, T. (1973) Chromatin: its isolation from cultured mammalian cells with particular reference to contamination by nuclear ribnucleoprotein particles. *Biochemistry 12*, 2766-2773

[13] Higashi T, Matsunaga S, Isobe K, Morimoto A, Shimada T, Kataoka S, Watanabe W, Uchiyama S, Itoh K, and Fukui K (2007) Histone H2A mobility is regulated by its tails and acetylation ofcore histone tails. *Biochem. Biophys. Res. Comm. 357*, 627-632.

[14] Kimura, H. & Cook, P. (2001) Kinetics of core histones in living human cells: little exchange of H3 and H4 and some rapid exchange of H2B, *J. Cell Biol. 153*, 1341–1353.

[15] Chen, D., Dundr, M., Wang, C., Leung, A., Lamond, A., Misteli, T., & Huang, S. (2005) Condensed mitotic chromatin is accessible to transcription factors and chromatin structural proteins. *J. Cell Biol. 168*, 41-54.

[16] Park, Y. J., Dyer, P. N., Tremethick, D. J., Luger, K. (2004) A new fluorescence resonance energy transfer approach demonstrates that the histone variant H2AZ stabilizes the histone octamer within the nucleosome. *J. Biol. Chem. 279*, 24274-82.

[17] Thakar, A., Gupta, P., Ishibashi, T., Finn, R., Silva-Moreno, B., Uchiyama, S., Fukui, K., Tomschik, M., Ausio, J., & Zlatanova, J. (2009) H2A.Z and H3.3 histone variants affect nucleosome structure: biochemical and biophysical studies. *Biochemistry 48*, 10852-10857.

[18] Chadwick, B. P., & Willard, H. F. (2002) Cell cycle-dependent localization of macroH2A in chromatin of the inactive X chromosome. *J. Cell Biol. 157*, 1113-23.

[19] Ichijima, Y., Sakasai, R., Okita, N., Asahina, K., Mizutani, S., Teraoka, H. (2005) Phosphorylation of histone H2AX at M phase in human cells without DNA damage response. *Biochem. Biophys. Res. Commun. 336*, 807-812.

[20] Ohsumi, K., Katagiri, C., & Kishimoto, T. (1993). Chromosome condensation in Xenopus mitotic extracts without histone H1. *Science 262*, 2033-2035

[21] Sirotkin, A. M., Edelmann, W., Cheng, G. Klein-Szanto, A., Kucherlapati, R., & Skoultchi, A.I. (1995) Mice develop normally without the H1(0) linker histone, *Proc. Natl. Acad. Sci. USA, 92*, 6434–6438

[22] Maresca T. J., Freedman, B.S. & Heald, R. (2005) Histone H1 is essential for mitotic chromosome architecture and segregation in Xenopus laevis egg extracts. *J. Cell Biol. 169*, 859–869.

[23] Fan, Y., Nikitina, T., Zhao, J., Fleury, T. J., Bhattacharyya, R., Bouhassira, E. E., Stein, A., Woodcock, C. L., & Skoultchi, A. I. (2005) Histone H1 depletion in mammals alters global chromatin structure but causes specific changes in gene regulation. *Cell, 123*, 1199–1212.

[24] Ramakrishnan, V., Finch, J. T., Graziano, V., Lee, P.L., & Sweet, R. M. (1993) Crystal structure of globular domain of histone H5 and its implications for nucleosome binding. *Nature 362*, 219–223

[25] Bednar, J., Horowitz, R. A., Grigoryev, S. A., Carruthers, L. M., Hansen, J. C., Koster, A. J., & Woodcock, C. L. (1998) Nucleosomes, linker DNA, and linker histone form a unique structural motif that directs the higher-order folding and compaction of chromatin. *Proc Natl Acad Sci U S A. 95*, 14173-14178.

[26] Takata, H., Matsunaga, S., Morimoto, A., Maniwa-Ono, R., Uchiyama, S., & Fukui, K. (2007) H1.X with different properties from other linker histones is required for mitotic progression. *FEBS Lett. 581*, 3783-8.

[27] Kouzarides, T (2007) Chromatin modifications and their function. *Cell 128*, 693-705.

[28] Kang, T. H., Park, D. Y., Choi, Y. H., Kim, K. J., Yoon, H. S., & Kim, K.T. (2007) Mitotic histone H3 phosphorylation by vaccinia-related kinase 1 in mammalian cells. *Mol. Cell. Biol. 27*, 8533-8546.

[29] Strahl, B. D. & Allis, C. D. The language of covalent histone modifications. (2000) *Nature 403*, 41-45.

[30] Smith, M. M. & Allis, C. D. (2000). Mitotic phosphorylation of histone H3 is governed by Ipl1/aurora kinase and Glc7/PP1 phosphatase in budding yeast and nematodes. *Cell 102*, 279- 291.

[31] Goto, H., Tomono, Y., Ajiro, K., Kosako, H., Fujita, M., Sakurai, M., Okawa, K., Iwamatsu, A., Okigaki, T., Takahashi, T., & Inagaki, M. (1999) Identification of a novel phosphorylation site on histone H3 coupled with mitotic chromosome condensation. *J. Biol. Chem. 274*, 25543-25549.

[32] Goto, H., Yasui, Y., Nigg, E. A., & Inagaki, M. (2002) Aurora-B phosphorylates Histone H3 at serine28 with regard to the mitotic chromosome condensation. *Genes Cells 7*, 11-17.

[33] Hendzel, M.J., Wei, Y., Mancini, M. A., van Hooser, A., Ranalli, T., Brinkley, B. R., Bazett-Jones, D. P., & Allis, C. D. (1997) Mitosis-specific phosphorylation of histone H3 initiates primarily within pericentromeric heterochromatin during G2 and spreads in an ordered fashion coincident with mitotic chromosome condensation. *Chromosoma 106*, 348-360.

[34] van Hooser, A., Goodrich, D. W., Allis, C. D., Brinkley, B. R., & Mancini, M. A. (1998) Histone H3 phosphorylation is required for the initiation, but not maintenance, of mammalian chromosome condensation. *J. Cell Sci. 111*, 3497-506.

[35] de la Barre, A. E., Angelov, D., Molla, A., & Dimitrov, S. (2001) The N-terminus of histone H2B, but not that of histone H3 or its phosphorylation, is essential for chromosome condensation. *EMBO J. 20*, 6383-6393.

[36] Hsu, J. Y., Sun, Z. W., Li, X., Reuben, M., Tatchell, K., Bishop, D. K., Grushcow, J.M., Brame, C. J., Caldwell, J. A., Hunt, D. F., & Lin, R., (2000) Mitotic phosphorylation of histone H3 is governed by Ipl1/aurora kinase and Glc7/PP1 phosphatase in budding yeast and nematodes. *Cell 102*, 279-291.

[37] Garcia, B. A., Barber, C. M., Hake, S. B., Ptak, C., Turner, F. B., Busby, S. A., Shabanowitz, J., Moran, R. G., Allis, C.D., & Hunt, D. F. (2005) Modifications of human histone H3 variants during mitosis. *Biochemistry 44*, 13202-13213.

[38] Fischle, W., Tseng, B. S., Dormann, H. L., Ueberheide, B. M., Garcia, B. A., Shabanowitz, J., Hunt, D. F., Funabiki, H., & Allis, C. D. (2005) Regulation of HP1-chromatin binding by histone H3 methylation and phosphorylation. *Nature 438*, 1116-11122.

[39] Sakabe, K., Wang, Z., & Hart, G. W. (2010) Beta-N-acetylglucosamine (O-GlcNAc) is part of the histone code. *Proc. Natl. Acad. Sci. U S A. 107*, 19915-19920.

[40] Sakabe, K., & Hart, G. W. (2010) O-GlcNAc transferase regulates mitotic chromatin dynamics. *J. Biol. Chem. 285*, 34460-34468.

[41] Sarg, B., Helliger, W., Talasz, H., Forg, B. & Lindner, H. H. (2006) Histone H1 phosphorylation occurs site-specifically during interphase and mitosis: identification of a novel phosphorylation site on histone H1. *J. Biol. Chem. 281*, 6573–6580.

[42] Roth, S. Y. & Allis, C. D. (1992) Chromatin condensation: does histone H1 dephosphorylation play a role? *Trends Biochem. Sci. 17*, 93–98.

In: Histones: Class, Structure and Function
Editor: Chang-Hui Shen
ISBN: 978-1-62100-274-1

Chapter 5

HISTONE MODIFICATIONS AND GENE ACTIVATION

***Roshini Nilupama Wimalarathna*[1,3,*] *and Chang-Hui Shen*[1,2,3]**
[1]Department of Biology, College of Staten Island, City University of New York, Staten Island, NY, US
[2]Institute for Macromolecular Assemblies, City University of New York, Staten Island, NY, US
[3]The Graduate Center, City University of New York, NY, US

ABSTRACT

Histones are subjected to post translational modifications such as acetylation, deacetylation, methylation, phosphorylation, ubiquitination, sumoylation, ADP ribosylation, glycosylation, biotinylation, and carbonylation. The first four post-translational modifications (acetylation, deacetylation, methylation, and phosphorylation) have been studied extensively in comparison with the other types of modification. Histone modification activity has been shown to participate in a variety of cellular processes including, but not limited to, gene activity. In this chapter, we will discuss current findings of histone modifications and its connection to gene activity.

1. INTRODUCTION

The nucleosome is the basic structural repeat unit of chromatin. A nucleosome contains 147 bp of DNA wrapped in 1.75 superhelical turns around a central octamer. This central octamer is composed of two of each of the four core histones H2A, H2B, H3 and H4. The chromatin structure helps to package a vast amount of DNA into the cellular nucleus, but this compaction of DNA creates a barrier in the accessibility of the information encoded within the DNA for many DNA template-based cellular activities, such as transcription, replication,

[*] Corresponding author. E-mail: Roshini.Wimalarathna@csi.cuny.edu.

recombination and repair. To overcome the repressive effects of chromatin on RNA polymerase II-mediated transcription, the chromatin structure must be perturbed at promoter regions prior to or during transcriptional activation [1]. There are two types of proteins that work on the nucleosomes, which allow for the access of the DNA by *trans-acting* factors. The first type is ATP-dependent chromatin remodelers, which typically form complexes with protein partners and use the free energy generated by ATP hydrolysis to remodel the chromatin structure [2, 3]. The other type of protein covalently modifies histone proteins by acetylation [4], deacetylation [5], methylation [6], phosphorylation [7], ubiquitynation [8], sumoylation [9], ADP ribosylation [10], biotinylation [11] glycosylation and carbonylation. Acetylation, deacetylation, methylation, and phosphorylation have been studied extensively, whereas relatively little is known about the other types of modification. In this chapter, we will discuss these post-translational modifications, as well as the connection of histone modification and gene activity.

2. Histone Acetylation

Acetylation of lysine residues on the core histones N-terminal tails is normally associated with transcriptionally active genes. In general, histone H3 Lys9, Lys14, Lys18 and Lys23, histone H4 Lys5, Lys8, Lys12 and Lys16, histone H2A Lys5 and Lys9, and histone H2B Lys5, Lys12, Lys15 and Lys20 are targets for acetylation [3, 12, 13]. Histone acetyltransferases (HATs) are the enzymes responsible for the transfer of an acetyl group to the ε-amine group of lysine residues. HATs consist of both large multi-subunit complexes and transcriptional co-activators, which are recruited to promoters by interactions with DNA-bound activator proteins [14]. HATs not only acetylate histone proteins but also modify non-histone proteins such as tumor suppressor protein p53 [15] and minichromosome maintenance 3 protein [16].

Based on sequence homology, HATs can be grouped into different families, such as Hat1p, Gcn5p/PCAF, p300/CBP, MYST, p160,CIITA, Atf2p, $TAF_{II}250$, TFIIC, Nut1p, Elp3p, CDY, Hpa2p, TFIIB, MCM3AP, Eco1p and Ard1p [17, 18; Table 1]. In terms of sub-cellular location, HATs can be categorized into two types, Type A and Type B. Type A HATs are normally found in the nucleus. This type of HAT acetylates histone in the chromatin context and may play important roles in the regulation of gene expression by functioning as transcriptional co-activators. PCAF, CBP, p300, $TAF_{II}250$, ACTR, Tip60p, Esa1p and Elp3p are type A HATs [4, 19-24]. Type B HATs can be found in the cytoplasm and are involved in the acetylation of nascent histones, H3 and H4. Hat1p is the only currently known type B HAT [19, 25-26]; however, recent evidence has shown that Hat1p may be present in multiple locations [19, 27], and thus may not precisely fit into this classification.

The Gcn5p/PCAF family is one of the nuclear HATs, which consists of Gcn5p, PCAF, and related proteins. Yeast Gcn5p contains a HAT domain and a bromodomain, which is highly homologous to the human PCAF and Gcn5p. Gcn5p was the first type A HAT discovered in *Tetrahymena thermophila* [28]. Later, homologs of Gcn5p were found in humans, mice, *Schizosaccharomyces pombe* and *Drosophila melanogaster*, suggesting that its function is highly conserved throughout eukaryotes [29- 32]. Gcn5p primarily acetylates histone H3 Lys14 and histone H4 Lys8 and Lys16 [33]. Yeast Gcn5p is required for normal progression through the G2/ M boundary [34] and mitotic gene expression [35]. The gene for PCAF, which is also referred to as P/CAF, was originally identified from a human cDNA database on the basis of its homology to Gcn5p. PCAF preferentially acetylates histone H3 Lys14 and weakly acetylates histone H4 Lys8 [36]. Although PCAF was originally characterized as a HAT, it can also acetylate various non-histone transcription related proteins, such as the chromatin proteins HMG17 and HMGI(Y), transcriptional activator p53 and MyoD, and general transcription factors TFIIE and TFIIF [15, 32].

The MYST family is the most diverse nuclear HAT family and is important in transcriptional activation, DNA repair, and gene silencing. Yeast Sas2p, Sas3p and Esa1p are members of the MYST family. Both Sas2p and Sas3p are involved in gene silencing [32, 37-38], and Esa1p is essential for cell cycle progression [39]. It has been shown that Esa1p can acetylate free histones H4 Lys5 with high efficiency, and acetylate H4 Lys8, Lys12 and Lys16 with moderate efficiency *in vitro* [40]. Histone H3 Lys14 and histone H2A Lys4 are potentially major targets, while histone H3 Lys4 and histone H2A Lys7 are potentially moderate targets. Furthermore, the *esa1* mutants succeed in replicating their DNA but fail to proceed through mitosis and cytokinesis, suggesting a role of Esa1p in cell cycle progression.

Both p300 and CBP (CREB binding protein) are transcriptional co-activators, which can acetylate both histones and non-histone proteins [17]. They are recruited to specific promoters through interactions with transcription factors, such as E1A and phosphorylated CREB [41, 42]. P300/CBP contain putative zinc finger domains, a bromodomain, a HAT domain, and at least two independent regions that interact with multiple transcription factors [42, 43]. Mutations of p300 are associated with certain cancers such as leukemia [17], colorectal carcinomas, and gastric carcinomas [44, 45]. The p300/CBP is able to acetylate all four histones within the nucleosome as well as free histones.

Yeast Hat1p is the only type B HAT, which is found in the cytoplasm and can acetylate newly synthesized histones H3 and H4. Hat1p preferentially acetylates H4 Lys5 and Lys12 [19, 46]. Hat2p is usually co-purified with Hat1p from the cytosolic extracts. Hat2p is not required for the catalytic activity of Hat1p but acts as a regulatory subunit when in complex with Hat1p. It has been shown that the catalytic activity of Hat1p increases 10 fold when Hat2p binds to Hat1p compared to the activity of the Hat1p alone [47]. Although Hat1p is mainly localized in the cytoplasm, a recent study has shown that Hat1p can be translocated into the nucleus for telomeric silencing. In contrast to other HATs, this process is mediated mainly through H4 Lys12. Hat1p is involved in transcriptional repression rather than gene activation [19].

It is commonly accepted that histone acetylation is important in transcriptional activation, although some exceptions do exist, such as the acetylation of histone H4 Lys12, which is a mark for the silent chromatin state as described above [19, 48, 49]. Histone acetylation can modulate gene transcription at two levels: global histone acetylation and promoter–specific histone acetylation. While global histone acetylation correlates with general transcription activity [50], promoter-specific acetylation is a critical mechanism for the control of specific gene activity [51].

Table 1. Classification of histone acetyltransferases [17, 18]

Family	HATs	Organism	Histone	Site	Function
Hat1	Hat1	*S.cerevisia*	H4	K5	Histone deposition
		to		K12	Gene silencing, Histone deposition
		mammals	H2A	K7	?
Gcn5/P	Gcn5	*S.cerevisia*	H2B	K11	Transcriptional activation
CAF		to		K16	Transcriptional activation
		mammals	H3	K9	Transcriptional activation
				K14	Transcriptional activation
				K18	Transcriptional activation
					DNA repair
				K27	Transcriptional activation
			H4	K8	Transcriptional activation
				K16	Transcriptional activation
	PCAF		H3	K14	Transcriptional activation
		Mammals	H4	K8	Transcriptional activation
p300/C	CBP	*C.elegans*	H2A	K5	Transcriptional activation
BP		to	H2B	K12	Transcriptional activation
		mammals		K15	
			H3	K18	DNA replication
				K23	Transcriptional activation
	P300		H2A	K5	Transcriptional activation
		Mammals	H2B	K12	Transcriptional activation
				K15	
				K20	Transcriptional activation
			H3	K14	Transcriptional activation
				K18	DNA replication
				K23	Transcriptional activation
MYST	Sas2	*S.cerevisia*	H3	K14	Gene silencing
			H4	K16	Gene silencing
	Sas3	*S.cerevisia*	H3	K14	Transcriptional activation,
					Elongation
			K23		DNA repair
	Eas1	*S.cerevisia*	H2A	K4	Transcriptional activation
				K7	Transcriptional activation
			H2B	K16	Transcriptional activation
			H3	K4	Transcriptional activation
				K14	Transcriptional activation,
					DNA repair
			H4	K5	Transcriptional activation,
					DNA repair
				K8	Transcriptional activation
				K12	Transcriptional activation,
					DNA repair
				K16	DNA repair

Family	HATs	Organism	Histone	Site	Function
	Mof	*Drosophila*	H4	K16	Gene dosage compensation
	TIP60	Mammals	H2A	K5	Transcriptional activation
			H3	K14	Transcriptional activation, DNA repair
			H4	K5	Transcriptional activation, DNA repair
				K8	Transcriptional activation, DNA repair
				K12	Transcriptional activation, DNA repair
	MOF	Mammals	H4		Transcriptional activation
	HBO1	Mammals	H3/H4		DNA replication, Transcriptional repressor
	MOZ	Mammals	H3/H4		Transcriptional activation
	MORF	Mammals	H3/H4		Transcriptional activation
p160	SCR-1	Mammals	H3	K9	Transcriptional activation
				K14	Transcriptional activation
	ACTR	Mammals	H3/H4		Transcriptional activation
CIITA	CIITA	Mammals	H4		Transcriptional activation
ATF2	ATF2	Mammals	H2B	K5	Transcriptional activation
				K12	Transcriptional activation
				K15	Transcriptional activation
			H4	K5	Transcriptional activation
				K8	Transcriptional activation
				K16	Transcriptional activation
TAF_{II} 250	$TAF_{II}230$	*Droso[hila*	H3/H4		Transcription initiation
	$TAF_{II}250$	Mammals	H3/H4		Transcriptional initiation, kinase and ubiquitin ligase
TFIIIC	hTFIIIC90	*S.cerevisia* to mammals	H3	K14	RNA polymerase III transcription
	TFIIIC subunits		H4		Transcriptional initiation
Nut1	Nut1	*S.cerevisia*	H3/H4		Transcriptional initiation
Elp3	Elp3	*S.cerevisia* to mammals	H3	K14	Transcriptional elongation
			H4	K8	Transcriptional elongation
CDY	CDY	Humans	H4		Histone to protamine transition during spermatogenesis
	CDYL	Mammals	H4		
Hpa2	Hpa2	*S.cerevisia*	H3	K4	Unknown
				K14	Unknown
			H4	K5	Unknown
				K12	Unknown
	Hpa3	*S.cerevisia*	H3/H4		Unknown

At the promoter-specific level, it has been shown that the induction of *CUP1* in *Saccharomyces cerevisiae* with copper resulted in targeted acetylation of both H3 and H4 at the *CUP1* promoter [52]. Nucleosomes containing upstream activating sequences of *CUP1* and sequences farther upstream were also the targets for acetylation. It has also been shown that the levels of H3 and H4 histone acetylation at the *INO1* promoter and *ORF* (open reading frame) increased under inducing conditions [53]. Such an increase of histone acetylation is not due to the increase of the nucleosome number; instead, it is because the acetylation level per nucleosome increased significantly. Therefore, these experiments demonstrated the direct connection between promoter-specific histone acetylation and gene activation.

At the global level, it has been reported that histone H4 Lys5 and Lys12 are preferentially bound by the TAFII250 bromodomains *in vitro*, and these interactions promote acetylation-stimulated global transcription [54].

3. Histone Deacetylation

Histone deacetylases (HDACs) oppose the effects of HATs by deacetylating acetylated lysine residues. Histone deacetylation reverses the charged, neutralizing effect of the acetyl group on histone tails to a charged tail state and promotes the interaction between DNA and histone tails. As such, histone deacetylation causes the condensation of chromatin fibers and promotes gene repression. HDACs are also called lysine deacetylases (KDAC). HDAC not only perform histone deacetylation, but can also deacetylate non-histone proteins such as α-tubulin [55].

HDACs can be divided into four classes based on their function and protein sequence similarity [Table 2]. The class I and class II HDACs are classical HDACs and their activities can be inhibited by trichostatin A (TSA), whereas the class III HDACs are a family of NAD+-dependent proteins which are not affected by TSA. HDAC1, HDAC2, HDAC3 and HDAC8 belong to class I HDACs. HDAC1, HDAC2 and HDAC8 are primarily found in the nucleus, whereas HDAC3 can be found in both the nucleus and cytoplasm. The yeast reduced potassium dependency 3 (Rpd3p) is also a class I HDAC.

HDAC4, HDAC5, HDAC6, HDAC7, HDAC9a, HDAC9b and HDAC10 are class II HDACs. Yeast histone deacetylase 1 (Hda1p) is also a class II HDAC [5,56]. Class II HDACs can be shuttled in and out of the nucleus in response to certain cellular signals [57-58]. HDAC6 is mainly localized in the cytoplasm and is responsible for deacetylation of α-tubulins.

Mammalian Sirtuins including SIRT1, SIRT2, SIRT3, SIRT4, SIRT5, SIRT6 and SIRT7 belong to class III HDACs. Yeast silent information regulator 2 (Sir2p) is also a class III HDAC [5, 58]. HDAC11 is the only class IV HDAC [55, 58-59] and is localized in the nucleus, although it has been shown that it can be co-precipitated with the cytoplasmically localized HDAC6 [60]. HDAC11 has a catalytic domain situated at the N-terminus, with proven HDAC activity that can be inhibited by trapoxin (a TSA analogue). There is no association of HDAC11 with any of the known HDAC complexes, suggesting a distinct biochemical function of HDAC11 [60].

Table 2. Classification of Histone Deacetylases [58]

Class I	Class II	Class III	Class IV
HDAC1, HDAC2, HDAC3, HDAC8	HDAC4, HDAC5, HDAC6, HDAC7, HDAC9a, HDAC9b, HDAC10	Sirtuins in mammals SIRT1, SIRT2, SIRT3, SIRT4, SIRT5, SIRT6, SIRT7	HDAC11
RPD3 in the yeast (*Saccharomyces cerevisiae*)	HDA1 in the yeast (*Saccharomyces cerevisiae*)	SIR2 in the yeast (*Saccharomyces cerevisiae*)	

4. HISTONE METHYLATION

Histone methylation was discovered over 35 years ago, however, only recently has this process been investigated. Methylation occurs at lysine and arginine residues on the histone tails of histones H3 and H4, and is associated with both transcriptional activation and repression. Histone methyltransferases (HMTs) are the enzymes responsible for these modifications. Based on their substrate specificity HMTs can be grouped into two families. Histone lysine methyltransferases (HKMTs) catalyze the methylation of lysine residues. On the other hand, protein arginine methyltransferases (PRMTs) catalyze the methylation of arginine residues. The PRMTs contain a conserved catalytic domain and an *S*-adenosyl methionine-binding region. Specificity with respect to the particular arginines methylated varies, as does the number of methyl groups attached. PRMTs can attach one or two methyl groups to arginine residues allowing for variation in the symmetry of these groups [6, 16, 61]. The HKMTs contain the evolutionarily conserved SET domain. The SET domain is a catalytic motif and is approximately 130 amino acids in length [62-66]. Based on the sequence homology within and surrounding the SET domain, SET-containing HKMTs can be divided into six different subfamilies including SET1, SET2, SUV39, EZH, SMYD, and PRDM [62-63, 67-68].

Table 3. Histone methyltransferases and their targeted residues [131]

HMTs	Histone	Site
PRMT1	H4	R3
PRMT5	H3 H4	R8 R3
NSD1	H3 H4	K36 K20
PR-SET7/SET8, SUV4-20H1, SUV4-20H2	H4	K20
EZH2	H3	K27
CARM1	H3	R26 R17
SET1, MLL/HRX/ALL1, MLL2/HRX2, MLL3/HALR, ASH1, SMYD3	H3	K4
SUV39H1, SUV39H2, G9a, EuHMTase, SETDB1/ESET, RIZ1	H3	K9
DOT1	H3	K79

Methylation of lysine residues is known to occur on histone H3 Lys4, Lys9, Lys14, Lys27, Lys36 and Lys79, and histone H4 Lys20. Histone H3 Lys4 methylation is a mark for transcriptionally active genes [69; Table 3]. There are usually three different states of H3 Lys4 methylation. The mono- and di-methylated states are associated with both transcriptionally active and inactive genes, respectively, whereas the tri-methylated state occurs exclusively at transcriptionally active genes [70].

Studies of fission yeast mating type loci have shown that a large stretch of silent heterochromatin (20kb) contains histones methylated at H3 Lys9 but is devoid of H3 Lys4 methylation. In contrast, the surrounding actively transcribed regions have H3 Lys4 methylation but not H3 Lys9. Removal of a boundary element results in the spreading of H3 Lys9 methylation into the active domain, resulting in depletion of H3 Lys4 methylation. This result suggests that methylation of H3 Lys9 is associated with gene silencing, while methylation of H3 Lys4 is associated with transcriptional activation [71-72].

The methylation of arginine residues is known to occur on histone H3 Arg 2, Arg 17 and Arg26, and histone H4 Arg26 [73-74]. Arginines can be either mono-methylated or di-methylated on the guanidino amine (or imine). The methylation of H3 Arg 17 has a positive influence on transcription [69].

5. Histone Phosphorylation

Histone phosphorylation is involved in transcriptional activation and chromatin condensation during mitosis. Histone phosphorylation occurs at serine, threonine and tyrosine residues. The ratio of phosphorylation for serine, threonine, and tyrosine residues is approximately 1000:100:1, respectively [75].

The phosphorylation of histone H3 Ser10 is conserved throughout eukaryotes and is crucial for chromosomal condensation and cell cycle progression during mitosis and meiosis [7]. In addition, it is important during interphase because H3 Ser10 phosphorylation enables an increase in gene transcription, which is activated as a consequence of a variety of cell-signaling events [76]. It has been shown that histone H3 phosphorylation is necessary for the initiation of chromosome condensation in *Tetrahymena thermophila* [7] and in mammalian cells [77]. Furthermore, it has been observed that phosphorylation of histone H3 Ser10 enhances gene expression after neuronal activation through the administration of agonists, which activate dopaminergic, muscarinic acetylcholine, and ionotropic glutamate receptors in hippocampal neurons [78].

It has been shown that H3 Ser28 phosphorylation occurred in chromosomes predominately during early mitosis and coincided with the initiation of mitotic chromosome condensation [79]. Furthermore, histone H3 phosphorylation at Ser28, together with Ser10 phosphorylation, is a conserved event and is likely to be involved in mitotic chromosome condensation. The phosphorylation of linker histones has also been related to mitotic chromosome condensation [80]. Cyclin-dependent kinases participate in cell cycle-dependent phosphorylation of linker histones [81]. Histone H2A from *Tetrahymena pyriformis* has been shown to be hyper-phosphorylated at its C-terminal tail [82], and this hyper-phosphorylation might be important for chromosome stability. H2A.X phosphorylation was shown to correlate with double stranded breakage [83].

6. Histone Ubiquitination

Ubiquitin is an 8.6-kDa, 76 amino acid protein that is highly conserved and is "ubiquitously" distributed throughout eukaryotes. It is a globular protein in which residues 1-72 are tightly folded and 73-76 are in a random coil configuration [84]. Ubiquitin is activated by an ATP-dependent reaction involving a ubiquitin activating enzyme, followed by its conjugation via a thioester bond to a cysteine residue in a ubiqutin-conjugating enzyme. In the final enzymatic step, ubiquitin is transferred from ubiqutin-conjugating enzyme to a targeted lysine residue in a particular substrate protein by ubiquitin-protein isopeptide ligase [85].

Ubiquitination occurs on the ε-amino group of lysine residues of H2A, H2B, H3, H4, H2A.Z and H1 [8, 86-90]. The C-terminal glycine residue of ubiquitin protein can form an iso-peptide bond with the ε-amino group of lysine. Most of the ubiquitination occurs in the cytosol and is associated with proteolytic breakdown and degradation of proteins within proteosomes. Studies in *Drosophila* and mammalian cells showed that the ubiquitinated forms of histones H2A and H2B are associated specifically with actively transcribed genes [91-92]. Histone H2A Lys119 and H2B Lys120 can be reversibly ubiquitinated within their C-terminal domains [93]. It has been shown that H2A and H2B are polyubiquitinated [87]. Although H2B is generally known to be mono-ubiquitinated at Lys123, a recent study has shown that H2B can be poly-ubiquitinated at Lys123 and it is dependent on the Rad6-Bre1 ubiquitination machinery [94].

Ubiquitination is a reversible histone modification like acetylation. In vertebrate cells, ubiquitinated histones have been found to be in a non-ubiquitinated state during mitosis specifically. Deubiquitination may be accomplished by a family of ubiquitin–specific proteases [86-87, 95-97]. In *Saccharomyces cerevisiase,* the deubiquitinating enzyme Ubp3p interacts with Sir4p, a protein involved in *MAT* silencing and heterochromatin formation [98]. In contrast, ubiquitinated histones are associated with transcriptionally active genes [91, 99]. Mono-ubiquitination of H2B is required for the optimal expression of two highly inducible yeast genes, *GAL1* and *SUC2* [99]. Furthermore, Ubp8p, a stable component of the SAGA complex, is capable of deubiquitinating H2B both *in vivo* and *in vitro*. Substitution of the histone H2B ubiquitination site or deletion of Ubp8p results in decreased transcription, which suggests that deubiquitination of H2B is required for optimal expression of *GAL1* [99].

7. Histone Sumoylation

Small ubiquitin related modifier (SUMO), a 100 amino acid poly-peptide is conserved in all eukaryotes and is a member of a growing family of ubiquitin-like proteins involved in post-translational modifications (sumoylation), which occur on lysine residues [100-101]. The three members of the SUMO family, which include SUMO-1, SUMO-2 and SUMO-3 are found in mammals [102-103]. It has been shown that histone H4 is modified by SUMO both *in vivo* and *in vitro*; this process requires two enzymes, SUMO-activating enzyme (E1) and SUMO-conjugating enzyme (E2). Histone H4 binds to SUMO-conjugating enzyme Ubc9, and can be sumoylated in an E1- and E2-dependent manner. Although the actual attachment sites have not been determined, it is believed that histone sumoylation mediates gene silencing through the recruitment of histone deacetylase and heterochromatin protein 1 [9].

It has been demonstrated that histone sumoylation occurs at all four core histones in *S. cerevisiae.* Histone H2B Lys6 and Lys7, histone H4 Lys5, Lys8, Lys12, Lys16 and Lys20, and histone H2A Lys126 are targets for sumoylation. Sumoylation may also have a negative effect on gene expression. It has been shown that the genes *GAL1, SUC2* and *TRP3* are negatively controlled by histone sumoylation [101]. It is believed that histone sumoylation may present in a dynamic interplay with histone acetylation and ubiquitination, as histone sumoylation may recruit HDACs to deacetylate neighboring sites [104].

8. Histone ADP-ribosylation

ADP-ribosylation is important in many physiological and pathological processes, including intercellular and intracellular signaling, transcription, DNA repair pathways, cell cycle regulation, mitosis, necrosis, and apoptosis [10]. ADP-ribosylation involves the attachment of a homopolymer of ADP-ribose (ADPr) units to a specific protein. These polymers can vary in length and can form either linear or branched structures, or a combination of both linear and branched structures. Polymers differ in both size and complexity. The short and linear structures are referred to as oligo (ADP-ribosyl) homopolymers. The longer and branched structures are referred to as poly (ADP-ribosyl) (pADPr) homopolymers that contain as many as 200 units *in vitro* [105]. The addition of the ADPr units is catalyzed by poly (ADP-ribose) polymerase, which is a 113-kDa enzyme and appears to be ubiquitous throughout eukaryotes, except for yeast [105-106].

In eukaryotes, histones are covalently modified by mono-ADPr. It has been shown that mono-ADP-ribosylation occurs at glutamate residues of H2A, arginine residues of H2A, H2B, H3, and H4, and the phosphate group of phosphoserine in histone H1, H2A, H2B, H3 and H4 [107-110]. Although all histones can be poly(ADP-ribosyl)ated, H1 and H2B are the main targets [111-113]. It has been shown that poly(ADP-ribosyl)ation can negatively regulate transcription. Mouse lymphoma cells treated with DNA-damaging agents have lower RNA polymerase activity, suggesting a decrease in mRNA and rRNA synthesis [114]. Reduction of RNA synthesis can be prevented by 3-aminobenzamide, a specific inhibitor of poly(ADP-ribose) synthetase, indicating the negative effect of poly-(ADP-ribosyl)ation in transcription. The transcriptional activity of the p53 tumor suppressor protein has also been shown to be regulated by poly(ADP-ribosyl)ation *in vitro* and *in vivo*[111,115].

9. Histone Biotinylation

Histone biotinylation occurs through covalent linkage of the vitamin, biotin, to histones. This covalent linkage is mediated by holocarboxylase synthetase and biotinidase [116-117]. Biotinidase belongs to the nitrilase superfamily [118] and promotes the process of biotinylation of histones through the hydrolytic cleavage of biocytin, followed by the transfer of the biotinyl residue to the free amine groups of histones [116].

The biotinylation of histones by holocarboxylase synthetase requires the presence of ATP and biotin [119]. Biotinylation of histones may play a role in diverse cellular processes including gene silencing [11], cell proliferation [119], DNA repair and apoptosis [11, 120]. Biotinylation sites have been identified within histones, such as histone H2A Lys9, Lys13, Lys125, Lys127 and Lys129 [121], histone H3 Lys4, Lys9 and Lys18 [122], and histone H4 Lys8 and Lys12 [123]. Biotinylation of histone H4 Lys12 is a mark for repeat regions and heterochromatin. Furthermore biotinylation of histone H4 Lys12 is co-localized with the repression marker of histone H3 Lys9 di-methylation [120, 124]. As such, biotinylation of histone H4 Lys12 may play a role in gene repression.

10. The Epigenetic Histone Code

The "histone code hypothesis" has been proposed in recent years and suggests that histone tail modifications constitute an epigenetic code to be read by other proteins. The epigenetic code occurs either sequentially or in combination as words in a language or symbols in a code. Consequently, the response to the histone code triggers downstream events resulting in a unique and specific biological outcome, such as gene expression, gene silencing, heterochromatin formation, DNA replication, and chromosome segregation.

It has been shown that yeast Gcn5p binds to histone H3, which is phosphorylated at Ser10 by Snf1p kinase. Subsequently, H3 Lys9 and Lys14 are acetylated, which is followed by *INO1* induction [76, 125-126]. Thus, phosphorylation of Ser10 facilitates the sequential acetylation of histone H3 Lys9 and Lys14. Here, phosphorylation and acetylation are targeted to the same histone through coordinated patterns of histone modification and as such, this histone code leads to gene activation [126]. Also histone H3 Lys4 methylation by an HKMT facilitates subsequent H3 and H4 acetylation through p300. Histone H3 Lys9 methylation, however, inhibits acetylation events as it impairs subsequent Ser10 phosphorylation, which leads to the acetylation of histone H3 Lys9 and Lys14 [127-128]. The PRMT1-mediated histone H4 Arg3 methylation is severely impaired by histone H4 acetylation, whereas histone H4 Lys8 and histone H4 Lys12 acetylation is elevated after methylation of histone H4 Arg3 [129]. Moreover, histone H4 Lys20 methylation and histone H4 Lys16 acetylation were found to preclude each other [130]. In yeast Rad6-mediated histone H2B Lys123 ubiquitination is required for subsequent histone H3 Lys4 and Lys79 methylation [131-132].

Conclusions

It has been established that histone acetylation is associated with gene activation, and such a connection can be regulated by an upstream epigenetic histone code as described above. However, the interplay between the epigenetic histone code and chromatin remodeling activity in the mechanism of gene activation still remains unclear. This is so, because it requires the detail of chromatin repositioning and the change of histone modification status. So far, relatively few genes have obtained detailed information as in the case of *CUP1* [52, 133], *PHO5* [134], *PHO8* [135], *HO* [136], and *INO1* [53]. These studies have demonstrated that different roles of histone acetylation in the mechanism of gene activation are possible.

For example, histone acetylation is not required until the very late stage when TATA-binding protein binds to the TATA box in *CUP1* induction whereas a histone acetylase is recruited by the chromatin remodeler at an early stage in *PHO8* and *HO* induction. Clearly, histone acetylation is needed for gene activation at different stages. As such, a comprehensive analysis of histone modification status and nucleosome repositioning pattern at a defined time and in a precise context would be helpful to further understand the mechanism of gene expression.

ACKNOWLEDGMENTS

We thank Mohammed Hassan Bhuiyan and Michelle Marie Esposito for critical reading of this chapter. This work was supported by an NSF Grant (MCB-0919218) and PSC-CUNY awards (62180-0040 and 64243-0042).

REFERENCES

[1] Kim, Y., McLaughlin, N., Lindstrom, K., Tsukiyama, T. & Clark, D. J. (2006). Activation of Saccharomyces cerevisiae HIS3 results in Gcn4p-dependent, SWI/SNF-dependent mobilization of nucleosomes over the entire gene. *Mol Cell Biol, 26,* 8607-8622.

[2] Smith, C. & Peterson, C. L. (2005). ATP-dependent chromatin remodeling. *Curr Top Dev, 65,* 115-148.

[3] Tsukiyama, T., Palmer, J., Landel, C.C., Shiloach, J. & Wu C. (1999). Characterization of the imitation switch subfamily of ATP-dependent chromatin-remodeling factors in Saccharomyces cerevisiae1. *Genes Dev, 13,* 686-697.

[4] Ogryzko, V. V., Schiltz, R. L., Russanova, V., Howard, B. H. & Nakatani, Y. (1996). The transcriptional coactivators p300 and CBP are histone acetyltransferases. *Cell, 87,* 953-959.

[5] Bjerling, P., Silverstein, R. A., Thon, G., Caudy, A.,Grewal, S. & Ekwall, K. (2002). Functional divergence between histone deacetylases in fission yeast by distinct cellular localization and in vivo specificity. *Mol Cell Biol, 22,* 2170-2181.

[6] O'Carroll, D., Scherthan, H., Peters, A. H., Opravil, S., Haynes, A. R., Laible, G., Rea, S., Schmid, M., Lebersorger, A., Jerratsch, M., Sattler, L., Mattei, M. G., Denny, P., Brown, S. D., Schweizer, D. & Jenuwein, T. (2000). Isolation and characterization of Suv39H2, a second histon H3 methyltransferase gene that displays testis-specific expression. *Mol Cell Biol, 20,* 9423-9433.

[7] Wei,Y., Mizzen, C. A., Cook, R. G., Gorovsky, M. A. & Allis, C. D. (1998). Phosphorylation of histone H3 at serine 10 is correlated with chromosome condensation during mitosis and meiosis. *Pro Natl Acad Sci U S A, 95,* 7180- 7484.

[8] Chen, H. Y., Sun, J. M., Zhang, Y., Davie, J. D. & Meistrch, M. L. (1998). Ubiquitination of Histone H3 in Elongating spermatids of Rat testes. *J Biol Chem, 273,* 13165-13169.

[9] Shiio, Y. & Eisenman, R. N. (2003). Histone sumoylation is as sociated with transcriptional repression. *Proc Natl Acad Sci, 10,* 13225-13230.

[10] Hassa, P. O., Haenni, S. S., Elser, M. & Hottiger, M. O. (2006). Nuclear ADP-Ribosylation reaction in mammalian cells: Where are we today and where are we going? *Microbiol Mol Biol Rev, 70,* 789-829.

[11] Peters, D. M., Griffin, J. B., Stanley, J. S., Beck, M. M. & Zempleni, J. (2002). Exposure to UV light causes increased biotinylation of histones in Jurkat cells. *Am J Physiol Cell Physiol, 283,* C878-C884.

[12] Marmorstein, R. (2001). Protein Modules that manipulate histone tails for chromatin regulation. *Nat Rev Mol Cell Biol, 2,* 422-431.

[13] Roth, S. Y., Denu, J. M. & Allis, C. D. (2001). Histone Acetyltransferases. *Annu Rev Biochem, 70*, 81-120.

[14] Utley, R. T., Ikeda, K., Grant, P. A., Cote, J., Steger, D. J., Eberharter, A., John, S. & Workman, J. L. (1998). Transcriptional activators direct histone acetyltransferase complexes to nucleosomes. *Nature, 98,* 498-502.

[15] Gu, W. & Roeder, R. G. (1997). Activation of p53 sequence-specific DNA binding by acetylation of the p53 C-terminal domain. *Cell, 90,* 595-606.

[16] Tachibana, M., Sugimoto, K., Fukushima, T. & Shinkai, Y. (2001). Set domain containing protein, G9a, is a novel lysine-preferring mammalian histone methyltransferase with hyperactivity and specific selectivity to lysines 9 and 27 of histone H3. *J Biol Chem, 276,* 25309-25317.

[17] Yang, X-J. (2004). The diverse super family of lysine acetyltransferases and their roles in leukemia and other diseases. *Nucleic Acid Res, 32,* 959-976.

[18] Peterson, C. L. & Laniel, M. C. (2004). Histones and histone modifications. *Curr Biol, 14,* R546- R551.

[19] Kelly, T. J., Qin, S., Gottschling, D. E. & Parthun, M. R. (2000). Type B histone acetyltransferase Hat1p participates in telomeric silencing. *Mol Cell Biol, 20,* 7051-7058.

[20] Yamamoto, T. & Horikoshi, M. (1997). Novel substrate specificity of the histone acetyltransferase activity of HIV-1-Tat interactive protein Tip60. *J Biol Chem, 272,* 30595-30598.

[21] Spencer, T. E., Jenster, G., Burcin, M. M., Allis, C. D., Zhou, J., Mizzen, C. A., McKenna, N. J., Onate, S. A., Tsai, S. Y., Tasai, M. J. & O'Maley, B. W. (1997). Steroid receptor coactivator-1 is a histone acetyltransferase. *Nature, 389,* 194-198.

[22] Mizzen, C. A., Yang, X. J., Kokubo, T., Brownell, J. E., Bannister, A. J., Owen-Hughes, T., Workman, J., Wang, L., Berger, S. L., Kouzarides, T., Nakatani, Y. & Allis, C. D. (1996). The TAF(II)250 subunits of TFIID has histone acetyltransferase activity. *Cell, 87,* 1261-1270.

[23] Clarke, A. S., Lowell, J. E., Jacobson, S. J. & Pilus, L. (1999). Esa1p is an essential histone acetyltransferase required for cell cycle progression. *Mol Cell Biol, 19,* 2515-2526.

[24] Chen, H., Lin, R. J., Schiltz, R. L., Chakravarti, D., Nash, A., Nagy, A., Nagy, L., Privalsky, M. L., Nakatani, Y. & Evans, R. M. (1997). Nuclear receptor coactivetor ACTR is a novel histone acetyltransferase and forms a multimeric activator compex with P/CAF and CBP/p300. *Cell, 90,* 569-580.

[25] Parthun, M. R. (2007). Hat1: the emerging cellular roles of a type B acetyltransferase. *Oncogene, 26,* 5319-5328.
[26] Brownell, J. E. & Allis, C. D. (1996). Special HATs for special occasions: linking histone acetylation to chromatin assembly and gene activation. *Curr Opin Genet Dev, 6,* 176-184.
[27] Poveda, A., Pamblanco, M., Rafrov, S., Tordera, V., Srernglanz, R. & Sendra, R. (2004). Hifl is component of yeast histone acetyltransferacse b, a complex mainly localized in the nucleus. *J Biol Chem, 279,* 16033-16043.
[28] Brownell, J. E., Zhou, J., Ranalli, T., Kobayashi, R., Edmondson, D. G., Roth, S. Y. & Allis, C. D. (1996). *Tetrahymena* histone acetyltransferase A: a transcriptional co-activator linking gene expression to histone acetylation. *Cell, 84,* 843-851.
[29] Candau, R., Moore, P.A., Wang, L., Barlev, N., Ying, C. Y., Rosen, C. A. & Berger, S. L. (1996). Identification of human proteins functionally conserved with the yeast putative adaptors ADA2 and GCN5. *Mol Cel Biol, 16,* 593-602.
[30] Xu, W., Edmondson, D. G. & Roth, S. Y. (1998). Mammalian GCN5 and P/CAF acetyltransferases have homologous amino-terminal domain important for recognition of nucleosomal substrates. *Mol Cell Biol, 18,* 5659-5669.
[31] Smith, E. R., Belote, J. M., Schiltz, R. L., Yang, X. J., Moore, P. A., Berger, S. L., Nakatani, Y. & Allis, C. D. (1998). Cloning of Drosophila GCN5: conserved features among metazoan GCN5 family members. *Nucleic Acids Res, 26,* 2948-2954.
[32] Sterner, D. E. & Berger, S. L. (2000). Acetylation of Histones and Transcription-Related Factors. *Microbio Mole Biol Rev, 64,* 435-459.
[33] Kuo, M.-H., Broenell, J. E. Sobel, R. E., Ranalli, T. A., Cook, R. G., Edmonoson, D. G., Roth, S. Y. & Allis, C. D. (1996). GCN5p, a yeast nuclear histone acetyltransferase, acetylates specific lysine in histone H3 and H4 that differ from deposition-related acetylation sites. *Nature, 383,* 269-272.
[34] Kreb, J. E., Fry, C. J., Samuels, M. L. & Peterson, C. L. (2000). Global role for chromatin remodeling enzymes in mitotic gene expression. *Cell, 102,* 587-598.
[35] Wenzheng Zhang, W., Bone, J. R., Edmondson, D. G., Turner, B. M. & Roth, S. Y. (1998). Essential and redundant functions of histone acetylation revealed by mutation of target lysines and loss of the Gcn5p acetyltransferase. *EMBO J, 17,* 3155-3167.
[36] Schiltz, R. L., Mizzen, C. A., Vassilev, A., Cook, C. D., Allis, C. D. & Nakatani, Y. (1999). Overlapping but distinc patterns of histone acetylation by the human coactivators p300 and PCAF within nucleosomal substrates. *J Biol Chem, 274,* 1189-1192.
[37] Carrozza, M. J., Utley, R. T., Workman, J. L. & Cote, J. (2003). The diverse functions of histone acetyltransferase complexes. *Trends Genet, 19,* 321-329.
[38] Xu, L., Lavinsky, R. M., Dasen, J. S., Flynn, S. E., McInerney, E. M., Mullen, T. M., Heinzel, T., Szeto, D., Koruzus, E., Kurokawa, R., Aggarwal, A. K., Rose, D. W., Glass, C. K. & Rosenfeld, M. G. (1998). Singal-specific co-activator domain requirements for Pit-1 activation. *Nature, 395,* 301-306.
[39] Smith, E. R., Eisen, A., Gu, W., Sattah, M. Pannuti, A., Zhou, R. G., Cook, J. C., Lucchesi, J. C. & Allis, C. D (1998). ESA1 is a histone acetyltrasferase that is essential for growth in yeast. *Proc Natl Sci USA, 95,* 3561-3565.

[40] Clarke, A. S., Lowell, J. E., Jacobson, S. J. & Pillus, L. (1999). Esa 1p is an essencial histone acetyltransferase required for cell cycle progression. *Mol Cell Biol, 19,* 2515-2526.

[41] Kwok R. P, Lundblad, J. R., Chrivia, J. C., Richards, J. P., Bachinger, H. P., Brennan, R. G., Roberts, S. G., Green, M. R. & Goodman, R. H. (1994). Nuclear protein CBP is a coactivator for the transcription factor CREB. *Nature, 370,* 223-226.

[42] Janknecht, R. & Hunter, T. (1996). Transcription. A growing coactivator network. *Nature*, 383, 22-23.

[43] Shikama N., Lyon, J. & La Thangue, N. B. (1997). The p300/CBP family: integrating signals with transcription factors and chromatin. *Trends Cell Biol, 7,* 230-236.

[44] Giles, R. H., Peter, D. J. & Breuning, M. H. (1998). Conjugation dysfunction:CBP/p300 in human disease. *Trends Genet, 14,* 178-183.

[45] Muraoka, M., Konishi, M., Kikuchi-Yanoshita, R., Tanaka, K., Shitara, N., Chong, J. M., Iwama, T. & Miyaki, M., (1996). p300 gene alterations in colorectal and gastric carcinomas. *Oncogene, 12,* 1565-1569.

[46] Kleff, S., Andrulis, E. D., Anderson, C. W. & Sternglanz, R. (1995). Identification of a gene encoding a yeast histone H4 acetyltransferase. *J Biol Chem, 270,* 24674-24677.

[47] Parthun, M. R., Widmon, J. & Gottschling, D. E., (1996). The Major Cytoplasmic Histone Acetyltransferase in Yeast: Links to Chromatin Replication and Histone Metabolism. *Cell, 87,* 85-94.

[48] Quana, A. S., Buschbeck, M. & Croce, L. Di. (2006). Chromatin structure and epigenetics. *Biochemical Pharmacology, 72,* 1563-1569.

[49] Lancher, M., O' Sulliva, R. J. & Jenuwein, T. (2003). An epigenetic road map for histone lysine methylation. *J Cell Sci, 116,* 2117-2124.

[50] Gottesfeld, J. M. & Forbes, D. J., (1997). Mitotic repression of the transcriptional machinery. *Trends Biochem Sci, 22,* 197-202.

[51] Vaissiere, T., Sawan, C. & Herceg, Z. (2008). Epigenetic interplay between histone modifications and DNA methylation in gene silencing. *Mutation Research, 659,* 40-48.

[52] Shen, C-H.,Leblanc, B. P., Neal, C., Akhaven, R. & Clark, D. J. (2002). Targeted histone acetylation at the yeast *CUP 1* promoter requires the transcriptional activator, the TATA boxes, and the putative histone acetylase encoded by *SPT10*. *Mol Cell Biol, 22,* 6406-6416.

[53] Esposito, M., Konarzewska, P., Odeyale, O. & Shen, C-H. (2010). Gene-wide histone acetylation at yeast *INO1* requires the transcriptional activator Ino2p. *Biochem Biophys Res Commun, 391,* 1285-1290.

[54] Jacobson, R. H., Ladumer, A. G., King, D. S. & Tjian, R. (2000). Structure and function of a human TAFII250 double bromodomain module. *Science, 288,* 1422-1425.

[55] Hubbert, C., Guardiola, A., Shao, R., Kawaguchi, Y., Ito, A., Nixon, A., Yoshida, M., Wang, X. F. & Yao, T. P. (2002). HDAC6 is a microtubule-associated deacetylase. *Nature, 417,* 455-458.

[56] Sengupta, N. & Seto, E., (2004). Regulation of histone deacetylase activities. *J Cell Biochem, 93,* 57-67.

[57] Longworth, M. S. & Laimins, L. A. (2006). Histone deacetylase 3 localizes to the plasma membrane and is a substrate of Src. *Oncogene, 25,* 4495-500.

[58] de Ruijter, A. J. M., van Gennip, A. H., Caron, H. N., Kemp, S. & van Kuilenburg, A. B. P. (2003). Histone deacetylases (HDACs): characterization of the classical HDAC family. *Biochem J, 370,* 737-749.

[59] Fischle, W., Dequiedt, F., Hendezel, M. J.,Guenther, M. G., Lazar, M. A., Voelter, W. & Verdin, E. (2002). Enzymatic activity associated with class II HDACs is dependent on a multiprotein complex containing HDAC3 and SMRT/N-CoR. *Mol Cell, 9,* 45-57.

[60] Gao, L., Cueto, M. A., Asselbergs, F. & Atadja, P. (2002). Cloning and functional characterization of HDAC11, a novel member of the human histone deacetylase family. *J Biol Chem, 277,* 25748-25755.

[61] Yang, L., Xia, L., Wu, D. Y., Wang, H., Chansky, H. A., Schubach, W. H., Hickstein, D. D. & Zhang, Y. (2002). Molecular cloning of ESET, a novel histone H3 specific methyltransferase that interacts with ERG transcription factor. *J Biol Chem, 277,* 148-152.

[62] Cheng, X., Collins, R. E. & Zhang, X. (2005). Structural and sequence motifs of protein (histone) methylation enzymes. *Annu Rev Biophys Biomol Struct, 34,* 267-294.

[63] Dillon, S. C., Zhang, X., Trievel, R. C. & Cheng, X. (2005). The SET-domain protein superfamily: protein lysine methyltransferases. *Genome Biol, 6,* 227.

[64] Qian, C. & Zhou, M. M. (2006). SET domain protein lysine methyltransferases: structure, specificity and catalysis. *Cell Mol Life Sci, 63,* 2755-2763.

[65] Couture, J. F. & Trievel, R. C. (2006). Histone-modifying enzymes: encrypting an enigmatic epigenetic code. *Curr Opin Struct Biol, 16,* 753-760.

[66] Cheng, X. & Zhang, X. (2007). Structural dynamics of protein lysine methylation and demethylation. *Mutat Res, 618,* 102-115.

[67] Volkel, P. & Angrand, P. O. (2007). The control of histone lysine methylation in epigenetic regulation. *Biochimie, 89,* 1-20.

[68] Upadhyay, A. K. & Cheng, X. (2011). Dynamics of Histone Lysine Methylation: Structures of Methyl Writers and Eraser. *Prog Drug Res, 67,* 107-124.

[69] Noma, K. I., Allis, C. D. & Grewal, S. I. S. (2001). Transitions in distinct histone H3 methylation patterns are the heterochromatin domain boundaries. *Science, 293,* 1150-1155.

[70] Santos-Rosa, H., Schneider, R., Bannister, A. J., Sherriff, J., Bernstein, B. E., Emre, N. C., Schreiber, S. L., Mellor, J. & Kouzarides, T. (2002). Active genes are tri-methylated at K4 of histone H3. *Nature, 419,* 407-411.

[71] Stec, I., Wright, T. J., van Ommen, G. J., de Boer, P. A., van Haeringen, A., Moorman, A. F., Altherr, M. R. & den Dunnen, J. T. (1998). WHSC1, a 90kb SET domain-containing gene, expressed in early development and homologous Drosophila dysmorphy gene maps in the Wolf-Hirschhorn syndrome critical region and is fused to IgH in t(4;14) multiple myeloma. *Hum Mol Genet, 7,* 1071-1082.

[72] Kouzarides, T. (2002). Histone methylation in transcriptional control. *Curr Opin Gene Devel, 12,* 198-208.

[73] McBrige, A. E. & Silver, P. A. (2001). State of the Arg: protein methylation at arginine comes of age. *Cell, 106,* 5-8.

[74] He, H. & Lehming, N. (2003). Global effect of histone modifications. *Briefings in functional genomics and proteomics, 2,* 234-243.

[75] Dieker, J. & Muller, S. (2010). Epigenetic histone code and Autoimmunity. *Clinic Rev Allerg Immunol, 39,* 78-84.

[76] Nowak, S. J. & Corces, V. G. (2004). Phosphorylation of histone H3: balancing act between chromosome condensation and transcriptional activation. *Trends Genet, 20,* 214-220.

[77] Van Hooser, A., Goodrich, D. W., Allis, C. D. & Brinkley, B. R. (1998). Histone H3 phosphorylation is required for the initiation, nut not maintenance, of mammalian chromosome condensation. *J Cell Sci, 111,* 3497-3506.

[78] Crosio, C., Heitz, E., Allis, C. D., Borelli, E. & Sassone-Corsi, P. (2003). Chromosome remodeling and neuronal response: multiple signaling pathways induce specific histone H3 modifications and early gene expression in hippocampal neurons. *J Cell Sci, 116,* 4905-4914.

[79] Goto, H., Tomono, Y., Ajiro, K., Kosako, H., Fujita, M., Sakurai, M., Okawa, K., Iwamatsu, A., Okigaki, T., Takahashi, T. & Inagaki, M. (1999). Identification of a novel phosphorylation site on histone H3 coupled with mitotic chromosome condensation. *J Biol Chem, 274,* 25543-25549.

[80] Bradbury, E. M. (1992). Reversible histone modifications and the chromosome cell cycle. *Bioessays, 14,* 9-16.

[81] Davie, J. R. & Chadee, D. N. (1998). Regulation and regulatory parameters of histone modifications. *J Cell Biochem, 30,* 203-213.

[82] Fusauchi, Y. & Iwai, K. (1984). Tetrahymena histone H2A. acetylation in the N-terminal sequence and phosphorelation in the C-terminal sequence. *J Biochem, 95,* 147-154.

[83] Hunter, N. & Borner, G. V. (2001). Gamma-H2AX illuminates meiosis. *Nat Genet, 27,* 236-238.

[84] Vijay-Kumar, S., Bugg, C. E., Wilkinson, K. D., Vierstra, R. D., Hatfeild, P. M. & Cook, W. J. (1987). Comparison of three dimensional structures of human, yeast, and oat ubiquitin. *J Biol Chem, 262,* 6396-6399.

[85] Weake, V. M. & Workman, J. L. (2008). Histone ubiquitination: Triggering gene activity. *Mol Cell, 29,* 653-663.

[86] Muller, R., Yasuda, H., Hatch, C., Bonner, W. & Bradbury, E. (1985). Identification of ubiquitinqted histones 2A and 2B in physarum polycephalum. Disappearance of these proteins at metaphase and reappearance at anaphase. *J. Biol Chem, 26,* 5147-5153.

[87] Nickel, B. E., Roth, S. Y., Cook, R. G., Allis, C. D. & Davie, J. R. (1987). Changes in the histone H2A variant H2A.Z and polyubiquitinated histone species in developing trout testis. *Biochemistry, 26,* 4417-4421.

[88] Robzyk, K., Recht, J. & Osley, M. A. (2000). Rad6-dependent ubiquitination of histone H2B in yeast. *Science, 287,* 501-504.

[89] Murantani, M. & Tansey, W. P. (2003). How the ubiquitin-proteasome system controls transcription. *Nat Rev Mol Cell Biol, 3,* 192-201.

[90] Kanda, F., Sykes, D. E., Yasuda, H., Sandberg, A. A. & Matsui, S. (1986). Substrate recognition of isopeptidase: specific cleavage of the epsilon-(alpha-glycyl)lysine linkage in ubiquitin-protein conjugates. *Biochim Biophys Acta, 870,* 64-75.

[91] Levinger, L. & Varshavsky, A. (1982). Selective arrangement of ubiquitinated and D1 protein-containing nucleosomes within the *Drosophila* genome. *Cell, 28,* 375-385.

[92] Huang, S. Y., Barnard, M. B., Xu, M., Matsui, S., Rose, S. M. & Garrard, W. T. (1986). The active immunoglobulin kappa chain gene is packaged by non-ubiquitin-conjugated nucleosomes. *Proc Natl Acad Sci USA, 83,* 3738-3742.

[93] Goldknopf, I. L. & Busch, H. (1977). Isopeptide linkage between non-histone and histone 2A polypeptides of chromosomal conjugated-protein A24. *Proc Natl Acad Sci U S A, 74,* 864-868.

[94] Geng, F. & Tansey, W. P. (2008). Polyubiquitylation of histone H2B. *Mol Biol Cell, 9,* 3616-3624.

[95] Kao, C.-F. & Osely, M. A. (2003). In vivo assays to study histone ubiquitylation. *Methods, 31,* 59-66.

[96] Cai, S. Y., Babbit, R. W. & Marchesi, V. T. (1999). A mutant deubiquitinating enzyme (Ubp-M) associates with mitotic chromosomes and blocks cell division. *Proc Natl Acad Sci USA, 96,* 2828-2833.

[97] Kerscher, O., Felberbaum, R. & Hochstrasser, M. (2006). Modification of protein by ubiquitin and ubiquitin like protein. *Annu Rev Cell Dev Biol, 22,* 159-180.

[98] Moazed, D. & Johnson, A. D. (1996). A deubiquitinating enzyme interacts with SIR4 and regulates silencing in S. cerevisiae. *Cell, 86,* 667-677.

[99] Henry, K. W., Wyce, A., Lo, W-S., Duggan, L. J., N.C. Tolga Emre, N. C. T., Kao, C-F., Pillus, L., Shilatifard, A., Osley, M. A. & Berger, S. L. (2003). Transcriptional activation via sequential histone H2B ubiquitylation and deubiquitylation, mediated by SAGA-associated Ubp8. *Genes Dev, 17,* 2648-2663.

[100] Johnson, E. S. & Gupta, A. A. (2001). An E3- like factor that promotes SUMO conjugation to the yeast septins. *Cell, 106,* 735-744.

[101] Nathan, D., Ingvarsdottir, K. & Sterner, D. E. (2006). Histone sumoylation is a negative regulator in *Saccharomyces cerevisiae* and shows dynamic interplay with positive-acting histone modifications. *Genes & Dev, 20,* 966-976.

[102] Picher, A., Knipscheer, P., Saithh, H., Sixma, T. K. & Melchior, F. (2004). The RanBP2 SUMO E3 ligase is neither HECT- nor RING-type. *Nature Stru Mole Biol, 11,* 984- 991.

[103] Tatham, M. H., Jaffray, E., Vaughan, O. A., Desterro, J. M. & Botting, C. H. (2001). Polmeric chains of SUMO-2 and SUMO-3 are conjugated to protein substrates by SAE1/SAE2 and Ubc9. *J Biol Chem, 276,* 35368-35374.

[104] Yang, S.-H. & Sharrocks, A. D. (2004). SUMO promotes HDAC-mediated transcriptional repression. *Mol Cell, 13,* 611-617.

[105] Diefenbach, J. & Burkle, A. (2005). Introduction to poly (ADP-ribose) metabolism. *Cell Mol Life Sci, 62,* 721-730.

[106] Burkle, A., (2005). Poly (ADP-ribose) The most elaborate metabolic of NAD+. *FEBS J, 18,* 4576-4589.

[107] Golderer, G. & Grobner, P. (1991). ADP-ribosylation of core histones and their acetylated subspecies. *Biochem J, 277,* 607-610.

[108] Ord, M. G. & Stocken, L. A. (1977). Adenosine diphosphate ribosylated histones. *Biochem J, 161,* 583-592.

[109] Smith, J. A. & Stoken, L. A. (1975). Chemical and metabolic properties of adenosine diphosphate ribose derivatives of nuclear proteins. *Biochem J, 147,* 523-529.

[110] Tanigawa, Y., Tsuchiya, M., Imai, Y. & Shimoyama, M. (1984). ADP-ribosyltransferase from hen liver nuclei. Purification and characterization. *J Biol Chem, 259,* 2022-2029.

[111] D' Amours, D., Desnoyers, S., D' Silva, I. & Poirier, G. G. (1999). Poly (ADP-ribosyl) ation reactons in the regulation of nuclear functions. *Biochem J, 342,* 249-268.

[112] Malanga, M., Atorino, L., Tramontano, F., Farina, B. & Quesada, P. (1998). Poly (ADP-ribose) binding properties of histone H1 variants. *Biochem Biophys Acta, 1399,* 154-160.

[113] Ausio, J., Abbott, D. W., Wang, X. & Moore, S. C. (2001). Histone variants and histone modifications: a structural perspective. *Biochem Cell Bio, 79,* 693-708.

[114] Taniguchi, T., Aggmori, M., Kameshita, I., Nishikimi, M. & Shiznta, Y.(1982). Participation of poly(ADP-ribosyl)ation in the depression of RNA synthesis caused by treatment of mouse lymphoma cells with methylnitrosourea. *J Biol Chem, 257,* 4027-4030.

[115] Wang, X., Ohnoshi, K., Takahashi, A. & Ohnishi, T. (1998). Poly(ADP-ribosyl)ation is required for p53-dependent signal transduction induced by radiation. *Oncogene, 17,* 2819-2825.

[116] Narang, M. A., Dumas, R., Ayer, L. M. & Gravel, R. A. (2004). Reduced histone biotinylation in multiple carboxylase deficiency patients: a nuclear role for holocarboxylase synthetase. *Hum Mol Genet, 13,* 15-23.

[117] Camporeale, G.,Giordano, E., Rendina, R., Zempleni, J. & Eisseenberg, J. C. (2006). Drosophila holocarboxylase synthetase is a chromosomal protein required for normal histone biotinylation, gene transxription patterns, lifespan and heat toleance. *J Nutr, 136,* 2735-2742.

[118] Brenner, C. (2002). Catalysis in the nitrilase superfamily. *Curr Opin Struct Biol, 12,* 775- 782.

[119] Zempleni, J. (2005). Uptake, localization and noncarboxylase role of biotin, *Annu Rev Nutr, 25,* 175-195.

[120] Kothapalli, N., Sarath, G. & Zempleni, J. (2005). Biotinylation of K12 in histone H4 decreases in response to DNA double-strand breaks in human JAr choriocarcinoma cells. *J Nutr, 135,* 2337-2342.

[121] Chew, Y. C., Camoreale, G., Kothapalli, N., Sarath, G. & Zempleni, J. (2006). Lysine residues in N-and C-terminal regions of human histone H2Aare targets for biotinylation by biotinidase. *J Nutr Biochem, 17,* 225-233.

[122] Kobaza, K., Comporeale, G., Rueckert, B., Kuch, A., Griffin, J. B., Sarath, G. & Zempleni, J. (2005). K4, K9 and K18 in human histone H3 are targets or biotinylation by biotinidase. *FEBS J, 272,* 4249-4259.

[123] Camporeale, G., Shubert, E. E., Sarath, G., Cerny, R. & Zempleni, J. (2004). K8 and K12 are biotinylated in human histone H4. *Eur J Biochem, 271,* 2257-2263.

[124] Camporeale, G., Oommen, A. M., Griffin, J. B., Sarath, G., Zempleni, J. (2007). K12-biotinylated histone H4 marks heterochromatin in human lymphoblastoma cells. *J Nutr Biochem, 18,* 760-768.

[125] Lo, W. S., Trievel, R. C., Rojas, J. R., Duggan, L., Hsu, J.-Y., Allis, C. D., Marmorsten, R. & Berger, S. L. (2000). Phosphorylation of serine 10 in histone H3 is functionally linked *in vitro* and *in vivo* to Ggcn5-mediated acetylation at lysine 14. *Mol Cell, 5,* 917-926.

[126] Lo, W. S., Duggan, L., Emre, T. N. C., Lane, Belotserkovsky, R., Lane, W. S., Shiekhsttar, R. & Berger, S. L. (2001). Snf1- a histone kinase that works in concert with the histone acetyltransferase Gcn5 to regulate transcription. *Science, 293,* 1142-1146.

[127] Wang, H. Cao, R., Xia, L., Erdjument-Bromzge, H., Borchres, C., Tempst, P. & Zhang, Y. (2001). Purification and Functional Characterization of a Histone H3-Lysine 4-Specific Methyltransferase . *Mol Cell, 8,* 1207-1217.

[128] Rea, S., Eisenhaber, F., O'Carrol, D., Strahl, B. D., Sun, Z. W., Schmid, M., Opravil, S., Mechtler, K., Ponting, C. P., Allis, C. D. & Jenuwein, T. (2000). Regulation of chromatin structure by site-specific histone H3 methyltransferases. *Nature, 406,* 593-599.

[129] Wang, H., Huang, Z. Q., Xia, L., Feng, Q., Erdjument-Bromzge, H., Strahl, B. D., Briggs, S. D., Allis, C. D., Wong, J., Tempst, P. & Zhang, Y.(2001). Methylationof histone H4 at arginine 3 facilitating transcriptional activation by nuclear hormone receptor. *Science, 298,* 853-857.

[130] Nishioka, K., Rice, J. C., Sarma, K., Erdjument-Bromzge, H., Werner, J., Wang, Y., Chuikov, S., Valenzuela, P., Tempst, P., Steward, R., Lis, J. T., Allis, C. D. & Reinberg, D. (2002). PR-Set7 is a nucleosome-specific methyltransferase that modifies lysine 20 of histone H4 and is associated with silent chromatin. *Mol Cell, 9,* 1201-1213.

[131] Margueron, R., Trojer, P. & Reinberg, D. (2005). The key to development: interpreting the histone code? *Curr Opin Genet Dev, 15,* 163-176.

[132] Ng, H. H., Xu, R. M., Zang, Y. & Struhl, K. (2002). Ubiquitination of histone H2B by Rad6 is required for Dot1-medated methylation of histone H3 lysine 79. *J Biol Chem, 277,* 34655-34657.

[133] Shen, C-H., Leblance, B. P., Alfieri, J.A. & Clark, D. J. (2001). Remodeling of yeast *CUP1* chromatin involves activator-dependent repositioning of nucleosomes over the entire gene and flanking sequences. *Mol Cell Biol, 21,* 534-547.

[134] Svaren, J. & Ho rs, W.(1997). Transcription factor vs. nucleosomes: regulation of the *PHO5* promoter in yeast. *Trends Biochem Sci, 22,* 93-97.

[135] Reinke, H., Gregory, P. D. & Horz, W. (2001). A transient histone hyper-acetylation signal marks nucleosomes for remodeling at the *PHO8* promoter *in vivo*. *Mol Cell, 7,* 529-538.

[136] Cosma, M. P., Tanaka, T. & Nasmyth, K. (1999). Ordered recruitment of transcription and chromatin remodeling factors to a cell cycle-and developmentally regulated promoter. *Cell, 97,* 299-311.

In: Histones: Class, Structure and Function
Editor: Chang-Hui Shen

ISBN: 978-1-62100-274-1

Chapter 6

HISTONE MODIFIERS IN CEREALS

Aliki Kapazoglou[1], Dimitra Papaefthimiou[1] and Athanasios Tsaftaris[1,2,*]

[1]Institute of Agrobiotechnology, CERTH, Thermi-Thessaloniki, Greece
[2]Department of Genetics and Plant Breeding, AUTH, Thessaloniki, Greece

ABSTRACT

Histone modifications play a crucial role in the epigenetic regulation of various aspects of eukaryotic development. Histone modifications such as methylation and acetylation of histone tails are regulated by histone modifiers which are specific enzymatic complexes including histone methyltransferases, histone demethylases, histone acetyltransferases and histone deacetylases. Histone modifications result in the modulation of chromatin structure, a process which leads to the optimal gene expression programmes that eukaryotic organisms require during development and in response to changing environments. The action of histone modifiers leads to open (active) or closed (inactive) forms of chromatin, which are associated with activation or silencing of genes, respectively. Whereas histone modifiers have been well investigated and reviewed in the model plant *Arabidopsis*, comprehensive reports regarding the study of histone modifiers in plants of high economic value, such as cereals, are limited. This review will present the current knowledge on epigenetic histone modifiers in cereal model plants and crops such as *Brachypodium*, rice, maize, and barley.

1. INTRODUCTION

In eukaryotes nuclear DNA is organized in chromatin, a tightly packaged higher order structure which permittts genomic DNA to fit within the nucleus. The fundamental unit of chromatin is the nucleosome which is composed of 147 base pairs of DNA that is wrapped almost twice around an octamer of histone proteins. The octamer consists of two copies of each of histone H2A, H2B, H3 and H4 [1].

[*] Corresponding author. Email: tsaft@certh.gr

Chromatin higher-order structure switches between condensed and relaxed states and plays a crucial role in the epigenetic regulation of gene expression [2]. These alterations in chromatin structure affect the accessibility of the transcriptional machinery (transcription factors, RNA polymerase) to the nucleosomal DNA and determine the levels of gene expression in response to developmental and environment stimuli. Chromatin modulation is achieved in part, through the following mechanisms: DNA methylation catalyzed by DNA cytosine methyltransferases, alterations in histone-DNA interactions facilitating nucleosome sliding catalyzed by chromatin remodeling complexes, and histone post-translational modifications catalyzed by a wide range of enzymes specific for each modification [3-4]. Histone post-translational modifications usually take place on the tails of histones extending out from their respective nucleosomes, and include methylation, acetylation, phosphorylation, ubiquitination, biotinylation, and sumoylation on specific lysine, arginine, serine and threonine residues [5-6]. A complex pattern of site-specific combinations of histone modifications on different residues, known as the 'epigenetic histone code', leads to specific chromatin states which regulate transcriptional activity and are inherited by daughter cells [7].

In plants, the best characterized post-translational histone modifications are histone methylations and histone acetylations. Generally, di-methylation of histone 3 on lysine 9 (H3K9me2) and tri-methylation of histone 3 on lysine 27 (H3K27me3) leads to a repressive chromatin state correlating with gene silencing whereas di-methylation (H3K4me2) or tri-methylation of histone 3 on lysine 4 (H3K4me3) leads to a permissive chromatin state correlating with gene activation [4]. In addition, the recently discovered demethylation of histones is a signal for gene activity [8]. Similarly, acetylation of histones H3 on lysines 9 and 14 (H3K9, 14ac) leads to gene activation whereas histone deacetylation results in gene silencing [9]. In plants, H3K9me2 is mostly associated with transcriptional silencing of transposons and repeats in centromeric/pericentromeric heterochromatin regions. Genome-wide analysis in Arabidopsis has determined the global H3 methylation patterns along chromatin and showed an agreement between the localization of H3 methylation marks and their proposed functions based on previous locus-specific studies [10-13]. The H3K4me3, which is associated with gene activation, was found mostly in endogenous genes but not in transposons, whereas H3K9me2 coincides almost exclusively with transposons and other repeats. H3K27me3 was correlated with transcriptional silencing of a large number of tissue-specific expressed genes, in euchromatic regions.

Despite the apparent universality behind the histone code, it is becoming more and more evident that epigenetic marks like histone modifications are species-specific, as well as gene-, tissue-, developmental-, and environmental-specific. This together with the cross-talk taking place between the different modifications makes the histone language even more complex than thought previously [6, 14]. It is important to understand the process of histone modifications and the genes and encoding proteins engaged in this epigenetic mechanism, for each individual organism, in the context of intrinsic and external signaling received and integrated by each one. Considering the significance of epigenetic chromatin modifications in the generation of epi-allelic variation which may impact agronomical characters the need to focus on crop species is evident [15-16].

In plants, the preponderance of research on histone modifications and the enzymes that perform them has been carried out in the model-plant Arabidopsis and includes both gene-specific and genome-wide studies. The technological advancement in genomic technologies, (next generation sequencing, ChIP-seq, bioinformatics tools) has driven the transition from epigenetics studies at small-scale to global studies of epigenomes (epigenomics) [16]. In the past several years the completion of the sequencing of the rice and maize genomes allowed for breakthrough studies which deciphered genome-wide epigenetic landscapes of histone modifications (as well as DNA methylation and small RNA patterns) in these two agronomically important monocot cereal crops [17-18]. In parallel, small scale locus-specific studies have shed light in the various histone modifications and the diverse histone modifying enzymes responsible for them.

The aim of this review is to summarize the current knowledge on the molecular and functional characterization of histone modifiers, giving emphasis on the most widely studied histone modifications and focusing mainly on the information concerning histone modifiers in monocot cereals.

2. Histone Methylation/Demethylation Modifiers

2.1. The Histone Methylation Mark and its Impact

Post-translational modifications involving methylation of plant histones have been intensively investigated in Arabidopsis where they were found to play a crucial role in the regulation of many developmental processes such as vegetative development, flowering, vernalization, seed development and stress responses. A large number of studies have been reported presenting results both from individual genes [reviewed in 19-25) and from large scale genome-wide analysis [10-13, 26].

In cereals, similar studies have been conducted both on a gene-specific and genome-wide level. For example, it was demonstrated that differential histone modifications were present on imprinted loci (ie loci with differential expression between the maternal and paternal allele), in maize. By performing Chromatin immunoprecipitation analysis (ChIP) with antibodies against specific histone marks it was shown that the silenced paternal alleles of the *Mez1, ZmFie1* and *Nrp1* genes were enriched for the H3K27me2 and H3K27me3 repressive marks whereas the expressed maternal alleles were enriched for the H3K4me2 (as well as the H3 and H4 acetylation) activating mark [27]. In another study, the light- induced activation of C4-specific Phosphoenolpyruvate (*C_4-Pepc*) and malic enzyme *(Me)* genes from maize were found to be associated with specific H3K4 modification changes in a cell-specific manner [28]. Similarly, the genes for the a-zein 22kd protein subfamily the major storage proteins in maize endosperm, were shown to be differentially methylated at their chromatin histones, in different tissues. The zein chromatin displayed high levels of H3K9me2 and H3K27me2 associated with transcriptional silencing in leaves. Conversely, a preactivation 'potentiation' state in endosperm tissue was found associated with decreased H3K9me2 and H3K27me2, and increased H3K4me2, whereas transcriptional activity was correlated with further increase of H3K4me2 in endosperm tissue [29]. Moreover, in barley, histone methylation at the barley *VERNALIZATION 1(HvVRN1)* locus that codes for a FRUITFULL-like MADS box transcription factor, was found to be associated with vernalization- induced flowering [30]. High levels of H3K27me3 at the *VRN1* chromatin were associated with repression of *HvVRN1* before vernalization. Upon vernalization the gene repressive H3K27me3 mark was decreased and H3K4 levels were increased. In this way, *HvVRN1* is activated and consequently downregulates the floral repressor *HvVRN2* which in turn induces the floral activator *HvVRN3* (encoding an FT-like protein) and the transition to flowering is achieved [31]. Finally, in the rice epigenetic mutant *epi-d1* which displays a dwarf phenotype due to the silencing of the *DWARF-1 (D1)* gene encoding a GTP-binding protein, differential histone modifications were detected at *D1* chromatin, between the mutant and the wild type rice plants. *D1* silencing in the *epi-d1*mutant was shown to correlate with increased levels of the repressive mark H3K9me2 (as well as decreased levels of the H3K9ac activating mark) along the coding region and 5'proximal to the ATG site [32].

The most comprehensive genome-wide profiling of histone modifications in cereals to date has been reported on the two fully sequenced genomes of maize and rice [17-18]. High throughput analysis of the genome of the B73 maize inbred line (widely used for F1 hybrid production) has revealed the epigenetic pattern of histone modifications as well as DNA methylation in root and shoot tissues, respectively [18]. Using Illumina/Solexa 1G parallel sequence after conducting ChIP analysis with antibodies specific for H3K4me3, H3K27me3, H3K36me3, a profile was unraveled that seems to be gene and tissue specific. In the maize epigenome of roots and shoots, the most prevalent histone methylation modification was the H3K4me3 activating mark. In general non-TE regions were more frequently modified by H3K4me3, H3K36me3, and H3K27me3, as compared to TE-rich regions and H3K27me3 were nearly twice as frequent in shoots as in roots. Unlike Arabidopsis, H3K4me3 was more prevalent at or near the transcriptional start site or the ATG site and less common in the gene body whereas H3K27me3 was most common along the gene body, as in Arabidopsis. H3K36me3 and was found mostly around the transcriptional start site or ATG in roots, but present throughout the gene body in shoots. In addition, the activating histone marks H3K4me3 and H3K36me3 correlated well with the highly expressed genomic regions, whereas H3K27me3 was found within genes having low expression levels in both tissues [18].

High resolution mapping in rice also provided important insights into the epigenetic landscape of this essential cereal crop on a genome-wide scale [17, 33]. Using tilling-path microarrays and ChIP analysis with antibodies specific for H3K4me2 and H3K4me3 the histone methylation profile for these marks was unraveled for chromosomes 4 and 10 [17]. In general euchromatic regions contain more H3K4me2 and H3K4me3 marks than heterochromatic regions and non-TE genes from both regions had elevated levels for both histone marks whereas TE genes had no H3K4me3. Genes marked predominately with H3K4me3 were actively transcribed whereas genes marked mainly with H3K4me2 were associated with moderately transcribed genes. As in maize, the H3K4me3 mark was mostly found near the 5' end of genes and both H3K4me3 and H3K4me2 were most common at about 150 bp upstream and 500 bp downstream from the ATG site. Genes marked with H3K4me3 on both promoter and the gene body (transcribed part) had the highest transcript abundance, whereas genes with H3K4me3 on just the body or just the promoter had ~ 40% less transcript levels, and very low transcript levels, respectively [17]. In another study, using high-throughput Illumina/Solexa 1G sequencing and ChIP-seq techiques for two rice parental lines and their reciprocal hybrids, He et al. (2010) unravelled the global transcription and epigenetic map of these genomes and demonstrated that differences in gene expression between parents and hybrids correlated with variation in histone modifications such as the H3K4me3 activation marker and the H3K4me27 silencing marker [33]. In particular, these two antagonizing markers were often found concurrently present on various chromosomal locations and the ratio between them was proposed to lead to fine-tuning of gene expression levels [33].

2.2. Histone Methyltransferases (HMTS): The "Writers" of the Methyl Marks

The different specificities of histone methylation modifications described above point to functionally different histone methyltransferases with different substrate specificities and enzymatic activities. Histone methyltransferases catalyze the methylation of specific lysines or arginine residues on the amino-terminal tails of histones as they protrude out of the nucleosomes. The most widely studied and best characterized ones are the histone lysine methyltransferases (HKMTs).

The HKMT proteins that have been implicated in the trimethylation of lysine 4 of Histone (H3K4me3), a histone mark associated with gene activation, belong to a group of proteins termed Trithorax Group (TrxG). On the other hand HKMT modifiers that perform the deposition of the silencing mark H3K27me3 belong to a group of proteins termed Polycomb group (PcG) which antagonizes the action of TrxG [21-22, 34]. PcG and TrxG histone modification systems have been extensively studied in animal systems, like human, mouse and Drosophila [34-35]. Regarding plants, numerous studies exist in *Arabidopsis* [21-22, 36-37] and in other plants like cereals studies on HKMTs are now emerging. Both PcG and TrxG proteins belong to a large superfamily known as the SET-domain proteins (or SET DOMAIN GROUP, SDG proteins) named after the founding members of the family in Drosophila: Suppressor of variegation, SU(V)3-9, Enhabcer of zestre, E(Z), and Trithorax , TRX [25]. These are mainly lysine methyltransferases and catalyze the mono-, di- and trimethylation of specific lysine residues on nucleosomal histone tails.

The SET domain is composed of approximately 130-150 amino acids and is responsible for the enzymatic activity of the HKMT proteins. Crystal structure resolution of human SET-domain proteins has revealed the catalytic mechanism and substrate specificity of the SET-domain and its flanking regions [38]. A highly conserved tyrosine in the SET- domain interacts through its hydroxyl group with the substrate and transfers a methyl group to the histone lysine residue using S-adenosylmethionine as the methyl-group donor [39].

In *Arabidopsis* there are 49 genes encoding putative SET-domain proteins (www.chromdb.org) [reviewed in 25]. Of those 31 have histone lysine methyltransferase activity and are divided in five classes based on the number and position of conserved domains and on substrate specificity differences. SET-domain HKMT encoding genes have also started to be characterized in maize and rice two monocots cereals with completed genome sequence [40-42] and also in barley [43-44]. Phylogenetic analysis has pointed to functional diversification prior to the divergence of monocots and dicots [25, 41]. The different HKMT classes along with representative Arabidopsis members and their homologues in maize, rice and barley are listed in Table 1. A detailed description of the different HKMTs with regard to their different specificities is presented below:

Table 1. Histone Lysine Methylatransferases (HKMTs)

Methylation Sites	Histone Lysine Mehtyltransferase (HKMT)			
	Arabidopsis	*Maize*	*Rice*	*Barley*
H3K4me1	ATXR7?			
H3K4me2 (activation)	***Trithorax group TrxG*** ATX2, SDG4, ATXR7	SDG106, SDG128, SDG115, SDG127	SDG721, SDG705, SDG717	HvTX1
H3K4me3 (activation)	ATX1, SDG8, ATXR7 SDG2 (ATXR3)	SDG108	SDG723, SDG701	
H3K27me1 (repression)	ATXR5, ATXR6	SDG139, SDG129	SDG720, SDG730	
H3K27me2 (repression)	SUVH2			
H3K27me3 (represssion)	***Polycomb group PcG***			
	PRC2 complex			
	AtMEA AtCLF **E(Z)** AtSWN	MEZ1 MEZ2, MEZ3	OsCLF (SDG711) OsEZ (SDG718), OSEZ1	HvE(Z)
	AtFIE	ZmFIE1 ZmFIE2	OsFIE1 OsFIE2	HvFIE
	AtFIS2 AtEMF2 **Su(z)12** AtVRN2	ZmEMF ZmEMF2-like (ZmVEF101)	OsEMF2a, OsEMF2b	HvSu(12a, HvSu(z)12 b
H3K36me1 (activation)	?			
H3K36me2 (activation)	SDG8			
H3K36me3 (activation)	SDG4, SDG7, SDG8, SDG24 SDG25, SDG26	SDG110, SDG102	SDG724, SD736, SDG708, SDG725	
H3K9me1 (repression)	SUVH2 , SUVH4 (KYP), SUVH5, SUVH6	SDG101, SDG105, SDG111, SDG113, SDG118, SDG136, SDG135, SDG137, SDG116, SDG103, SDG104, SDG119, SDG107,SDG117, SDG131	SDG704, SDG713, SDG728, SDG709, SDG733, SDG734, SDG714, SDG727, SDG726, SDG715, SDG729, SDG703, SDG710, SDG712, SDG706	
H3K9me2 (respressio)	SUVH2, SUVH4 (KYP), SUVH6, SUVR4			
H3K9me3 (repression)	?			

2.2.1. HKMTs Leading to Gene Activation

H3K4me Methyltransferases

Plant histone methyltransferases performing H3K4 methylation belong in the TRITHORAX subfamily of SDG proteins (termed after the Drosophila Trithorax protein). The molecular and functional characterization of H3K4 methyltransferases has been the subject of intensive research, in Arabidopsis [reviewed in 25]. They contain a SET domain and a post-SET domain, as well as a PHD (plant homeodomain), and a PWWP (proline-tryptophan-tryptophan-proline) domain [21-22]. The PHD domain is believed to interact with H3Kme3, whereas the PWWP domain interacts with both histones and DNA [25]. In Arabidopsis, there are seven Trithorax–like proteins (ATX1-5, ATXR3, and ATXR7) and biological function has been assigned to five of them. ATX1 and ATX2 are involved in H3K4me3 and H3K4me2, respectively, and both play important roles in proper flowering time and the flowering process. Specifically, they mediate H3K4 methylation of the key flowering regulator *FLOWERING LOCUS (FLC)*, a repressor of the floral transition, resulting in *FLC* gene activation and subsequent flowering repression [36, 45]. Two other Arabidopsis proteins have been classified in this group, with only a SET and a post-SET domain, named ATXR3 and ATXR7. ATXR7 has been implicated mainly in H3K4me3 but also in H3K4me2 and H3K4me1 at *FLC* chromatin leading to *FLC* activation and consequent repression of flowering [46-47]. Very recently, ATXR3, also known as SDG2, has been characterized as the main H3K4me3 methyltransferase in Arabidopsis, as *sdg* mutants exhibit severe loss of H3K4me3 marking and display pleitoropic phenotypes throughout the plant's vegetative, flowering and reproductive development [48-49]. Significant changes in the expression of a wide range of genes were observed in the *sdg* mutant, including severe repression of the *FLC* gene and of the egg and endosperm specific genes, *EC1* and *MEA,* respectively [48] as well as downregulation of genes specific for sporophytic or gametophytic development [49].

Maize and rice homologues of H3K4 methyltransferases have been identified [41; Table 1] but detailed molecular characterization is lacking, as yet. In barley, a gene encoding a trithorax-like H3K4 methyltransferase, *HvTX1,* that is inducible by drought has been recently identified and characterized by our group [44].

H3K36 Methyltransferases

The deposition of the H3K36me histone mark found in regions of decondensed chromatin associated with gene activation is deposited by another class of SDG HKMTases [21, 25, and 50]. In Arabidopsis, the *SDG8* gene has been implicated in H3K36 di- and trimethylation as *sdg8* mutants display a decrease in H3Kme2 and H3Kme3 on *FLC* chromatin, leading to reduced expression of the *FLC* gene and an early flowering phenotype. The Arabidopsis *SDG4* gene was reported to mediate H3K36 trimethylation as well as H3K4 dimethylation, and plays a role in stamen and pollen development [51-52]. H3K36me homologues have been identified in maize and rice [41] but detailed characterization has not been reported as yet.

2.2.2. HKMTs Leading to Gene Repression

H3K27 Methyltransferases

The repressive histone mark H3K27 associated with gene silencing is deposited by another class of SDG proteins homologous to the Drosophila E(Z) protein [53].

The most extensively studied H3K27 HKMTs are the ones that perform H3K27 trimethylation which belong to the Polycomb Repressive Complex 2 (PRC2), a major repressive complex and one of two Polycomb group (PcG) complexes in Drosophila and mammals [34, 53]. The PRC2 complex was first identified and characterized in Drosophila and consists of four essential subunits: a SET–domain protein, termed ENHANCER OF ZESTE E(Z) [E(Z)], responsible for H3K27me3 methyltransferase activity, two WD40 proteins, named EXTRA SEX COMBS (ESC), and P55 containing the WD40 domain involved in protein-protein interactions, and a zinc finger protein, SUPRESSOR OF ZESTE 12 [Su(z)12] [37, 54]. The PRC2 complex is an essential epigenetic regulator that determines cell fate and proper organ formation. Loss of PcG function leads to cell overproliferation, loss of organ identity, and abnormal organ development [37, 54].

In Arabidopsis three homologues of E(Z), namely MEDEA (MEA), SWINGER (SWN) and CURLY LEAF (CLF) are found associated with specific variants of the Polycomb Repressive Complex 2 (PRC2). All three contain a SET domain, two EZ domains, a SANT domain and a CXC (cysteine rich) domain. The different variants of the PRC2 complex are associated with different functions during plant development. The E(Z)-MEA-FIE-MS1 regulates the initiation of seed development [37, 54]. It consists of FIE, MEA, FIS2 and MSI1, which are homologous to the Drosophila ESC, E(Z), Su(z)12, and P55, respectively [55-58]. Mutants of all four subunits *fie*, *mea*, *fis2*, *msi1* result in improper seed development [59-60]. In Arabidopsis all PcG complexes are predicted to contain the FIE and MSI1 subunits. Depending on cell type and function the different PcG complexes contain one of the three E(Z) homologues, MEA, CURLY LEAF (CLF) or SWINGER (SWN), and one of the three homologues of the Drosophila Su(z)12 protein, EMBRYONIC FLOWER 2 (EMF2), FERTILIZATION INDEPENDENT SEED 2 (FIS2), and VERNALIZATION2 (VRN2), respectively. The EMF2-CLF/SWN-FIE-MSI1 complex has been suggested to play a role in suppressing the transition from vegetative development to flowering and flower organ formation [61-63] and the VRN2-CLF/SWN-FIE-MSI1 has been implicated in the regulation of the process of vernalization [61, 64-65].

PRC2 gene homologues have been studied in maize and rice [40, 66-67] and more recently the respective PRC2 homologues in barley, HvE(Z), HvFIE, HvSu(z)12a and HvSu(z)12b were identified and characterized by our group [43; Table 1]. The maize and rice putative protein E (Z) homologues cluster in two clades: the so-called SWN-like clade which contains ZmMEZ2, ZmMEZ3 and OsiEZ1, and the CLF-like clade which includes ZmMEZ1 and OsjEZ (also named OsCLF). The single barley homologue HvE(Z) identified, so far, groups out with the SWN-like clade. The barley HvE(Z) homologue [43], the maize ZmMEZ1, ZmMEZ2 and ZmMEZ3 [40] and the rice OsiEZ1 and OsCLF (same as OsjEZ) homologues [67] are expressed during vegetative, reproductive and seed development implying a widespread function throughout plant growth. In addition, all four PRC2 barley homologues were mapped on barley chromosomes and co-localized with regions responsible for malting-quality traits. Furthermore, the barley E(Z) homologue was shown to have differential expression in different barley cultivars during seed development and it is strongly induced by ABA, implying an association of a putative barely H3K27me3 histone methyltransferase with different cultivar traits, ABA signalling, and the stress response [43].

Aside from the well studied H3K27me3 histone mark and H3K27me3 methyltransferases studies on H3K27 methyltransferases with other specificities are coming into sight. A recent report described a study on the H3K27me1 mark which was shown to be associated with gene repression in heterochromatic regions (unlike H3K27me3 that mediates gene silencing in euchromatic regions). The ATXR5 and ATXR6 HKMTs were characterized as novel H3K27 histone lysine methylatransferases mediating H3K27me1 in Arabidopsis [68], however, no such homologues have been characterized in cereals, to date.

H3K9 Methyltransferases

H3K9 methylation is carried out by the so called SUVH and SUVR [SU(VAR)3-9 RELATED] proteins [69]. SUVH proteins have a SET- domain, a pre-SET domain, a post-SET domain and the SET and RING finger associated (SRA) domain, whereas SUVR proteins lack the SRA domain [25]. Arabidopsis SUVH proteins mediate mono-, di-, and trimethylation of histones and mostly associate with highly condensed heterochromatic regions and silencing of repetitive DNA (TEs, centromeric repeats and ribosomal RNA repeats), but also with gene inactivation in euchromatic regions [11]. In addition they interact with DNA methylation in heterochromatic regions and there exists a dynamic interplay between the two repressive marks. For example mutations in the SUVH4/KRYPTONITE protein caused a loss of DNA methylation conferred by the Chromomethylase 3 (CMT3) cytosine methyltransferase. Conversely, in a DNA methyltransferases triple mutant drm1 drm2 cmt3 with defects in the DNA methyltransferases DRM1, DRM2, CMT3, the H3K9me2 levels were significantly reduced [70-71]. Genome-wide profiling also revealed a significant correlation between these two repressive marks, in Arabidopsis [11]. SUVH proteins often act synergistically, as is the case for AtSUVH4, AtSUVH5 and AtSUVH6 which were shown to act together to silence inverted repeats [72-73]. Furthermore it was demonstrated that the triple mutant suvh2 suvh4 suvh9, exhibits altered expression of the SUPRESSOR OF DRM1 DRM2 CMT3 gene which results in a curly leaf phenotype [71]. Moreover, overexpression of AtSUVH2 induced an increase in repressive histone marks, such as H3K9me2, H3K27me2, H3K27me3 and H4K20me2 [74] and resulted in delayed leaf senescence [75]. Little is known about SUVR proteins. AtSUVR4 protein has H3K9me1 and H3K9me2 activity [76]. AtSUVR5 interacts with the remodeling factor SWP1and is associated with H3K9me2 and H3K27me2 marking at the FLC promoter and with reduced FLC expression [77].

Very recently, 12 SUVH rice homologues were identified and functionally characterized through RNAi silencing or overexpression analysis after transformation of rice plants with the appropriate constructs [42]. An obvious phenotype was not apparent for most transgenic plants, implying functional redundancy among rice SUVH proteins. Interestinlgly, OsSDG728 RNAi plants displayed a decrease in seed size and weight. OsSDG728 overexpressing lines were shown to have increased levels of H3K9me2 and H3K9me3 histone marks, whereas a decrease in both modifications was observed in the OsSDG728 RNAi plants. Similarly, OsSDG713 RNAi plants exhibited a decrease in H3K9me2. Since H3K9me2 is strongly associated with heterochromatic regions in condensed chromatin these results suggest that rice SUVH proteins are involved in heterochromatin formation. In addition, it was also observed that reduced expression of OsSDG714, OsSDG727, and

OsSDG710 resulted in partial complementation of the mutant phenotype (abnormal panicle development) of the JMJ706 gene encoding a heterochromatin-associated histone H3K9 demethylase (described in the section 2.3 below), supporting a functional role for SUVH in heterochromatin formation. Furthermore, H3K9 methylation at the specific target loci Tos17, a copia-like retrotransposon was examined by ChIP analysis of OsSDG728 in overexpressors and RNAi transgenic plants. Both H3K9me2 and H3K9me3 were increased across the Tos17 locus in the overexpressors whereas both histone marks were reduced in the RNAi plants suggesting a role for rice SUVH in silencing retrotransposons.

2.3. Histone Demethylases (HDMs): The "Erasers" of Methyl Marks

The discovery of the first histone demethylase KDM1 (LSD1), capable to directly reverse the state of methylation filled a gap in the world of epigenetic modifiers revealing the antagonists of methyltransferases [78]. Specific demethylation of mono- and di-methylated H3K4, leading to transcriptional activation, was now attributed to the function of LSD1. Indeed, histone methylation is a highly dynamic process that clearly benefits from the existence of a counteracting system, required for the effective fine-tuning of complex epigenetic signals [79-80]. More recently another group of histone demethylases were detected in animals, that contain the conserved Jumonji C (JmjC) domain and are able to remove all three methylation states from H3K4, K9, K27 and K36 [81]. The jumonji demethylases include several other conserved modules in addition to the enzymatically active cupin metalloenzyme JmjC, such as PHD, zinc fingers and ARID [82]. On the basis of domain architecture, non-plant jumonji proteins were classified into seven families (JHDM1, PHF2/PHF8, JARID, JHDM3/JMJD2, UTX/UTY, JHDM2 and JmjC-domain only) [82]. Soon after the discovery of the nuclear amine oxidase LSD1 conserved from yeast to humans [78], plant histone demethylases were gradually uncovered. While plant KDM1 family histone demethylases had probably arisen from the same ancestor as their non-plant homologues, distinct evolutionary patterns were revealed for the jumonji family members [8].

Four LSD1-type histone demethylases were found in Arabidopsis, three of which are the flowering regulators FLOWERING LOCUS D (FLD), LSD1-LIKE 1 (LDL1) and LSD1-LIKE 2 (LDL2)[20]. Moreover, plant analogues for five out of the seven human JmjC domain containing (jumonji) protein groups were mined by a bioinformatics study in Arabidopsis and rice [83; Table 2].

The plant jumonji proteins studied so far in Arabidopsis have shown specificity for demethylation of di- and tri-methylated H3K4 and K9 [84]. In total, four LSD1-like and 20 JmjC domain containing genes have been identified in rice [84], whereas information from other cereals is scarce. Important functions in plants implicating histone demethylases include circadian systems [85] and plant growth development [86] including flowering control [87]. The control of flowering time has been studied extensively and involves a large number of protein regulators such as the FLOWERING LOCUS C (FLC) that prevents the transition to flowering in response to short days [87]. Repression of FLC and transition from vegetative to reproductive plant development is achieved by a complex mechanism that involves H3K4 demethylation by FLD, and LDL1 [20]. Directly implicated with FLC repression is also the putative plant jumonji demethylase REF6 [86].

Assumptions regarding functionality of the putative rice jumonji demethylases [83] were drawn by sequence similarity [Table 2]. The sequence with ChromDB accession no JMJ706 was studied further experimentally revealing that it encodes a JHDM3 (JMJD2) family protein [Table 2] capable of removing the di- and tri-methyl states from H3K9 and possibly H3K36me2 [88]. This rice JmjC containing protein is actively involved in the regulation of a subset of genes involved in flower development. In the cereal crop barley, one putative plant-specific PKDM7 subfamily histone demethylase was characterised and was shown to be significantly induced by drought stress (Papaefthimiou and Tsaftaris, submitted). As histone demethylases have just been discovered and due to their high significance in development, one should be expecting an increasing amount of reports about histone demethylases for agronomically important cereal crop species.

3. Histone Acetylation/Deacetylation Modifiers

3.1. The Histone Acetylation Mark and its Impact

Reversible histone acetylation constitutes a major and dynamic epigenetic switch leading to activation of many plant genes, including light-response integrators [89], major flowering regulators [20] plant development modulators [9] and stress-response activators [90-91]. In contrast to non-plant eukaryotic organisms, the most extensively acetylated histone in plants is H3, instead of H4 which in plants occurs in mono- to penta-acetylated forms in a progressive manner [9] from the histone body to the N-terminus [92-93]. Histone acetylation sites identified in *A. thaliana* include H2AK5 and K144; H2BK6, K11, K27 and K32; H3K9, K14, K18, K23 and K56; and H4K5, K8, K12, K16 and K20 [5]. Acetylation of H3K9 and K14 and hyperacetylation of H4 lysines occur at the promoters of active ribosomal RNA [92], while promoters of highly expressed *Arabidopsis* genes contain acetylated H3K56 [12]. Similarly in maize, H4K16 is controlled independently of other acetylation sites and is constitutively acetylated in euchromatin [93]. Moreover, H3K9, K23 and H4K5, K16 acetylation in maize is selectively induced by light, whereas H3K14, K18 acetylation is light independent, showing differential regulation of individual acetylation sites [93-94]. In addition, UVB radiation tolerance in maize is achieved by inducing acetylation on N-terminal tails of histones H3 and H4, and on promoter and transcribed regions of UVB–regulated genes [95]. Generally, a histone acetylation code becomes obvious in the case of the maize phosphoenolpyruvate carboxylase (*Pepc*) gene where multiple HATs and HDACs targeting different sites on the gene are employed in a complex regulatory acetylation network involving external and endogenous stimuli [93-94]. There is also high frequency of dominant H3K9 acetylation mark on 5' regions of highly expressed genes, and H3K9ac and H3K27me3 coexist in much higher proportion in maize shoots and roots, indicating tissue specificity [18]. Accordingly in rice, elevated patterns of H3K9 and H4K12 acetylation were detected on active genes [84], also involved in floral transition [96], and interplay between H3 acetylation and H3K4 methylation occur at submergence-inducible genes [84]. The dynamics of H4 acetylation in barley involve pair-wise acetylation of K12 and K5 in nucleolar regions and of K8 and K16 in gene-rich regions, thus regulating the structure and function of the chromosome throughout the cell cycle [97]. In addition, H3 acetylation is fairly constant throughout the cell cycle and involves low acetylation intensity at K9, K18 and K14 and more intense uniform acetylation on the remaining chromatin [98].

3.2. Histone Acetyltransferases (HATs): The "Writers" of Acetyl Marks

HATs are often subunits of large protein complexes and have been classified with respect to their intracellular location and substrate specificity into two types: type A involved in promoting transcriptional activation by acetylation of nucleosomal histones; and type B proteins acetylating newly synthesised free histones. Phylogenetic and domain analyses revealed that plant HATs are essentially divided into three families. These are the GNAT/MYST superfamily acetylating small molecules and non-histone proteins; the CBP (CREB-binding protein) family involved in gene regulation for cell cycle control, differentiation and apoptosis; and the TAFII250 family, which are TATA binding protein-associated factors (TAFs) [99]. Several uncharacterised HAT sequences (GCN5, MYST, ESA, and HAT-B) are present in plant genome databases [100]. In the Arabidopsis GNAT family, single homologues of each of the GCN5 (required for H3K9 acetylation and up-regulation of many genes), ELP3 and HAT1 subfamilies were detected, indicating conservation of their functions [101]. However, substantial evolutionary diversification was identified for the other HAT families. Three distinct types of CBP family proteins have evolved in plants, all of which were shown to be involved in flowering time regulation in Arabidopsis [92]. In total, over fifteen HAT proteins have been detected in Arabidopsis so far [24], whereas in cereals significant work has been done in maize. Phylogenetic analysis revealed high conservation among the different cereal GNAT/MYST family members, while only rice proteins representing the other HAT-A families, CBP and TAFII250, were extracted by data mining and included in the analysis [102].

Three distinct HAT activities were identified in maize, HAT-A1, HAT-A2 and HAT-B [103]. One maize HAT type B (HAT-B) protein was detected and characterized, consisting of two catalytic subunits, p45 and p50 [104]. HAT-B is highly specific for histone H4, selectively mono and di-acetylating lysines 12 and 5 in a sequential manner [105], while it also acetylates H2A to a lesser extent [103]. Interestingly, the maize HAT-B protein is accumulating during embryogenesis and HAT-B p50 mRNA expression sharply increases between 15-22 h after germination, inducing specific transitions in chromatin structure [104]. During germination in maize, mainly HAT-B promotes a fluctuating enhancement of total histone acetyltransferase activity with two peaks of highest activity at 40 and 72 h [106]. Among type A HATs isolated from maize is the single copy gene *ZmGCN5*, which is constitutively expressed in dividing cells [107]. ZmGCN5 is a putative transcriptional co-activator that acetylates the core histones H2A, H2B, H3 and H4 by interacting with its counterpart ZmADA2 in GCN5-adapter complexes [107]. They both form a SAGA-like complex by interacting with the endosperm-specific bZIP-type transcription activator Opaque2 (O2) and inducing an ordered sequence of changes in maize endosperm development that promote accumulation of histone marks at specific O2 targets [29].

Three single homologues were described from the barley GNAT/MYST family (HvGCN5, HvELP3 and HvMYST) [102]. Significant quantitative differences of the three barley transcripts between seed size, developmental stage and cultivar, suggested their association with barley seed development [102]. *HvGCN5* transcript accumulation was more induced at initial stages of seed development, while *HvELP3* and *HvMYST* mRNAs were up-regulated later in the development [103], coinciding with down-regulation of metabolic genes and endosperm reserve accumulation [108]. In addition, a possible regulation of these genes by the stress-related phytohormone abscisic acid (ABA) during barley seed development, germination and stress response was indicated by HAT transcript induction after ABA treatment [102].

3.3. Histone Deacetylases (HDACs): The "Erasers" of Acetyl Marks

As mentioned above, acetylation of histones is often associated with increased gene activation, whereas deacetylation of histones is correlated with transcriptional repression and gene silencing [9]. This is achieved through the action of specific enzymes. Histone acetyltransferases (HATs) transfer an acetyl moiety ("writers") to the ε-amino group of highly conserved lysines in the N-terminal extensions of core histones, thereby neutralizing the positive charge of lysines and resulting in less affinity to the negatively charged DNA molecules. Highly ordered condensed chromatin is then unfolded and accessibility of promoters to transcription factors is increased. Histone deacetylases (HDACs) remove ("erasers") the acetyl moieties from lysines resulting in restablishment of the positive charge and compact chromatin configuration [9, 109].

Eukaryotic HDACs can be grouped into three major families based on their primary homology to the yeast HDACs, HDA1, and SIR2 (sirtuins). These are: 1) the HDA1 family, 2) the SIR2 family and 3) the plant specific HD2 family [99]. The presence of a unique HDAC protein family in plants, HD2, possibly reflects the divergence of developmental processes and environmental responses of plant organisms from those of yeast and mammals. SIR2 proteins differ from the other HDACs in that they depend on nicotinamide adenine dinucleotide for enzymatic activity and have no structural similarity to other HDACs. Sequence analysis by Pandey et al. (2002) revealed that the first family of histone acetyltransferases, HDA1, can be subdivided further into three major clades, Class I, Class II, and Class III [99]. Sequence and phylogenetic analysis of the rice genome [110] identified the HDA1 family in rice, and showed that rice HDACs belonging in this family can also be divided in Class I, Class II and Class III. In addition sequence analysis in rice revealed a fourth clade, Class IV. Expression analyses of 18 rice *HDAC* genes from HDA1, SIR2 and HD2 families demonstrated distinct spatial expression patterns and differential responses to environmental stresses and hormones [110] (A phylogenetic tree of different HDACs from different organisms is shown in Fig.1).

Characterization of HDAC mutants in Arabidopsis revealed functional diversification among the members of different HDAC families and classes [111]. *AtHDAC6* gene (*HDA1*-Class I) for instance was shown to be involved in transgenic and repetitive DNA silencing and in the nuclear organization of rDNA loci [112-113]. *AtHDAC19,* another member of the *HDA1*-Class I type was shown to be important for proper vegetative development as *hdac19* mutants displayed various developmental abnormalities [114-115]. Mutations in the *AtHDA18* gene (*HDA1*-Class II) resulted in alterations in the cellular patterning in the root epidermis in Arabidopsis [116] and *HDA6* and *HDA19* were shown to be involved in repression of embryonic-specific gene functions during germination [117]. Similar to the *HDA1* family, the plant specific *HD2* family has been suggested to be crucial for proper reproductive development. Silencing of the *AtHD2A* gene resulted in aborted seed development [118] and overexpression of *AtHD2A* resulted in repression of a number of genes involved in seed development and maturation [119].

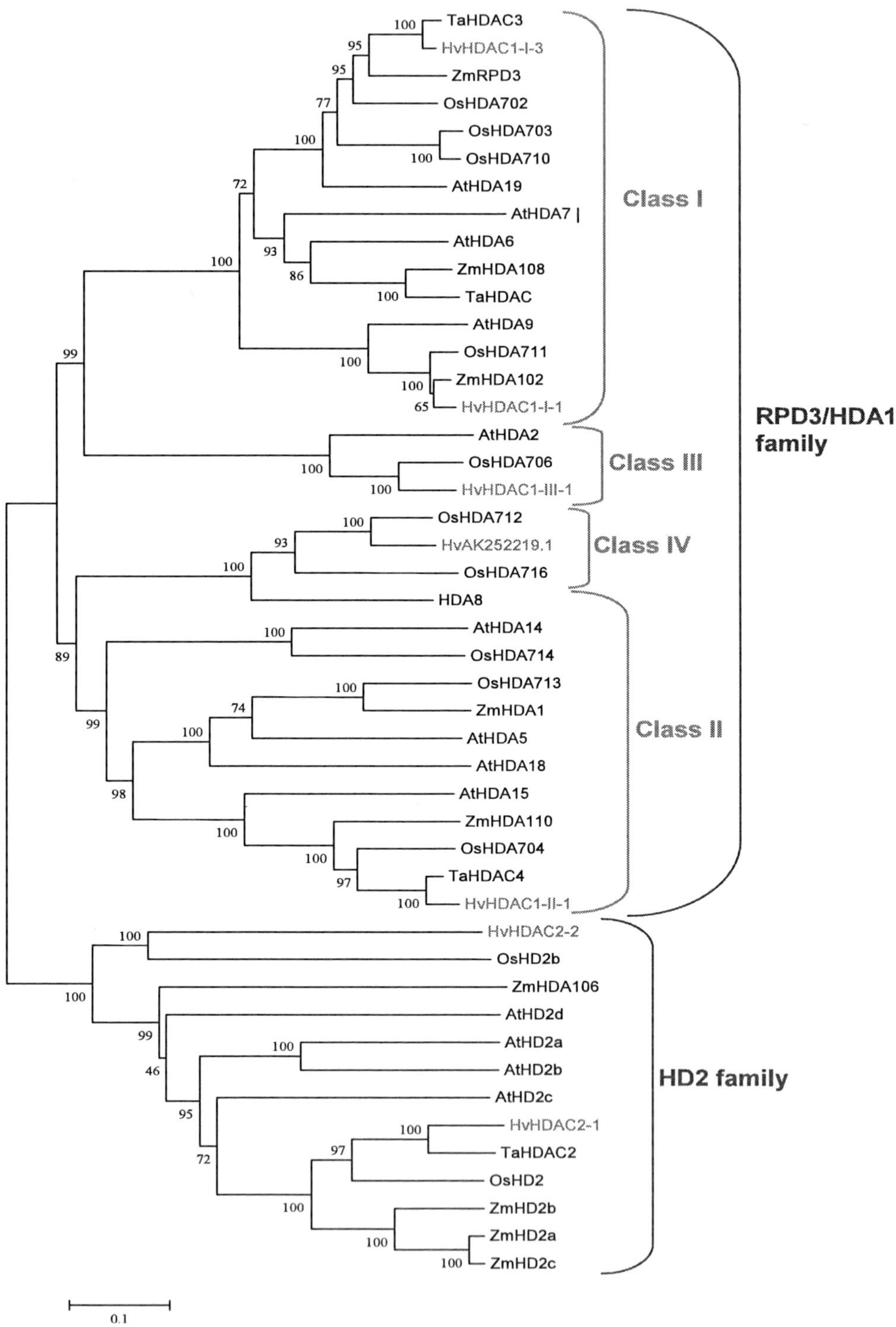

Figure 1.

Furthermore, HDAC genes from both families have been associated with the response to biotic and abiotic stress. *AtHD2C* was proposed to play a role in abscisic acid (ABA) signaling and abiotic stress [120]. *AtHDAC19* was suggested to mediate jasmonic acid (JA) and ethylene signaling during pathogen defense [114-115] and *AtHDA6* was shown to be required for JA response, senescence and flowering, in *Arabidopsis* [121]. Moreover, *AtHDA6* was shown to be involved in the deposition of H3K9, 14 ac but also of H3K4me3 histone activating markers in response to ABA and salt-stress [122].

In maize, 15 HDAC genes have been identified (10 HDA1, 1 SIR2, and 4 HD2-like genes; www.chromdb.org) and a number of HDA1 members have been biochemically characterized [123-125]. In addition, upregulation and downregulation of the maize *HDA101* gene (the *HDAC19* homologue in maize), resulted in numerous morphological and developmental defects including plant morphology, flower architecture and kernel size. In addition to its role in vegetative and reproductive development, maize *HDA101* was demonstrated to mediate other chromatin modifications such as histone methylations and phosphorylations [126]. Furthermore, mutants upregulating or downregulating the maize *HDA101*gene exhibited changes in the metabolic profiles of developing maize kernels, indicating a role for HDA101 in the modulation of metabolic pathways in seeds [127].

Recently, functional analysis using rice overexpression or RNAi transgenic lines has revealed the potential roles of some rice HDACs. For example the *OsHDAC1* gene was demonstrated to regulate seedling root growth in rice by controlling the expression of the *OsNAC6* gene (a NAC transcription factor). Specifically, *OsHDAC1* repressed *OsNAC6* activity by deacetylating K9, K14, and K18 on H3 and K5, K12, K16 on H4 at the *OsNAC6* promoter [128]. Furthermore, using artificial miRNAs to induce gene silencing, Hu *et al.* (2009) showed that the rice homologues of the *AtHDA19* gene, *OsHDA702, OsHDA710*, and *OsHDA703*, play a role in determining plant stature, leaf and stem morphology, and grain size [129].

In barley, the characterization of histone deacetylase genes belonging to the *HDA1* and *HD2* families was reported recently by our group [130-131; Figure 1]. *HvHDAC2-1* and *HvHDAC2-2* of the HD2 family were isolated, mapped on barley chromosomes and their expression investigated during seed development and in two cultivars with varying seed size. Differential expression during seed development was observed, suggesting possible functional specificity. *HvHDAC2-1*, exhibited significantly increased levels of expression in a small-seed cultivar, as compared to a large-seed cultivar and both genes displayed an induction upon JA treatment and altered expression in response to external ABA, suggesting an association with stress response. Similarly, the barley HDA1 genes were induced by JA implying an association of barley HDAC genes with JA-mediated stress response processes.

4. "Readers" of Histone Marks and Histone Cross-talk

Effector molecules have been identified in animals that have the ability to recognize and "read" the histone modification marks to produce functional outcomes. The chromodomain containing HETEROCHROMATIN PROTEIN 1, HP1, was the first protein identified to bind histone H3K9me3 [132]. Other effectors-readers were discovered in animals during the past several years recognizing histone methylation marks, including the tudor domain, PWWP (proline-tryptophan-tryptophan-proline) domain, the Agelent module, the PHD (plant homeodomain finger), as well as WD40 repeats [133]. The Origin Recognition Complex (ORC) was identified as a nucleosome ineractor recruited cooperatively by DNA methylation and histone methylation at H3K4, H3K9 and H3K27 sites [134]. In addition, a genome-wide interaction proteomics analysis assigned reader effectors to specific complexes. The human SAGA complex -via a Tudor domain- was found to bind H3K4me3 and the ORC was shown to bind H39me27 [135].

In Arabidopsis, several readers were identified, such as, LIKE HETEROCHROMATIN PROTEIN 1-LHP1/TFL2, ORC1, AtING, AL, and WDR5, which can bind methylated histones in vitro and in vivo. The LHP1/TFL2 has similar domains to the animal HP1, but unlike HP1 it binds H3K27me3 and co-localizes with PRC2 complexes [26, 136]. In this manner it behaves like the Pc subunit of PRC1 complex in animals that binds and co-localizes with PRC2-mediated H3K27me3 modifications. The chromodomain of LHP1/TFL2 was shown to be essential for the binding of H3K27me3 and to be required for the silencing of the *FLC* gene when *FLC* chromatin is marked by H3K27me3 [137-138]. Furthermore, it was demonstrated that Arabidopsis ORC1 interacts with H3K4me3 via the PHD finger domain, and this interaction is a requirement for the activation of target genes [139]. Moreover, the PHD domain containing proteins, AtING and AL, were identified as H3Kme3/2 interactors in vitro [14]. Given the current rapid progress of histone modification studies in cereals it is expected that reports on the cereal counterparts of histone mark "readers" will emerge soon.

Growing evidence has suggested that histone modifiers act combinatorialy either in a synergistic or antagonistic manner. For example, the phosphorylation of serine 10 on histone H3 induces GCN5 to perform H3K14 [140] and H2B monoubiquitination is required for H3K4 methylation, a process well conserved in eukaryotes [141]. Conversely, HP1 protein binding of H3K9me3 is inhibited when the adjacent serine H3S10 is phosphorylated [142]. Additionally, cross-talk has been detected between methylation/acetylation/deacetylation signals. H3K4me and H3K36me gene activation marks have been shown to provide sites where both HATs (activators) and HDACs (repressors) are recruited [14].

Locus-specific and large-scale epigenomic studies in Arabidopsis maize and rice (described previously) support the cooperative action of activating histone modifiers by the simultaneous existence of activating signals like H3K4me3, H3K4me2 and H3K4ac on actively transcribed chromatin. On the other hand, bivalent chromatin states have been proposed where activating and repressing histone modification marks coexist and "poise" a locus for transcription later in development [143]. The chromatin at the Arabidopsis flowering locus *AGAMOUS (AG)* is concurrently marked with the H3K4me3 activating marker and the H3K27me3 silencing marker, when in its silent state [144]. Histone modification cross-talk and interplay is well exemplified in Arabidopsis, for the *FLC* locus. The expression of *FLC* is under the control of a multitude of both positive and negative chromatin regulatory pathways both endogenous (developmental) and exogenous (environmental) [20]. Upon vernilization, *FLC* is repressed by a series of histone modifications including H3K27me2, H3K27me3, H3K9me and H4R3sme2 (histone 4 arginine 3 symmetric dimethylation), which subsequently leads to flowering. It was shown that the PHD-domain proteins VIN3, VIL1/VRN5, and VERNALIZATION5/VIN3 LIKE 1 (VEL1) interact with a PRC2-like PcG repressive complex consisting of the four PRC2 subunits, SWN, VERNALIZATION 2 (VRN2), FIE, and MSI1 and that this complex is required for *FLC* repression [20 and references therein]. On the other hand H3K4me3, histone

H2B mononoubiquination, and deposition of the histone H2A.Z variant by an homologue of the yeast SWR1c complex, were shown to be associated with *FLC* activation and consequent flowering repression [16 and references therein].

Genome-wide studies in rice showing co-localization of the activating H3K4me3 and silencing H3K27me3 markers, at the same chromatin loci [33] lend support to the notion of complex histone modification interplay in cereals as well. Future studies should provide detailed information on the particular interactions in monocots and the enzymes involved.

Conclusions

The area of histone modifications has been under intensive research in animals and in the past several years major advances have been achieved in plants as well. In particular, distinct lysine and arginine residues in histone tails have been found to be modified by specific "writers" and "erasers" of the methyl and acetyl histone marks. The "writers" comprise the enzymes histone methyltransferases (HMTs) [the best characterized ones being the lysine methyltransferases (HKMTs] and histone acetyltransferases (HATs), whereas the "erasers" comprise the enzymes histone demethylases (HDMs) and histone deacetylases (HDACs). Once the mark has been deposited by the "writers" or removed by the "erasers" the particular histone modification is recognized by specific "readers" that perform translating of the specific modification into a functional outcome. Much genetic and biochemical research conducted in Arabidopsis lead to the identification and characterization of many of the genes and encoding proteins responsible for "writing" and "erasing" specific histone marks, and studies about the potential "readers" of the histone code have been reported too. In cereals, significant progress has been made in recent years on the identification and molecular and functional characterization of the homologues of "writers" and "erasers" of histone marks: HMKTs, HATs, and HDMs and HDACs, respectively. However a great deal of effort is still needed to identify more members of histone modifiers and fully characterize the components of each group, especially of the recently discovered histone demethylases for which information is presently obscure. Furthermore, the complex combinations between histone modifications and how these lead to specific gene expression patterns is an area where little is known in Arabidopsis and even less in cereals. Genetic and biochemical studies are required to complement the emerging information from global epigenomic studies in order to shed light on the complex interplay between different histone modifications and the way it influences chromatin folding and subsequent gene activation and/or repression. Elucidating the molecular mechanisms underlying histone modifications, chromatin structure, and gene expression in cereals, will further our knowledge on how epigenetic regulation affects the development of economically important monocot crops, and may have a substantial impact in agriculture.

References

[1] Luger, K., Mader, A. W., Richmond, R. K., Sargent, D. F. & Richmond, T. J. (1997). Crystal structure of the nucleosome core particle at 2.8 A resolution. *Nature, 389*, 251-60.

[2] Goldberg, A. D., Allis, C. D. & Bernstein, E. (2007). Epigenetics: a landscape takes shape. *Cell, 128*, 635 - 38.

[3] Pfluger, J. & Wagner, D. (2007). Histone modifications and dynamic regulation of genome accessibility in plants. *Curr. Opin. Plant Biol, 10*, 645-652.

[4] Kouzarides, T. (2007). Chromatin modifications and their function. *Cell, 128*, 693-705.

[5] Zhang, K., Sridhar, V. V., Zhu, J., Kapoor, A. & Zhu, J.-K. (2007). Distinctive Core Histone Post-Translational Modification Patterns in *Arabidopsis thaliana. PLoS ONE, 2*, e1210.

[6] Berger, S. L. (2007). The complex language of chromatin regulation during transcription. *Nature, 447*, 407-412.

[7] Strahl, B. D. & Allis, C. D. (2000). The language of covalent histone modifications. *Nature, 403*, 41-45.

[8] Zhou, X. & Ma, H. (2008). Evolutionary history of histone demethylase families: distinct evolutionary patterns suggest functional divergence. *BMC Evol. Biol, 8*, 294.

[9] Chen, Z. J. & Tian, L. (2007). Roles of dynamic and reversible histone acetylation in plant development and polyploidy. *Biochim. Biophys. Acta, 1769*, 295-307.

[10] Zhang, X., Clarenz, O., Cokus, S., Bernatavichute, Y. V., Pellegrini, M., Goodrich, J. & Jacobsen, S. E. (2007). Whole-genome analysis of histone H3 lysine 27 trimethylation in *Arabidopsis*. *PLoS Biol, 5*, e129.

[11] Bernatavichute, Y. V., Zhang, X., Cokus, S., Pellegrini, M. & Jacobsen, S. E. (2008). Genome-wide association of histone H3 lysine nine methylation with CHG DNA methylation in *Arabidopsis thaliana. PLoS One, 3*, e3156.

[12] Roudier, F., Teixeira, F. K. & Colot, V. (2009). Chromatin indexing in Arabidopsis: an epigenomic tale of tails and more. *Trends Genet, 25*, 511-517.

[13] Zhang, X., Bernatavichute, Y., Cokus, S., Pellegrini, M. & Jacobsen, S. (2009). Genome-wide analysis of mono-, di- and trimethylation of histone H3 lysine 4 in Arabidopsis thaliana. *Genome Biol, 10*, R62.

[14] Lee, J.-S., Smith, E. & Shilatifard, A. (2010). The Language of Histone Crosstalk. *Cell, 142*, 682-685.

[15] Tsaftaris, A., Polidoros, A., Kapazoglou, A., Tani, E., Kovacevic, N. (2008). Epigenetics and Plant Breeding. *Plant Breeding Reviews, 30*, 49-179.

[16] Tsaftaris, A. S., Kapazoglou, A., Darzentas, N. Epigenetics, Epigenomics, and Implications in Plant Breeding *in "Plant biotechnology and agriculture: Prospects for the 21st century"* (in press).

[17] Li, X. Y., Wang, X. F., He, K., Ma, Y. Q., Su, N., He, H., Stolc, V., Tongprasit, W., Jin, W. W., Jiang, J. M., Terzaghi, W., Li, S. G. & Deng, X. W. (2008). High-resolution mapping of epigenetic modifications of the rice genome uncovers interplay between DNA methylation, histone methylation, and gene expression. *Plant Cell, 20*, 259-276.

[18] Wang, X., Elling, A. A., Li, X., Li, N., Peng, Z., He, G., Sun, H., Qi, Y., Liu, X. S. & Deng, X. W. (2009). Genome-Wide and Organ-Specific Landscapes of Epigenetic

Modifications and Their Relationships to mRNA and Small RNA Transcriptomes in Maize. *Plant Cell, 21*, 1053-1069.

[19] Chinnusamy, V. & Zhu, J. K. (2009). Epigenetic regulation of stress responses in plants. *Curr. Opin. Plant Biol, 12*, 133 - 139.

[20] He, Y. (2009). Control of the Transition to Flowering by Chromatin Modifications. *Mol. Plant, 2*, 554-564.

[21] Avramova, Z. (2009). Evolution and pleiotropy of TRITHORAX function in Arabidopsis. *Int. J. Dev. Biol, 53*, 371.

[22] Alvarez-Venegas, R. (2010). Regulation by Polycomb and Trithorax Group Proteins in Arabidopsis. *The Arabidopsis Book, 8*, 1-20.

[23] Amasino, R. M. (2010). Seasonal and developmental timing of flowering. *Plant J, 61*, 1001-1013.

[24] Chen, M., Lv, S. & Meng, Y. (2010). Epigenetic performers in plants. *Development, Growth & Differentiation, 52*, 555-566.

[25] Pontvianne, F., Blevins, T., Pikaard, C. S. & Jean-Claude Kader and Michel, D. (2010). Arabidopsis Histone Lysine Methyltransferases. *Advances in Botanical Research, 53*, 1-22.

[26] Turck, F., Roudier, F. O., Farrona, S., Martin-Magniette, M.-L., Guillaume, E., Buisine, N., Gagnot, S. v., Martienssen, R. A., Coupland, G. & Colot, V. (2007). *Arabidopsis* TFL2/LHP1 Specifically Associates with Genes Marked by Trimethylation of Histone H3 Lysine 27. *PLoS Genet, 3*, e86.

[27] Haun, W. J. & Springer, N. M. (2008). Maternal and paternal alleles exhibit differential histone methylation and acetylation at maize imprinted genes. *The Plant J, 56*, 903-912.

[28] Danker, T., Dreesen, B., Offermann, S., Horst, I. & Peterhänsel, C. (2008). Developmental information but not promoter activity controls the methylation state of histone H3 lysine 4 on two photosynthetic genes in maize. *The Plant J, 53*, 465-474.

[29] Locatelli, S., Piatti, P., Motto, M. & Rossi, V. (2009). Chromatin and DNA Modifications in the Opaque2-Mediated Regulation of Gene Transcription during Maize Endosperm Development. *Plant Cell, 21*, 1410-1427.

[30] Oliver, S. N., Finnegan, E. J., Dennis, E. S., Peacock, W. J. & Trevaskis, B. (2009). Vernalization-induced flowering in cereals is associated with changes in histone methylation at the VERNALIZATION1 gene. *PNAS, 106*, 8386-8391.

[31] Distelfeld, A., Li, C. & Dubcovsky, J. (2009). Regulation of flowering in temperate cereals. *Curr. Opin. Plant Biol, 12*, 178-184.

[32] Miura, K., Agetsuma, M., Kitano, H., Yoshimura, A., Matsuoka, M., Jacobsen, S. E. & Ashikari, M. (2009). A metastable DWARF1 epigenetic mutant affecting plant stature in rice. *PNAS, 106*, 11218-11223.

[33] He, G., Zhu, X., Elling, A. A., Chen, L., Wang, X., Guo, L., Liang, M., He, H., Zhang, H., Chen, F., Qi, Y., Chen, R. & Deng, X.-W. (2010). Global Epigenetic and Transcriptional Trends among Two Rice Subspecies and Their Reciprocal Hybrids. *Plant Cell, 22*, 17-33.

[34] Schwartz, Y. B. & Pirrotta, V. (2008). Polycomb complexes and epigenetic states. *Curr. Opin. Cell Biol, 20*, 266-273.

[35] Papp, B. & Müller, J. (2006). Histone trimethylation and the maintenance of transcriptional ONand OFF states by trxG and PcG proteins. *Genes Dev, 20*, 2041-2054.

[36] Pien, S., Fleury, D., Mylne, J. S., Crevillen, P., Inze, D., Avramova, Z., Dean, C. & Grossniklaus, U. (2008). ARABIDOPSIS TRITHORAX1 Dynamically Regulates FLOWERING LOCUS C Activation via Histone 3 Lysine 4 Trimethylation. *Plant Cell, 20*, 580-588.

[37] Kohler, C. & Villar, C. B. R. (2008). Programming of gene expression by Polycomb group proteins. *Trends Cell Biol, 18*, 236 - 243.

[38] Marmorstein, R. (2003). Structure of SET domain proteins: a new twist on histone methylation. *Trends Biochem. Sci, 28*, 59-62.

[39] Couture, J.-F. & Trievel, R. C. (2006). Histone-modifying enzymes: encrypting an enigmatic epigenetic code. *Curr. Opin. Struc. Biol, 16*, 753-760.

[40] Springer, N. M., Napoli, C. A., Selinger, D. A., Pandey, R., Cone, K. C., Chandler, V. L., Kaeppler, H. F. & Kaeppler, S. M. (2003). Comparative Analysis of SET Domain Proteins in Maize and Arabidopsis Reveals Multiple Duplications Preceding the Divergence of Monocots and Dicots. *Plant Physiol, 132*, 907-925.

[41] Ng, D. W. K., Wang, T., Chandrasekharan, M. B., Aramayo, R., Kertbundit, S. & Hall, T. C. (2007). Plant SET domain-containing proteins: Structure, function and regulation. *BBA - Gene Structure and Expression, 1769*, 316-329.

[42] Qin, F.-J., Sun, Q.-W., Huang, L.-M., Chen, X.-S. & Zhou, D.-X. (2010). Rice SUVH Histone Methyltransferase Genes Display Specific Functions in Chromatin Modification and Retrotransposon Repression. *Mol. Plant, 3*, 773-782.

[43] Kapazoglou, A., Tondelli, A., Papaefthimiou, D., Ampatzidou, H., Francia, E., Stanca, M., Bladenopoulos, K. & Tsaftaris, A. (2010). Epigenetic chromatin modifiers in barley: IV. The study of barley Polycomb group (PcG) genes during seed development and in response to external ABA. *BMC Plant Biology, 10*, 73.

[44] Papaefthimiou, D., & Tsaftaris, A.S. Characterization of a drought inducible trithorax-like H3K4 methyltransferase from barley (in press).

[45] Saleh, A., Alvarez-Venegas, R., Yilmaz, M., Le, O., Hou, G., Sadder, M., Al-Abdallat, A., Xia, Y., Lu, G., Ladunga, I. & Avramova, Z. (2008). The Highly Similar Arabidopsis Homologs of Trithorax ATX1 and ATX2 Encode Proteins with Divergent Biochemical Functions. *The Plant Cell, 20*, 568-579.

[46] Berr, A., Xu, L., Gao, J., Cognat, V., Steinmetz, A., Dong, A. & Shen, W.-H. (2009). SET DOMAIN GROUP25 Encodes a Histone Methyltransferase and Is Involved in FLOWERING LOCUS C Activation and Repression of Flowering. *Plant Physiol, 151*, 1476-1485.

[47] Tamada, Y., Yun, J.-Y., Woo S. c. & Amasino, R. M. (2009). ARABIDOPSIS TRITHORAX-RELATED7 Is Required for Methylation of Lysine 4 of Histone H3 and for Transcriptional Activation of FLOWERING LOCUS C. *Plant Cell, 21*, 3257-3269.

[48] Guo, L., Yu, Y., Law, J. A. & Zhang, X. (2010). SET DOMAIN GROUP2 is the major histone H3 lysine 4 trimethyltransferase in Arabidopsis. *PNAS, 107*, 18557-18562.

[49] Berr, A., McCallum, E. J., Menard, R., Meyer, D., Fuchs, J., Dong, A. & Shen, W.-H. (2010). Arabidopsis SET DOMAIN GROUP2 Is Required for H3K4 Trimethylation and Is Crucial for Both Sporophyte and Gametophyte Development. *Plant Cell, 22*, 3232-3248.

[50] Lee, J.-S. & Shilatifard, A. (2007). A site to remember: H3K36 methylation a mark for histone deacetylation. *Mutat. Res, 618*, 130-134.

[51] Cartagena, J. A., Matsunaga, S., Seki, M., Kurihara, D., Yokoyama, M., Shinozaki, K., Fujimoto, S., Azumi, Y., Uchiyama, S. & Fukui, K. (2008). The Arabidopsis SDG4 contributes to the regulation of pollen tube growth by methylation of histone H3 lysines 4 and 36 in mature pollen. *Dev. Biol, 315*, 355-368.

[52] Thorstensen, T., Grini, P., Mercy, I., Alm, V., Erdal, S., Aasland, R. & Aalen, R. (2008). The Arabidopsis SET-domain protein ASHR3 is involved in stamen development and interacts with the bHLH transcription factor ABORTED MICROSPORES (AMS). *Plant Mol. Biol, 66*, 47-59.

[53] Hennig, L. & Derkacheva, M. (2009). Diversity of Polycomb group complexes in plants: same rules, different players? *Trends Genet, 25*, 414-423.

[54] Schatlowski, N., Creasy, K., Goodrich, J. & Schubert, D. (2008). Keeping plants in shape: Polycomb-group genes and histone methylation. *Seminars in Cell & Dev. Biol, 19*, 547 - 553.

[55] Grossniklaus, U., Vielle-Cazalda, J. P., Hoepner, M. A. & Gagliano, W. (1998). Maternal Control of Embryogenesis by MEDEA, a Polycomb Group Gene in *Arabidopsis thaliana. Science, 280*, 448 - 449.

[56] Ohad, N., Yadegari, R., Margossian, L., Hannon, M. & Michaeli, D. (1999). Mutations in FIE, a WD polycomb group gene, allow endosperm development without fertilization. *Plant Cell, 11*, 407-416.

[57] Luo, M., Bilodeau, P., Koltunow, A., Dennis, E. S., Peacock, W. J. & Chaudhury, A. M. (1999). Genes controlling fertilization-independent seed development in Arabidopsis thaliana, *PNAS, 96*, 296-301

[58] Kohler, C., Hennig, L., Bouveret, R., Gheyselinck, J., Grossniklaus, U. & Gruissem, W. (2003). Arabidopsis MSI1 is a component of the E(Z)/FIE Polycomb group complex and required for seed development. *EMBO J, 22*, 4804 - 4814.

[59] Gehring, M., Choi, Y. & Fischer, R. L. (2004). Imprinting and seed development. *Plant Cell, 16 Suppl*, S203-13.

[60] Huh, J. H., Bauer, M. J., Hsieh, T. F. & Fischer, R. L. (2008). Cellular Programming of Plant Gene Imprinting. *Cell, 132*, 735 - 744.

[61] Chanvivattana, Y., Bishopp, A., Schubert, D., Stock, C., Moon, Y.-H., Sung, Z. R. & Goodrich, J. (2004). Interaction of Polycomb-group proteins controlling flowering in *Arabidopsis. Development, 131*, 5263-5276.

[62] Schubert, D., Clarenz, O. & Goodrich, J. (2005). Epigenetic control of plant development by Polycomb-group proteins. *Curr. Opin. Plant Biol, 8*, 553-561.

[63] Schubert, D., Primavesi, L., Bishopp, A., Roberts, G., Doonan, J., Jenuwein, T. & Goodrich, J. (2006). Silencing by plant Polycomb-group genes requires dispersed trimethylation of histone H3 at lysine 27. *EMBO J, 25*, 4638-4649.

[64] Sung, S. & Amasino, R. M. (2004). Vernalization in Arabidopsis thaliana is mediated by the PhD finger protein VIN3. *Nature, 427*, 159 - 164.

[65] Wood, C. C., Robertson, M., Tanner, G., Peacock, W. J., Dennis, E. S. & Helliwell, C. A. (2006). The Arabidopsis thaliana vernalization response requires a polycomb-like protein complex that also includes VERNALIZATION INSENSITIVE 3. *PNAS, 103*, 14631 - 14636.

[66] Thakur, J. K., Malik, M. R., Bhatt, V., Reddy, M. K. & Sopory, S. K. (2003). A POLYCOMB group gene of rice (*Oryza sativa L.* subspecies indica), OsiEZ1, codes for

a nuclear-localized protein expressed preferentially in young seedlings and during reproductive development, *Gene*, *314*, 1-13.

[67] Luo, M., Platten, D., Chaudhury, A., Peacock, W. J. & Dennis, E. S. (2009). Expression, imprinting, and evolution of rice homologues of the Polycomb group genes. *Mol. Plant, 2*, 711 - 723.

[68] Jacob, Y., Feng, S., LeBlanc, C. A., Bernatavichute, Y. V., Stroud, H., Cokus, S., Johnson, L. M., Pellegrini, M., Jacobsen, S. E. & Michaels, S. D. (2009). ATXR5 and ATXR6 are H3K27 monomethyltransferases required for chromatin structure and gene silencing. *Nat. Struct. Mol. Biol, 16*, 763-768.

[69] Rea, S., Eisenhaber, F., O'Carroll, D., Strahl, B. D., Sun, Z.-W., Schmid, M., Opravil, S., Mechtler, K., Ponting, C. P., Allis, C. D. & Jenuwein, T. (2000). Regulation of chromatin structure by site-specific histone H3 methyltransferases. *Nature, 406*, 593-599.

[70] Johnson, L. M., Bostick, M., Zhang, X., Kraft, E. & Henderson, I. (2007). The SRA methyl-cytosine-binding domain links DNA and histone methylation. *Curr. Biol, 17*, 379.

[71] Johnson, L. M., Law, J. A., Khattar, A., Henderson, I. R. & Jacobsen, S. E. (2008). SRA-domain proteins required for DRM2-mediated de novo DNA methylation. *PLoS Genet, 4*, e1000280.

[72] Ebbs, M. L., Bartee, L. & Bender, J. (2005). H3 Lysine 9 Methylation Is Maintained on a Transcribed Inverted Repeat by Combined Action of SUVH6 and SUVH4 Methyltransferases. *Mol. Cell. Biol, 25*, 10507-10515.

[73] Ebbs, M. L. & Bender, J. (2006). Locus-Specific Control of DNA Methylation by the Arabidopsis SUVH5 Histone Methyltransferase. *Plant Cell, 18*, 1166-1176.

[74] Naumann, K., Fischer, A., Hofmann, I., Krauss, V. & Phalke, S. (2005). Pivotal role of AtSUVH2 in heterochromatic histone methylation and gene silencing in Arabidopsis. *EMBO J, 24*, 1418.

[75] Ay, N., Irmler, K., Fischer, A., Uhlemann, R., Reuter, G. & Humbeck, K. (2009). Epigenetic programming via histone methylation at WRKY53 controls leaf senescence in Arabidopsis thaliana. *The Plant J, 58*, 333-346.

[76] Thorstensen, T., Fischer, A., Sandvik, S. V., Johnsen, S. S. & Grini, P. E. (2006). The Arabidopsis SUVR4 protein is a nucleolar histone methyltransferase with preference for monomethylated H3K9. *Nucleic Acid Res, 34*, 5461.

[77] Krichevsky, A., Gutgarts, H., Kozlovsky, S. V., Tzfira, T., Sutton, A., Sternglanz, R., Mandel, G. & Citovsky, V. (2007). C2H2 zinc finger-SET histone methyltransferase is a plant-specific chromatin modifier. *Dev. Biol, 303*, 259 - 269.

[78] Shi, Y., Lan, F., Matson, C., Mulligan, P., Whetstine, J. R., Cole, P. A., Casero, R. A. & Shi, Y. (2004). Histone Demethylation Mediated by the Nuclear Amine Oxidase Homolog LSD1. *Cell, 119*, 941-953.

[79] Nightingale, K. P., Gendreizig, S., White, D. A., Bradbury, C., Hollfelder, F. & Turner, B. M. (2007). Cross-talk between Histone Modifications in Response to Histone Deacetylase Inhibitors. *J. Biol. Chem, 282*, 4408-4416.

[80] Feng, S., Jacobsen, S. E. & Reik, W. (2010). Epigenetic Reprogramming in Plant and Animal Development. *Science, 330*, 622-627.

[81] Zhou, D.-X. (2009). Regulatory mechanism of histone epigenetic modifications in plants. *Epigenetics, 4*, 15-18.

[82] Klose, R. J., Kallin, E. M. & Zhang, Y. (2006). JmjC-domain-containing proteins and histone demethylation. *Nat. Rev. Genet, 7*, 715-727.

[83] Lu, F., Li, G., Cui, X., Liu, C., Wang, X.-J. & Cao, X. (2008). Comparative Analysis of JmjC Domain-containing Proteins Reveals the Potential Histone Demethylases in *Arabidopsis* and Rice. *J. Integr. Plant Biol, 50*, 886-896.

[84] Zhou, D.-X. & Hu, Y. (2010). Regulatory Function of Histone Modifications in Controlling Rice Gene Expression and Plant Growth. *Rice, 3*, 103-111.

[85] Jones, M. A., Covington, M. F., DiTacchio, L., Vollmers, C., Panda, S. & Harmer, S. L. (2010). Jumonji domain protein JMJD5 functions in both the plant and human circadian systems. *PNAS, 107,* 21623-21628.

[86] Clouse, S. D. (2008). The molecular intersection of brassinosteroid-regulated growth and flowering in Arabidopsis. *PNAS, 105*, 7345-7346.

[87] Amasino, R. M. & Michaels, S. D. (2010). The Timing of Flowering. *Plant Physiol. 154*, 516-520.

[88] Sun, Q. & Zhou, D.-X. (2008). Rice jmjC domain-containing gene JMJ706 encodes H3K9 demethylase required for floral organ development. *PNAS, 105*, 13679-13684.

[89] Benhamed, M., Bertrand, C., Servet, C. & Zhou, D.-X. (2006). *Arabidopsis* GCN5, HD1, and TAF1/HAF2 Interact to Regulate Histone Acetylation Required for Light-Responsive Gene Expression. *Plant Cell, 18*, 2893-2903.

[90] Bharti, K., von Koskull-Doring, P., Bharti, S., Kumar, P., Tintschl-Korbitzer, A., Treuter, E. & Nover, L. (2004). Tomato Heat Stress Transcription Factor HsfB1 Represents a Novel Type of General Transcription Coactivator with a Histone-Like Motif Interacting with the Plant CREB Binding Protein Ortholog HAC1. *Plant Cell, 16*, 1521-1535.

[91] Mao, Y., Pavangadkar, K. A., Thomashow, M. F. & Triezenberg, S. J. Physical and functional interactions of *Arabidopsis* ADA2 transcriptional coactivator proteins with the acetyltransferase GCN5 and with the cold-induced transcription factor CBF1. *BBA - Gene Structure and Expression, 1759*, 69-79.

[92] Earley, K. W., Shook, M. S., Brower-Toland, B., Hicks, L. & Pikaard, C. S. (2007). *In vitro* specificities of Arabidopsis co-activator histone acetyltransferases: implications for histone hyperacetylation in gene activation. *The Plant J, 52*, 615-626.

[93] Offermann, S., Dreesen, B., Horst, I., Danker, T., Jaskiewicz, M. & Peterhansel, C. (2008). Developmental and Environmental Signals Induce Distinct Histone Acetylation Profiles on Distal and Proximal Promoter Elements of the C4-Pepc Gene in Maize. *Genetics, 179*, 1891-1901.

[94] Horst, I., Offermann, S., Dreesen, B., Niessen, M. & Peterhansel, C. (2009). Core promoter acetylation is not required for high transcription from the phosphoenolpyruvate carboxylase promoter in maize. *Epigenetics & Chromatin, 2*, 17-17.

[95] Casati, P., Campi, M., Chu, F., Suzuki, N., Maltby, D., Guan, S., Burlingame, A. L. & Walbot, V. (2008). Histone Acetylation and Chromatin Remodeling Are Required for UV-B-Dependent Transcriptional Activation of Regulated Genes in Maize. *Plant Cell, 20*, 827-842.

[96] Komiya, R., Ikegami, A., Tamaki, S., Yokoi, S. & Shimamoto, K. (2008). Hd3a and RFT1 are essential for flowering in rice. *Development, 135*, 767-774.

[97] Wako, T., Murakami, Y. & Fukui, K. (2005). Comprehensive analysis of dynamics of histone H4 acetylation in mitotic barley cells. *Genes & Genetic Systems, 80*, 269-276.

[98] Jasencakova, Z., Meister, A. & Schubert, I. (2001). Chromatin organization and its relation to replication and histone acetylation during the cell cycle in barley. *Chromosoma, 110*, 83-92.

[99] Pandey, R., Muller, A., Napoli, C. A., Selinger, D. A., Pikaard, C. S., Richards, E. J., Bender, J., Mount, D. W. & Jorgensen, R. A. (2002). Analysis of histone acetyltransferase and histone deacetylase families of Arabidopsis thaliana suggests functional diversification of chromatin modification among multicellular eukaryotes. *Nucleic Acids Res, 30*, 5036-55.

[100] Lusser, A., Kölle, D. & Loidl, P. (2001). Histone acetylation: lessons from the plant kingdom. *Trends Plant Sci, 6*, 59-65.

[101] Pandey, R., Muller, A., Napoli, C. A., Selinger, D. A., Pikaard, C. S., Richards, E. J., Bender, J., Mount, D. W. & Jorgensen, R. A. (2002). Analysis of histone acetyltransferase and histone deacetylase families of *Arabidopsis thaliana* suggests functional diversification of chromatin modification among multicellular eukaryotes. *Nucl. Acids Res, 30*, 5036-5055.

[102] Papaefthimiou, D., Likotrafiti, E., Kapazoglou, A., Bladenopoulos, K. & Tsaftaris, A. (2010). Epigenetic chromatin modifiers in barley: III. Isolation and characterization of the barley GNAT-MYST family of histone acetyltransferases and responses to exogenous ABA. *Plant Physiol. Biochem, 48*, 98-107.

[103] López-Rodas, G., Georgieva, E. I., Sendra, R. & Loidl, P. (1991). Histone acetylation in Zea mays.I. Activities of histone acetyltransferases and histone deacetylases. *J. Biol. Chem, 266*, 18745-18750.

[104] Lusser, A., Eberharter, A., Loidl, A., Schramel, M. G., Horngacher, M., Haas, H. & Loidl, P. (1999). Analysis of the histone acetyltransferase B complex of maize embryos. *Nucl. Acids Res, 27*, 4427-4435.

[105] Kölle, D., Sarg, B., Lindner, H. & Loidl, P. (1998). Substrate and sequential site specificity of cytoplasmic histone acetyltransferases of maize and rat liver. *FEBS Letters, 421*, 109-114.

[106] Georgieva, E. I., López-Rodas, G., Sendra, R., Gröbner, P. & Loidl, P. (1991). Histone acetylation in Zea mays. II. Biological significance of post-translational histone acetylation during embryo germination. *J. Biol. Chem, 266*, 18751-18760.

[107] Bhat, R. A., Riehl, M., Santandrea, G., Velasco, R., Slocombe, S., Donn, G., Steinbiss, H.-H., Thompson, R. D. & Becker, H.-A. (2003). Alteration of GCN5 levels in maize reveals dynamic responses to manipulating histone acetylation. *The Plant J, 33*, 455-469.

[108] Sreenivasulu, N., Usadel, B., Winter, A., Radchuk, V., Scholz, U., Stein, N., Weschke, W., Strickert, M., Close, T. J., Stitt, M., Graner, A. & Wobus, U. (2008). Barley Grain Maturation and Germination: Metabolic Pathway and Regulatory Network Commonalities and Differences Highlighted by New MapMan/PageMan Profiling Tools. *Plant Physiol, 146*, 1738-1758.

[109] Liu, C., Lu, F., Cui, X. & Cao, X. (2010). Histone Methylation in Higher Plants. *Annual Review of Plant Biol, 61*, 395-420.

[110] Fu, W., Wu, K. & Duan, J. (2007). Sequence and expression analysis of histone deacetylases in rice. *Biochem. Biophys. Res. Commun, 356*, 843-50.

[111] Hollender, C. & Liu, Z. (2008). Histone Deacetylase Genes in Arabidopsis Development. *Journal of Integrative Plant Biol, 50*, 875-885.

[112] Aufsatz, W., Mette, M. F., van der Winden, J., Matzke, M. & Matzke, A. J. (2002). HDA6, a putative histone deacetylase needed to enhance DNA methylation induced by double-stranded RNA. *EMBO J, 21*, 6832-41.

[113] Probst, A. V., Fagard, M., Proux, F., Mourrain, P., Boutet, S., Earley, K., Lawrence, R. J., Pikaard, C. S., Murfett, J., Furner, I., Vaucheret, H. & Mittelsten Scheid, O. (2004). Arabidopsis histone deacetylase HDA6 is required for maintenance of transcriptional gene silencing and determines nuclear organization of rDNA repeats. *Plant Cell, 16*, 1021-34.

[114] Tian, L., Fong, M. P., Wang, J. J., Wei, N. E., Jiang, H., Doerge, R. W. & Chen, Z. J. (2005). Reversible histone acetylation and deacetylation mediate genome-wide, promoter-dependent and locus-specific changes in gene expression during plant development. *Genetics, 169*, 337-45.

[115] Zhou, C., Zhang, L., Duan, J., Miki, B. & Wu, K. (2005). HISTONE DEACETYLASE19 is involved in jasmonic acid and ethylene signaling of pathogen response in *Arabidopsis*. *Plant Cell, 17*, 1196-204.

[116] Xu, C.-R., Liu, C., Wang, Y.-L., Li, L.-C., Chen, W.-Q., Xu, Z.-H. & Bai, S.-N. (2005). Histone acetylation affects expression of cellular patterning genes in the Arabidopsis root epidermis. *PNAS, 102*, 14469-14474.

[117] Tanaka, M., Kikuchi, A. & Kamada, H. (2008). The *Arabidopsis* histone deacetylases HDA6 and HDA19 contribute to the repression of embryonic properties after germination. *Plant Physiol, 146*, 149-61.

[118] Wu, K., Tian, L., Malik, K., Brown, D. & Miki, B. (2000). Functional analysis of HD2 histone deacetylase homologues in *Arabidopsis thaliana*. *Plant J, 22*, 19-27.

[119] Zhou, C., Labbe, H., Sridha, S., Wang, L., Tian, L., Latoszek-Green, M., Yang, Z., Brown, D., Miki, B. & Wu, K. (2004). Expression and function of HD2-type histone deacetylases in Arabidopsis development. *Plant J, 38*, 715-24.

[120] Sridha, S. & Wu, K. (2006). Identification of AtHD2C as a novel regulator of abscisic acid responses in Arabidopsis. *Plant J, 46*, 124-33.

[121] Wu, K., Zhang, L., Zhou, C., Yu, C. W. & Chaikam, V. (2008). HDA6 is required for jasmonate response, senescence and flowering in Arabidopsis. *J. Exp. Bot, 59*, 225-34.

[122] Chen, L.-T., Luo, M., Wang, Y.-Y. & Wu, K. (2010). Involvement of Arabidopsis histone deacetylase HDA6 in ABA and salt stress response. *J Exp Bot, 61*, 3345-3353.

[123] Lusser, A., Kolle, D. & Loidl, P. (2001). Histone acetylation: lessons from the plant kingdom. *Trends Plant Sci, 6*, 59-65.

[124] Rossi, V., Locatelli, S., Lanzanova, C., Boniotti, M. B., Varotto, S., Pipal, A., Goralik-Schramel, M., Lusser, A., Gatz, C., Gutierrez, C. & Motto, M. (2003). A maize histone deacetylase and retinoblastoma-related protein physically interact and cooperate in repressing gene transcription. *Plant Mol Biol, 51*, 401-13.

[125] Varotto, S., Locatelli, S., Canova, S., Pipal, A., Motto, M. & Rossi, V. (2003). Expression Profile and Cellular Localization of Maize Rpd3-Type Histone Deacetylases during Plant Development, *Plant Physiol, 133*, 606-617.

[126] Rossi, V., Locatelli, S., Varotto, S., Donn, G., Pirona, R., Henderson, D. A., Hartings, H. & Motto, M. (2007). Maize histone deacetylase hda101 is involved in plant

development, gene transcription, and sequence-specific modulation of histone modification of genes and repeats. *Plant Cell,* 19, 1145-62.

[127] Castro, C., Motto, M., Rossi, V. & Manetti, C. (2008). Variation of metabolic profiles in developing maize kernels up- and down-regulated for the hda101 gene. *J. Exp. Bot, 59*, 3913-3924.

[128] Chung, P. J., Kim, Y. S., Jeong, J. S., Park, S.-H., Nahm, B. H. & Kim, J.-K. (2009). The histone deacetylase OsHDAC1 epigenetically regulates the OsNAC6 gene that controls seedling root growth in rice. *The Plant J, 59*, 764-776.

[129] Hu, Y., Qin, F., Huang, L., Sun, Q., Li, C., Zhao, Y. & Zhou, D.-X. (2009). Rice histone deacetylase genes display specific expression patterns and developmental functions. *Biochem. Biophys. Res. Commun, 388*, 266-271.

[130] Demetriou, K., Kapazoglou, A., Bladenopoulos, K. & Tsaftaris, A. (2010). Epigenetic Chromatin Modifiers in Barley: II. Characterization and Expression Analysis of the HDA1 Family of Barley Histone Deacetylases during Development and in Response to Jasmonic Acid. *Plant Mol. Biol. Rep, 28*, 9-21.

[131] Demetriou, K., Kapazolgou, A., Tondelli, A., Francia, E., Stanca, M. S., Bladenopoulos, K. & Tsaftaris, A. S. (2009). Epigenetic chromatin modifiers in barley: I. Cloning, mapping and expression analysis of the plant specific HD2 family of histone deacetylases from barley, during seed development and after hormonal treatment. *Physiologia Plantarum, 136*, 358 - 368.

[132] Bannister, A. J., Zegerman, P., Partridge, J. F., Miska, E. A., Thomas, J. O., Allshire, R. C. & Kouzarides, T. (2001). Selective recognition of methylated lysine 9 on histone H3 by the HP1 chromo domain. *Nature, 410*, 120-124.

[133] Taverna, S. D., Li, H., Ruthenburg, A. J., Allis, C. D. & Patel, D. J. (2007). How chromatin-binding modules interpret histone modifications: lessons from professional pocket pickers. *Nat. Struct. Mol. Biol, 14*, 1025-1040.

[134] Bartke, T., Vermeulen, M., Xhemalce, B., Robson, S. C., Mann, M. & Kouzarides, T. Nucleosome-Interacting Proteins Regulated by DNA and Histone Methylation. *Cell, 143*, 470-484.

[135] Vermeulen, M., Eberl, H. C., Matarese, F., Marks, H., Denissov, S., Butter, F., Lee, K. K., Olsen, J. V., Hyman, A. A., Stunnenberg, H. G. & Mann, M. (2010). Quantitative Interaction Proteomics and Genome-wide Profiling of Epigenetic Histone Marks and Their Readers. *Cell, 142*, 967-980.

[136] Zhang, X., Germann, S., Blus, B. J., Khorasanizadeh, S., Gaudin, V. & Jacobsen, S. E. (2007). The Arabidopsis LHP1 protein colocalizes with histone H3 Lys27 trimethylation. *Nat Struct Mol Biol, 14*, 869-871.

[137] Exner, V., Aichinger, E., Shu, H., Wildhaber, T., Alfarano, P., Caflisch, A., Gruissem, W., Kohler, C. & Hennig, L. (2009). The Chromodomain of LIKE HETEROCHROMATIN PROTEIN 1 Is Essential for H3K27me3 Binding and Function during Arabidopsis Development. *PLoS ONE, 4*, e5335.

[138] Mylne, J. S., Barrett, L., Tessadori, F., Mesnage, S. p., Johnson, L., Bernatavichute, Y. V., Jacobsen, S. E., Fransz, P. & Dean, C. (2006). LHP1, the Arabidopsis homologue of HETEROCHROMATIN PROTEIN1, is required for epigenetic silencing of FLC. *PNAS, 103*, 5012-5017.

[139] de la Paz Sanchez, M. & Gutierrez, C. (2009). Arabidopsis ORC1 is a PHD-containing H3K4me3 effector that regulates transcription. *PNAS, 106*, 2065-2070.

[140] Lo, W. S. & *et al*. (2000). Phosphorylation of serine 10 in histone H3 is functionally linked in vitro and in vivo to Gcn5-mediated acetylation at lysine 14. *Mol Cell, 5*, 917-926.

[141] Kim, J., Guermah, M., McGinty, R. K., Lee, J.-S., Tang, Z., Milne, T. A., Shilatifard, A., Muir, T. W. & Roeder, R. G. (2009). RAD6-Mediated Transcription-Coupled H2B Ubiquitylation Directly Stimulates H3K4 Methylation in Human Cells. *Cell, 137*, 459-471.

[142] Mateescu, B., Bourachot, B., Rachez, C., Ogryzko, V. & Muchardt, C. (2008). Regulation of an inducible promoter by an HP1β-HP1γ switch. *EMBO Rep, 9*, 267-272.

[143] Bernstein, B. E., Mikkelsen, T. S., Xie, X., Kamal, M., Huebert, D. J., Cuff, J., Fry, B., Meissner, A., Wernig, M., Plath, K., Jaenisch, R., Wagschal, A., Feil, R., Schreiber, S. L. & Lander, E. S. (2006). A Bivalent Chromatin Structure Marks Key Developmental Genes in Embryonic Stem Cells. *Cell, 125*, 315-326.

[144] Saleh, A., Al-Abdallat, A., Ndamukong, I., Alvarez-Venegas, R. & Avramova, Z. (2007). The Arabidopsis homologs of trithorax (ATX1) and enhancer of zeste (CLF) establish "bivalent chromatin marks" at the silent AGAMOUS locus. *Nucleic Acids Res, 35*, 6290-6296.

In: Histones: Class, Structure and Function
Editor: Chang-Hui Shen
ISBN: 978-1-62100-274-1

Chapter 7

HISTONE PHOSPHORYLATION AND THE CELL DIVISION CYCLE

Geoffrey Woodard*

Department of Medicine, Uniformed Services University of the Health Sciences, Bethesda, MD, US

ABSTRACT

Chromatin condensation and subsequent decondensation are processes required for proper execution of various cellular events including cell division and DNA replication. During mitosis, chromatin compaction is at its highest, whereas relaxation of chromatin is necessary for DNA replication, repair, recombination, and gene transcription. Since histone proteins are directly complexed with DNA in the form of a nucleosome, great emphasis is put on deciphering histone post-translational modifications that control the chromatin condensation state. Histone H3 phosphorylation is a mark present in mitosis, where chromatin condensation is necessary, and in transcriptional activation of genes, when chromatin needs to be decondensed. There are four characterized phospho residues within the H3 N-terminal tail during mitosis: Thr3, Ser10, Thr11, and Ser28. Interestingly, H3 phosphorylated at Ser10, Thr11, and Ser28 has been observed on genomic regions of transcriptionally active genes. Therefore, H3 phosphorylation is involved in processes requiring opposing chromatin states. The level of H3 phosphorylation is mediated by opposing actions of specific kinases and phosphatases during mitosis and gene transcription. The cellular contexts under which specific residues on H3 are phosphorylated in mitosis and interphase are known to some extent. However, the functional consequences of H3 phosphorylation are still unclear. Likewise the role of epigenetic signaling events, regarding methylation and acetylation of histone proteins, that regulate histone modification during mitosis is still actively being explored. Several converging and reinforcing signals, including transcription factors, noncoding RNAs, DNA methylation, and histone modifications orchestrate complex epigenetic states. Although all of these pathways modulate transcription from chromatin in vivo, the mechanism by which epigenetic information is transmitted through cell division remains unclear.

* E-mail: Geoffrey.Woodard.CTR@usuhs.edu.

1. INTRODUCTION

Nuclear DNA is packaged into nucleosomes, the basic repeating structural units in chromatin. The nucleosome consists of a histone octamer, arranged as a $(H3\text{-}H4)_2$ tetramer and two H2A-H2B dimers, around which DNA is wrapped. The core histones (ie, H2A, H2B, H3 and H4) have a similar structure, with a basic N-terminal domain, a globular domain organized by the histone fold and a C-terminal tail. The N-terminal tails emanate from the nucleosome in all directions and are available to interact with linker DNA, nearby nucleosomes or with other proteins. Histone H1 binds to the nucleosome and the linker DNA, which joins nucleosomes together. Histone H1 and variants of this linker histone are phosphorylated, with levels of phosphorylation increasing throughout interphase and peaking at mitosis [1]. Recent studies demonstrate a role of H1 phosphorylation in facilitating transcription [2,3]. Core histones undergo a variety of reversible post-translational modifications (PTMs), including acetylation, methylation and phosphorylation. Core histone PTMs function to alter chromatin structure and/or provide a 'code' for the recruitment or occlusion of non-histone chromosomal proteins to chromatin [4-6]. This section provides an overview of the biological roles of core histone phosphorylation in mammalian cells particularly during cell division.

2. PHOSPHORYLATION OF HISTONE H3

There are several known residues (eg, Tyr3, Tyr6, Ser10, Tyr11, Ser28, Tyr41, Tyr45 and Tyr118) within the H3 molecule that are phosphorylated in response to different cellular processes. Each phospho-H3 modification has a specific kinase that mediates the phosphorylation of a particular residue within the H3 protein (Table 1) [7]. Some phospho-H3 modifications are well characterized (eg, Ser10 and Ser28), while others have only been recently described (eg, Tyr6, Tyr41 and Tyr45). Phosphorylation of Tyr118 was identified by mass spectrometry [8], and may function to increase nucleosome mobility [9].

The best-characterized H3 phosphorylation residue is Ser10 (H3S10ph). Extensive research has implicated H3S10ph in processes during mitosis and the transcriptional activation of specific genes in interphase [7]. During mitosis in mouse and human cells, H3S10ph is first evident in pericentromeric chromatin of late G2-phase cells. The distribution of H3S10ph subsequently spreads throughout the chromosomes and peaks during prophase; however, H3S10ph is still evident at the metaphase plate [10]. Aurora B is the major mitotic kinase responsible for H3S10ph [11]. Dephosphorylation of H3S10 by protein phosphatase 1 starts in late anaphase and is completed by early telophase, before detectable post-mitotic decondensation of chromosomes [7,10]. The function of H3S10ph during mitosis has been correlated with the initiation of chromosome condensation, rather than the maintenance of mitotic chromatin condensation [7].

H3S10ph also has a role in the initiation of localized chromatin decondensation associated with the transcriptional activation of genes, such as immediate-early genes [7]. This H3S10ph event is targeted to a small proportion of H3 molecules within the chromatin of mouse and human interphase cells during the activation of signaling pathways (eg, Ras-MAPK) [7]. For example, during phorbol ester-induced Ras-MAPK signaling, MAPK and stress-activated protein kinase-1 (MSK1) phosphorylate H3S10 at the regulatory nucleosomes of immediate-early genes and initiate the binding of 14-3-3to H3S10ph, which can further recruit chromatin modifiers and remodelers (eg, SWI/SNF) to aid in the initiation of transcription [12]. Different kinases (eg, IKKα and PIM1 responding to a specific stimulus can phosphorylate H3 and elicit a similar mechanism at the regulatory nucleosomes of the responsive genes [7,13-15]. Overall, H3S10ph has two conflicting roles: one in the initiation of mitotic chromosome condensation and the other in localized chromatin decondensation leading to gene expression. Similar to H3S10ph, H3 phosphorylation at Ser28 (H3S28ph)occurs during mitosis and interphase in mouse fibroblasts [7,16]. The mitotic distribution and timing of H3S28ph is similar to that of H3S10ph, and Aurora B is the kinase responsible for this phosphorylation event [7,16]. H3S28ph coincides with the onset of chromosome condensation, similar to H3S10ph. During interphase in mouse fibroblasts, H3S28ph is established by stimulus-specific activated kinases on the regulatory nucleosomes of specific genes in response to the activation of various signaling cascades. Similar to H3S10ph, the establishment of H3S28ph is involved in recruiting 14-3-3ε and 14-3-3γ and, subsequently, in chromatin modifying and enzyme remodeling, in order to alter the local chromatin environment and allow transcriptional initiation [7,12]. H3S10ph and H3S28ph do not alter chromatin structure, but have a role in attracting and repelling proteins binding to the H3 N-terminal domain. 14-3-3 proteins bind to these H3 PTMs. H3S10ph repels heterochromatin protein 1γ (HP1-gamma) from its binding site at the neighboring trimethylated Lys9 [17,18]. This molecular event is referred to as the 'switch' or 'binary-switch' hypothesis [19,20]. H3S10ph also repels the binding of the SR-rich RNA splicing factors SRp20 and ASF/SF2 from mitotic chromosomes [21]. Phosphorylation of the H3 residues Thr3, Ser10 and Thr11 prevents the binding of the DNA methyltransferase DNMT3a to the H3 N-terminal tail domain, resulting in reduced DNA methyation at chromatin regions with phosphorylated H3 [22] Upon stimulation of the MAPK pathway in 10T1/2 and Ciras3 fibroblasts, numerous small foci of H3S10ph and H3S28ph appear in the nuclei of these cells. Surprisingly, the majority of H3S10ph foci do not coincide with H3S28ph foci, indicating that H3S10ph and H3S28ph do not coexist on the same histone tail or even adjacent nucleosomes [23,24]. In addition, chromatin immunoprecipitation (ChIP) assays revealed that strings of nucleosomes of two to three long had H3S10ph or H3S28ph [24]. Thus, these two phosphorylation events at Ser10 and Ser28 are independent and act separately to promote gene expression. In sequential ChIP assays, it was demonstrated that the upstream promoter regions of immediate-early genes were associated with H3S10ph or H3S28ph [12]. These results suggest that there may be parental allele-specific association of H3S10ph and H3S28ph with the regulatory region of the immediate-early genes.

The phosphorylation of H3 at its threonine residues has been receiving increasing attention. Two mitotic marks, H3 phosphorylation at Thr3 (H3T3ph) and Thr11 (H3T11ph) have been characterized [25,26] and, recently, H3 phosphorylation at Thr6 (H3T6ph) was observed ininterphase [27]. H3T3ph appears in the G2 phase of the cell cycle as a speckle along chromosomes in HeLa cells and, by late prophase, H3T3ph has spread throughout the chromosome arms. The kinase responsible for this H3 phosphorylation event is Haspin [28]. The most intense immunofluorescence staining for H3T3ph occurs at the inner centromeric regions [25]. Levels of H3T3ph start to decline prior to chromosome decondensation in telophase. The distribution of H3T3ph overlaps with that of cohesion protein subunits (Rad21/Scc1 and SA2), indicating a possible role in chromosome cohesion [29]. Similar to

H3S10ph, H3T11 phosphorylation has been reported to occur both during mitosis and in interphase in rat and human cells [26,30]. H3T11ph is evident from early prophase to early anaphase, and its localization is largely restricted to centromeric regions; thus, H3T11ph may play a role in kinetochore recognition [30]. Although H3T11ph is considered a mitotic mark, this modification differs in its spatial and temporal characteristics from the other well-established mitotic marks H3S10ph and H3S28ph, which originate from pericentromeric regions and spread across chromosome arms, as H3T11ph is observed mostly at centromeric regions. Mitotic H3T11ph is mediated by death-associated protein-like kinase (DLK) or zipper interaction protein kinase (ZIP) [30]; however, its role has not been determined.

H3T11ph also occurs in interphase cells. Checkpoint protein 1 (Chk1)-mediated H3T11ph has been associated with DNA damage-induced gene repression in mouse fibroblasts [31]. Interestingly, PKC-related kinase 1-mediated H3T11ph was determined to be associated with the expression of androgen receptor target genes in prostate carcinoma cell lines [32]. Therefore, depending on the stimulus and cellular context, H3T11ph can be involved in opposing functions during interphase.

Table 1. An overview of histone phosphorylation

Histone	Amino acid	Kinase	Phosphatase	Function
H3	Ser^{10}	MSK1, MSK2, IKK-α and PIM1 Aurora B	PP1	TS Mitosis
	Ser^{28}	MSK1, MSK2 and Aurora B	PP1	TS Mitosis
	Thr^{3}	Haspin	ND	Mitosis
	Thr^{11}	DLK Chk1	ND	Mitosis TS
	Thr^{6}	PKCβ$_{I}$	ND	TS
	Tyr^{41}	JAK2	ND	TS
	Thr^{45}	PKCδ	ND	Apoptosis DNA replication
H4	Ser^{1}	CK2	ND	TS repression DNA repair Mitosis
	His^{18}	His H4 kinase	ND	ND
H2A	H2A.1 Ser^{1}	Aurora B RSK2 MSK1	ND	Mitosis TS repression
	H2A.X Ser^{139} (γ-H2A.X)	ATM ATR DNA-PK	PP1 PP2A PP4 Wip1	DNA repair Mitosis Apoptosis
	H2A.X Tyr^{142}	WSTF	EYA1/3	DNA repair
	MacroH2A Thr^{128}	ND	ND	Chromatin condensation
	MacroH2A Ser^{137}	Cdks (in vitro)	ND	Chromatin condensation Senescence
	H2A.Z Tyr[illegible]	ND	ND	ND
H2B	Ser^{14}	Mst1 and PKCδ ND	ND ND	Apoptosis DNA repair

ND - Not determined, TS - transcription.

A novel H3 phosphorylation modification at Thr6 has been recently described in interphase cells. H3T6ph, and the corresponding kinase PKCβ1, were observed in a LNCaP human prostate cancer cell line [27]. This study provided evidence that androgen-dependent signaling can lead to the establishment of the H3T6ph mark, which prevents the removal of methyl marks from H3 Lys4, mediated by histone lysine-specific demethylase 1 (LSD1) during androgen receptor-induced gene expression [27]. Metzger *et al* also demonstrated that the increased levels of PKC-β1 and H3T6ph positively correlated with a high Gleason score of prostate carcinomas. Furthermore, inhibition of PKC-β1 blocked androgen receptor-induced tumor cell proliferation *in vitro* and the cancer progression of tumor xenografts *in vivo* [27]. This study provided insight into the mechanistic outcome of H3T6ph. Nevertheless, further studies examining the molecular role of H3T6ph in other cell types/tissues and cellular contexts (eg, mitosis) are needed.

Two recently characterized H3 phosphorylation sites have been reported at Tyr41 (H3Y41ph) and Thr45 (H3T45ph). H3Y41ph was identified as an event during chromatin regulation in human PBMCs [33]. JAK2-mediated H3Y41ph prevents the binding of HP1-alpha to a region close to H3T45 Tyr41 at the nucleosomes of specific genes (eg, *LMO2*). The exclusion of HP1-alpha binding may promote a more open configuration of localized chromatin. Given that H3Y41 is located within a histone core region that affects nucleosome remodeling, H3Y41ph may have an impact on chromatin dynamics at specific gene promoters [33].

H3T45ph was recently identified with the use of mass spectrometry [34], and occurs within the histone core, unlike most other histone H3 PTMs, which are located on the N-terminal tail. The kinase responsible for H3T45ph is PKC-DELTAδ, and this modification is associated with apoptosis in human promyelocytic leukemia cells around the time of DNA nicking [34]. As H3T45 is located within the histone core, phosphorylation of this residue may mediate structural changes within the nucleosome to facilitate DNA nicking and/or fragmentation during apoptosis [34]. Furthermore, H3T45ph was recently reported to be a replication-associated modification in budding yeast [35] and, therefore, can be observed during interphase and DNA replication.

Mammalian cells have three variants of histone H3, H3.1, H3.2 and H3.3 that are highly conserved in their amino acid composition [36-38]. The replication-independent variant H3.3 differs in only five amino acids from the canonical H3, with one of the amino acid differences being at position 31 (a proline residue in H3), which is a serine in H3.3 (H3.3 Ser31) and an alanine in H3.1 and H3.2 [36]. Interestingly, H3.3 Ser31 is phosphorylated (H3.3S31ph) during mitosis [39]. Although Ser10 and Ser28 can be phosphorylated on H3.3, the timing and distribution of H3S10ph and H3S28ph differs from H3.3S31ph [40]. H3.3S31ph is first evident during late prometaphase and peaks in metaphase, with its localization predominantly at pericentromeric regions in mitotic HeLa cells [39]. Currently, the role of mitotic H3.3S31ph is unknown, and further studies need to be conducted on this unusual phosphorylation event.

Another histone H3 variant, centromere protein A (CENP-A), is located exclusively at centromeres and is phosphorylated at Ser7 during mitosis. In HeLa cells, mitotic CENP-A Ser7 phosphorylation occurs at all centromeres, albeit after the spreading of H3 phosphorylation to the chromosome arms [41,42]. CENP-A Ser7 phosphorylation is implicated in the proper localization of Aurora B and INCENP, as well as in normal daughter cell separation [41]; however, the exact role of this PTM has not been fully resolved.

3. Phosphorylation of Histone H4

Histone H4 is phosphorylated at several residues (eg, Ser1, His18, Ser47, Tyr51, Tyr72 and Tyr88), with phosphorylation at Ser1 (H4S1ph) being the best characterized (Table 1). Phosphorylation at Ser47, Tyr51, Tyr72 and Tyr88 were identified by mass spectrometry [8,43], and the function of modification at these sites is unknown [9].

In yeast, H4S1ph plays a pivotal role in chromatin condensation and segregation. During yeast sporulation, H4S1ph has a transcriptional repressive function and is localized at the transcription start site of genes [44]. H4S1ph mediated by casein kinase 2 (CK2) has also been demonstrated to have a role in DNA repair in yeast [45]. In HeLa cells, H4S1ph has a biphasic pattern, with high levels of phosphorylation at the G1/S border and in early S-phase that subsequently decline to baseline levels during mid/late S-phase/early G2 before peaking again during mitosis [46].

The phosphorylation of H4 at His18 (H4H18ph) is a rare event that occurs during DNA synthesis. The lack of knowledge of this H4 PTM is attributed to the acid-labile nature of the phosphoramidate bond of phosphohistidine. Histidine H4 kinase activity increases when growth is stimulated by partial hepatectomy and, when the liver is damaged, histidine H4 kinase activity has also been demonstrated to be induced [47].

4. Phosphorylation of Histone H2A

The core histone H2A has the largest family of macroheterogeneous-described variants, which are mainly distinguished from the major H2A by the sequence and length of their N- and C-termini [48]. As a result of its position close to the dyad axis, histone H2A phosphorylation is likely to impact chromatin structure [49]. In humans, 12 H2A gene family members have been identified [50], some of which are conserved throughout evolution (eg, H2A.Z) [51], while others, such as macroH2A [52] and H2A-Bbd [53], are restricted to vertebrates or mammals [54]. In mammals, the H2A.Z variant represents approximately 10% of the H2A complement, H2A.X represents 2 to 25%, and H2A.1-H2A.2 represents the remaining 65% [55].

The phosphorylation of H2A.1 at Ser1 (H2AS1ph) is the most abundant PTM in histone H2A. This modification has been observed in asynchronous growing cells, greatly increasing during mitosis, and is conserved in the worm, fly and mammalian cells. Together with H4S1ph, H2AS1ph is initiated during early prophase in HeLa cells and remains at high levels throughout anaphase, declining upon the completion of telophase. The onset, duration and subcellular localization of H2AS1ph is therefore similar to H3S10ph and H3S28ph, and is proposed to have a compensatory or redundant function in chromosome condensation/segregation [46]. Aurora B, RSK2 and MSK1 are the kinases that have been demonstrated to modify H2AS1 *in vitro*. H2AS1ph occurs outside of mitosis, and MSK1-mediated H2AS1ph was determined to inhibit transcription [56].

The phosphorylation of histone H2A.X at Ser139 (γ-H2A.X) occurs in response to DNA double-strand breaks (DSBs) and spreads across megabase chromatin domains flanking the lesion [57,58]. The kinases responsible for this modification are members of the PI3K family, including ataxia telangiectasia mutated protein (ATM), ataxia telangiectasia and Rad3-related protein (ATR) and DNA-dependent protein kinase, in mutually exclusive pathways depending on the underlying insult. ATM induces H2A.X phosphorylation in response to γ-irradiation, whereas ATR induces phosphorylation in response to UV damage or replicative stress [59]. However, some overlapping activities among these kinases have been reported. The role of γ-H2A.X may involve recruiting some, but probably not all, mediators/repair proteins to DNA lesions and stabilizing the repair focus to recruit late factors. The major mediator of γ-H2A.X recognition is MDC1, which binds γ-H2A.X through its tandem BRCA1 C-terminus (BRCT) domain [60]. The subsequent recruitment and concentration of ATM/ATR would allow further phosphorylation of adjacent H2A.X molecules. This DNA-damage signal amplification would concentrate several mediator/repair proteins, including BRCA1 and 53BP1, which, in turn, would facilitate downstream checkpoint activation, chromatin remodeling activities and other histone modifications, such as acetylation, that appear to also have important functions in the cellular response to DSBs [61].

A role for γ-H2AX outside the context of DNA repair can be inferred from the presence of hundreds of γ-H2A.X sites throughout the genome of normally growing mammalian cells. A small proportion of these sites is similar to the foci induced by ionizing irradiation and likely represents 'naturally occurring' DNA DSBs. However, another population of abundant, smaller domains, equally present both in euchromatin and heterochromatin, is not associated with the recruitment of DSB repair proteins [59]. The conversion of single-strand breaks by advancing replication forks has been proposed to be an alternative source of DSBs. Both ATM and ATR mediate H2A.X phosphorylation in these numerous foci, although novel kinases may also be involved.

The steady-state phosphorylation of histone H2A.X is under cell-cycle regulation. The level of γ-H2A.X is at a minimum during G1 and increases as cells progress through the S phase. This increase correlates with the DNA DSBs that occur during DNA replication as a result of stalled replication forks. However, γ-H2A.X signal intensities do not plateau in the S phase, but rather continue to increase until reaching maximal levels in G2/M [59]. The phosphorylation cycle mirrors the pattern reported for H3 phosphorylation, indicating a redundant/compensatory role, and suggesting that mitotic defects may also arise from alterations in ATM-dependent mitotic phosphorylations during the chromosome segregation process [59].

The mechanism via which γ-H2A.X is attenuated and sister chromatids are segregated following repair remains unknown, but several hypotheses have been proposed. A histone phosphatase complex containing protein phosphatase 4 (PP4) has been demonstrated to be responsible for dephosphorylating γ-H2A.X, but only after its release from chromatin [62,63]. PP2A has also been reported to bind directly to and dephosphorylate γ-H2A.X. In addition, a histone exchange mechanism has been proposed, as H2A.X phosphorylation by DNA-PK induces FACT-mediated H2A.X replacement by H2A [64]. Chromatin-bound Wip1 phosphatase can also dephosphorylate γ-H2A.X in response to DNA damage [65]. SNF2-family ATPase complexes have been implicated in the removal of γ-H2A.X by histone exchange [66-69]. However, the mobility of H2A.X in live mammalian nuclei appears to be low, indicating that turnover by histone exchange may not be the predominant mechanism.

One possible interpretation is that a chromatin remodeling complex near the DSB would remove γ-H2A.X and alter the chromatin structure, allowing repair factors to access the DSB, while more distally, γ-H2A.X would be eliminated by direct dephosphorylation. Future studies may help to define the mechanism of γ-H2A.X elimination and its role in DSB repair.

γ-H2A.X formation that is dependent on caspase-activated DNase activation occurs during apoptosis in different cellular systems, as well as in caspase-independent programmed necrosis [70]. JNK kinases are also involved in γ-H2A.X-mediated, UVA-induced apoptosis. Therefore, DNA repair and apoptosis may share common initial chromatin-related steps. In the case of DNA repair, γ-H2A.X would mediate the maintenance of the two DNA ends in close proximity, whereas in the case of apoptosis, γ-H2A.X might facilitate the packaging of fragmented DNA into apoptotic bodies

The C-terminus of H2A.X can be phosphorylated at Tyr142 (H2A.XY142ph), and this modification exists in metazoans, but not in unicellular eukaryotes [71]. The function of Tyr142 appears to be related to DSB repair. In mouse embryonic fibroblast cells, Tyr142 is constitutively phosphorylated under normal growth conditions and becomes gradually dephosphorylated during the DNA damage response, as the levels of γ-H2A.X increase. Mutation of Tyr142 greatly reduced Ser139 phosphorylation levels and foci formation [71]. Interestingly, the tyrosine kinase activity of Williams syndrome transcription factor (WSTF), a member of the BAZ/WAL family of chromatin remodeling factors, targets H2A.XY142ph [71]. H2A.XY142ph may alter the local chromatin structure for the subsequent maintenance of Ser139 phosphorylation. The subsequent dephosphorylation of Tyr142 after DNA damage may act to enhance MDC1 and ATM recruitment to extend and maintain γ-H2A.X. EYA1/3 has been identified as the phosphatase that effectively removes the phosphotyrosine mark from H2A.X, acting as a switch between the apoptotic and DNA repair pathways [72].

MacroH2A is an unusual histone variant, with one-third of the protein (at the N-terminus) sharing > 60% amino-acid sequence identity with the canonical H2A. The C-terminus of macroH2A contains a highly conserved large non-histone domain, the macro domain [73]. MacroH2A1.1 and macroH2A1.2 are enriched on the facultative heterochromatin of the inactive X chromosome (Xi) and in senescent cells, suggesting a role in the repression of transcription. The phosphorylation of Thr128 and Ser137 (both located in the hinge region of macroH2A) has been reported. Thr128 phosphorylation may modulate macroH2A.1 localization, as it resides in the region that binds SPOP, which plays a role in the correct localization of this variant histone [74]. Different roles related to chromatin structure have been proposed for macroH2A Ser137 phosphorylation. In an Xi-related context, while macroH2A is enriched on the Xi, the Ser137 phosphorylated form is not. Furthermore, macroH2A Ser137 phosphorylation occurs throughout the cell cycle, and increases during mitosis. Xi-independent roles, including during senescence and mitosis, have been proposed, although further studies are necessary to clarify the dynamics and functions of this modification [75,76].

H2A.Z is a highly conserved H2A variant that has a role in gene regulation and in antagonizing the spreading of silent heterochromatin. One phosphorylation site for H2A.Z, Tyr61, has been reported [77]; however, the role of H2A.Z phosphorylation awaits further investigation.

5. PHOSPHORYLATION OF HISTONE H2B

H2B is modified by several PTMs in its N-terminal tail region, resulting in a myriad of biological outcomes. The phosphorylation of human histone H2B occurs primarily at Ser14 (H2BS14ph), with additional, less well-known phosphorylation sites at Ser32, Tyr40 and Tyr42. There are also several H2B variants that have individual predicted sites of phosphorylation, although the biological roles of these variant phosphorylation sites have yet to be determined.

H2BS14ph has a role in DNA repair and apoptosis [78], and occurs at the sites of DNA DSBs during irradiation-induced DNA damage in mice and during camptothecin-induced death of human neuroblastoma cells [79,80]. H2BS14ph has been observed during B-cell-specific cell death; both this process and the phosphorylation of H2BS14 are catalyzed by PKC-δ [81]. During apoptosis, γ-H2A.X coincides with H2BS14ph [82]. In HL60 cells, caspase-3-activated mammalian sterile 20-like kinase (Mst1) can also phosphorylate H2B at Ser14 [78].

Histone H2B has 15 variants in humans, 11 of which are phosphorylated at several predicted and curated sites. Tyrosine phosphorylation at residues 40 and 42 in H2B; 37, 40 and 42 in H2B1D, H2B1H, H2B1K and H2B1L; 38, 41 and 43 in H2B1M; 38, 41 and 43 in H2B1N; 43 in H2B2E; 38 and 41 in H2B2E; and 37, 40 and 42 in H2BFM were detected in a global survey of phosphotyrosine signaling in lung cancer [43]. Several of these residues, with the exception of Tyr37, have also been detected in the signaling networks of oncogenic EGFR and c-Met [83], and histone H2B variants have also been reported to be upregulated in colon cancer [84]. However, these H2B variant phosphorylation sites have not been confirmed by methods other than mass spectrometry and, thus, there remains much more to be resolved regarding the biological role of H2B phosphorylation.

CONCLUSIONS

Histone phosphorylation may disrupt chromatin structure and/or provide a 'code' for the recruitment or occlusion of non-histone chromosomal proteins to chromatin. The well-characterized H3S10ph, H3S28ph and γ-H2A.X, which are involved in transcription or DNA repair, function by recruiting or repelling chromatin-binding proteins. H3S10ph and H3S28ph recruit 14-3-3 proteins, and H3S10ph repels HP1 association, while γ-H2A.X has a critical role in recruiting DNA repair proteins to DSBs. Much remains to be learned about the function of the many other core histone phosphorylations in altering chromatin structure and/or regulating the binding of non-histone chromosomal proteins. Although this review focuses on histone phosphorylation, it is important to note that the phosphorylation event often occurs in the presence of other histone PTMs, such as acetylation and methylation. Together, phosphorylation events and the surrounding PTMs determine the net impact of the PTMs on regulating non-histone chromatin proteins and modulating the structure and function of chromatin.

REFERENCES

[1] Parseghian, M.H., Henschen, A.H., Krieglstein, K.G., & Hamkalo, B.A. (1994). A proposal for a coherent mammalian histone H1 nomenclature correlated with amino acid sequences. *Protein Sci, 3(4),* 575-587.

[2] Zheng, Y., John, S., Pesavento, J.J., Schultz-Norton, J.R., Schiltz, R.L., Baek, S., Nardulli, A.M., Hager, G.L., Kelleher, N.L., & Mizzen, C.A. (2010). Histone H1 phosphorylation is associated with transcription by RNA polymerases I and II. *J Cell Biol, 189(3)*, 407-415.

[3] O'Brien, S.K., Cao, H., Nathans, R., Ali, A., & Rana, TM (2010). P-TEFB kinase complex phosphorylates histone H1 to regulate expression of cellular and HIV-1 genes. *J Biol Chem, 285(39),* 29713-20.

[4] Jenuwein. T., & Allis, C.D. (2001). Translating the histone code. *Science, 293(5532),* 1074-1080.

[5] Hake, S.B., Xiao, A., & Allis, C.D. (2004). Linking the epigenetic 'language' of covalent histone modifications to cancer. *Br J Cancer, 90(4)*, 761-769.

[6] Taverna, S.D., Li, H., Ruthenburg, A.J., Allis, C.D., & Patel, D.J. (2007). How chromatin-binding modules interpret histone modifications: Lessons from professional pocket pickers. *Nat Struct Mol Biol, 4(11),* 1025-1040.

[7] Pérez-Cadahía, B., Drobic, B., & Davie, J.R. (2009). H3 phosphorylation: Dual role in mitosis and interphase. *Biochem Cell Biol, 87(5),* 695-709.

[8] Freitas, M.A., Sklenar, A.R., & Parthun, M.R. (2004). Application of mass spectrometry to the identification and quantification of histone post-translational modifications. *J Cell Biochem, 92(4),* 691-700.

[9] Cosgrove, M.S., Boeke, J.D., & Wolberger, C. (2004). Regulated nucleosome mobility and the histone code. *Nat Struct Mol Biol, 11(11),* 1037-1043.

[10] Hendzel, M.J., Wei, Y., Mancini, M.A., Van Hooser, A., Ranalli, T., Brinkley, B.R., Bazett-Jones, D.P., & Allis, C.D. (1997). Mitosis-specific phosphorylation of histone H3 initiates primarily within pericentromeric heterochromatin during G2 and spreads in an ordered fashion coincident with mitotic chromosome condensation. *Chromosoma, 106(6),* 348-360.

[11] Crosio, C., Fimia, G.M., Loury, R., Kimura, M., Okano, Y., Zhou, H., Sen, S., Allis, C.D., & Sassone-Corsi, P. (2002). Mitotic phosphorylation of histone H3: Spatio-temporal regulation by mammalian Aurora kinases. *Mol Cell Biol, 22(3),* 874-885.

[12] Drobic, B., Pérez-Cadahía, B., Yu, J., Kung, S.K., & Davie, J.R. (2010). Promoter chromatin remodeling of immediate-early genes is mediated through H3 phosphorylation at either serine 28 or 10 by the MSK1 multi-protein complex. *Nucleic Acids Res, 38(10)*, 3196-3208.

[13] Anest, V., Hanson, J.L., Cogswell, P.C., Steinbrecher, K.A., Strahl, B.D., & Baldwin, A.S. (2003). A nucleosomal function for IκB kinase-α in NF-κB-dependent gene expression. *Nature, 423(6940),* 659-663.

[14] Yamamoto, Y., Verma, U.N., Prajapati, S., Kwak, Y.T., & Gaynor, R.B. (2003). Histone H3 phosphorylation by IKK-α is critical for cytokine-induced gene expression. *Nature, 423(6940)*, 655-659.

[15] Zippo, A., De Robertis, A., Serafini, R., & Oliviero, S. (2007). PIM1-dependent phosphorylation of histone H3 at serine 10 is required for MYC-dependent transcriptional activation and oncogenic transformation. *Nat Cell Biol, 9(8),* 932-944.

[16] Goto, H., Tomono, Y., Ajiro, K., Kosako, H., Fujita, M., Sakurai, M., Okawa, K., Iwamatsu, A., Okigaki, T., Takahashi, T., & Inagaki, M. (1999). Identification of a novel phosphorylation site on histone H3 coupled with mitotic chromosome condensation. *J Biol Chem, 274(36),* 25543-25549.

[17] Davie, J.R,, Drobic, B., Perez-Cadahia, B., He, S., Espino, P.S., Sun, J.M., Chen H.Y., Dunn, K.L., Wark, L., Mai, S., Khan, D.H., Davie, S.N., Lu, S., Peltier, C.P., & Delcuve, G.P. (2010). Nucleosomal response, immediate-early gene expression and cell transformation. *Adv Enzyme Regul, 50(1),* 135-145.

[18] Vicent, G.P., Ballaré, C., Nacht, A.S., Clausell, J., Subtil-Rodríguez, A., Quiles, I., Jordan, A., Beato, M. (2006). Induction of progesterone target genes requires activation of Erk and Msk kinases and phosphorylation of histone H3. *Mol Cell, 24(3),* 367-381.

[19] Schreiber, S.L., & Bernstein, B.E. (2002). Signaling network model of chromatin. *Cell, 111(6),* 771-778.

[20] Dormann, H.L., Tseng, B.S., Allis, C.D., Funabiki, H., & Fischle, W. (2006). Dynamic regulation of effector protein binding to histone modifications: The biology of HP1 switching. *Cell Cycle, 5(24),* 2842-2851.

[21] Loomis, R.J., Naoe, Y., Parker, J.B., Savic, V., Bozovsky, M.R., Macfarlan, T., Manley, J.L., & Chakravarti, D. (2009). Chromatin binding of SRp20 and ASF/SF2 and dissociation from mitotic chromosomes is modulated by histone H3 serine 10 phosphorylation. *Mol Cell, 33(4),* 450-461.

[22] Zhang, Y., Jurkowska, R., Soeroes, S., Rajavelu, A., Dhayalan, A., Bock, I., Rathert, P., Brandt, O., Reinhardt, R., Fischle, W., & Jeltsch, (2010). A. Chromatin methylation activity of Dnmt3a and Dnmt3a/3L is guided by interaction of the ADD domain with the histone H3 tail. *Nucleic Acids Res, 38(13),* 4246-4253.

[23] Dunn K.L., & Davie, J.R. (2005). Stimulation of the Ras-MAPK pathway leads to independent phosphorylation of histone H3 on serine 10 and 28. *Oncogene, 24(21),* 3492-3502.

[24] Dyson, M.H., Thomson, S., Inagaki, M., Goto, H., Arthur, S.J., Nightingale, K., Iborra, F.J., & Mahadevan, L.C. (2005). MAP kinase-mediated phosphorylation of distinct pools of histone H3 at S10 or S28 via mitogen- and stress-activated kinase 1/2. *J Cell Sci, 118(Pt 10),* 2247-2259.

[25] Dai, J., Sultan, S., Taylor, S.S., & Higgins, J.M. (2005). The kinase haspin is required for mitotic histone H3 Thr 3 phosphorylation and normal metaphase chromosome alignment. *Genes Dev, 19(4),* 472-488.

[26] Houben, A., Demidov, D., Rutten, T., & Scheidtmann, K.H. (2005). Novel phosphorylation of histone H3 at threonine 11 that temporally correlates with condensation of mitotic and meiotic chromosomes in plant cells. *Cytogenet Genome Res, 109(1-3),* 148-155.

[27] Metzger, E., Imhof, A., Patel, D., Kahl, P., Hoffmeyer, K., Friedrichs, N., Müller J.M., Greschik, H., Kirfel, J., Ji. S., Kunowska, N., Beisenherz-Huss, C., Günther, T., Buettner, R., & Schüle, R. (2010). Phosphorylation of histone H3T6 by PKC-β1 controls demethylation at histone H3K4. *Nature, 464(7289),* 792-796.

[28] Dai, J., & Higgins J.M. (2005). Haspin – A mitotic histone kinase required for metaphase chromosome alignment. *Cell Cycle, 4(5),* 665-668.

[29] Dai, J., Sullivan, B.A., & Higgins, J.M. (2006). Regulation of mitotic chromosome cohesion by Haspin and Aurora B. *Dev Cell, 11(5),* 741-750.

[30] Preuss, U., Landsberg, G., & Scheidtmann, K.H. (2003). Novel mitosis-specific phosphorylation of histone H3 at Thr11 mediated by Dlk/ZIP kinase. *Nucleic Acids Res, 31(3),* 878-885.

[31] Shimada, M., Niida, H., Zineldeen, D.H., Tagami, H., Tanaka, M., Saito, H., & Nakanishi, M. (2008). Chk1 is a histone H3 threonine 11 kinase that regulates DNA damage-induced transcriptional repression. *Cell, 132(2),* 221-232.

[32] Metzger, E., Yin, N., Wissmann, M., Kunowska, N., Fischer, K., Friedrichs, N., Patnaik, D., Higgins, J.M., Potier, N., Scheidtmann, K.H., Buettner, R. & Schüle R. (2008). Phosphorylation of histone H3 at threonine 11 establishes a novel chromatin mark for transcriptional regulation. *Nat Cell Biol, 10(1),* 53-60.

[33] Dawson, M.A., Bannister, A.J., Göttgens, B., Foster, S.D., Bartke, T., Green, A.R., & Kouzarides, T. (2009). JAK2 phosphorylates histone H3Y41 and excludes HP1-alpha from chromatin. *Nature, 461(7265),* 819-822.

[34] Hurd, P.J., Bannister, A.J., Halls, K., Dawson, M.A., Vermeulen, M., Olsen, J.V., Ismail, H., Somers, J., Mann, M., Owen-Hughes, T., Gout, I. & Kouzarides, T. (2009). Phosphorylation of histone H3 Thr-45 is linked to apoptosis. *J Biol Chem, 284(24),* 16575-16583.

[35] Baker, S.P., Phillips, J., Anderson, S., Qiu, Q., Shabanowitz, J., Smith, M.M., Yates, J.R. 3rd, Hunt, D.F., & Grant, P.A. (2010). Histone H3 Thr**45** phosphorylation is a replication-associated post-translational modification in *S. cerevisiae*. *Nat Cell Biol, 12(3),* 294-298.

[36] Marzluff, W.F., Gongidi, P., Woods, K.R., Jin, J., & Maltais, L.J. (2002). The human and mouse replication-dependent histone genes. *Genomics, 80(5),* 487-498.

[37] Hake, S.B., & Allis, C.D. (2006). Histone H3 variants and their potential role in indexing mammalian genomes: The 'H3 barcode hypothesis'. *Proc Natl Acad Sci USA, 103(17),* 6428-6435.

[38] Hake, S.B., Garcia, B.A., Duncan, E.M., Kauer, M., Dellaire, G., Shabanowitz, J., Bazett-Jones, D.P., Allis, C.D., & Hunt, D.F. (2006). Expression patterns and post-translational modifications associated with mammalian histone H3 variants. *J Biol Chem, 281(1),* 559-568.

[39] Hake, S.B., Garcia, B.A., Kauer, M., Baker, S.P., Shabanowitz, J., Hunt, D.F., & Allis, C.D. (2005). Serine 31 phosphorylation of histone variant H3.3 is specific to regions bordering centromeres in metaphase chromosomes. *Proc Natl Acad Sci USA, 102(18),* 6344-6349.

[40] Garcia, B.A., Barber, C.M., Hake, S.B., Ptak, C., Turner, F.B., Busby, S.A., Shabanowitz, J., Moran, R.G., Allis, C.D., & Hunt D.F. (2005). Modifications of human histone H3 variants during mitosis. *Biochemistry, 44(39),* 13202-13213.

[41] Zeitlin, S.G., Barber, C.M., Allis, C.D., & Sullivan, K.F. (2001). Differential regulation of CENP-A and histone H3 phosphorylation in G2/M. *J Cell Sci, 114(Pt 4)*, 653-661.

[42] Slattery, S.D., Moore, R.V., Brinkley, B.R., & Hall, R.M. (2008). Aurora-C and Aurora-B share phosphorylation and regulation of CENP-A and Borealin during mitosis. *Cell Cycle, 7(6),* 787-795.

[43] Rikova, K., Guo, A., Zeng, Q., Possemato, A., Yu, J., Haack, H., Nardone, J., Lee, K., Reeves, C., Li, Y., Hu, Y. Tan, Z., Stokes, M., Sullivan, L., Mitchell, J., Wetzel, R., Macneill, J., Ren, J.M., Yuan, J., Bakalarski, C.E., Villen, J., Kornhauser, J.M., Smith, B., Li, D., Zhou, X., Gygi, S.P., Gu, T.L., Polakiewicz, R.D., Rush, J., & Comb, M.J. (2007). Global survey of phosphotyrosine signaling identifies oncogenic kinases in lung cancer. *Cell, 131(6),* 1190-1203.

[44] Govin, J., Schug, J., Krishnamoorthy, T., Dorsey, J., Khochbin, S., & Berger, S.L. (2010). Genome-wide mapping of histone H4 serine-1 phosphorylation during sporulation in *Saccharomyces cerevisiae. Nucleic Acids Res, 38(14),* 4599-4606.

[45] Utley, R.T., Lacoste, N., Jobin-Robitaille, O., Allard, S., & Côté, J. (2005). Regulation of NuA4 histone acetyltransferase activity in transcription and DNA repair by phosphorylation of histone H4. *Mol Cell Biol, 25(18),* 8179-8190.

[46] Barber, C.M., Turner, F.B., Wang, Y., Hagstrom, K., Taverna, S.D., Mollah, S., Ueberheide, B., Meyer, B.J., Hunt, D.F., Cheung, P., & Allis, C.D. (2004). The enhancement of histone H4 and H2A serine 1 phosphorylation during mitosis and S-phase is evolutionarily conserved. *Chromosoma, 112(7),* 360-371.

[47] Attwood, P.V., Ludwig, K., Bergander, K., Besant, P.G., Adina-Zada, A., Krieglstein, J., & Klumpp, S. (2010). Chemical phosphorylation of histidine-containing peptides based on the sequence of histone H4 and their dephosphorylation by protein histidine phosphatase. *Biochim Biophys Acta, 1804(1),* 199-205.

[48] Ismail, I.H., & Hendzel, M.J. (2008). The γ-H2A.X: Is it just a surrogate marker of double-strand breaks or much more? *Environ Mol Mutagen, 49(1),* 73-82.

[49] Talbert, P.B., & Henikoff, S. (2010). Histone variants – Ancient wrap artists of the epigenome. *Nat Rev Mol Cell Biol, 11(4),* 264-275.

[50] Boyne, M.T. 2nd, Pesavento, J.J., Mizzen, C.A., & Kelleher, N.L. (2006). Precise characterization of human histones in the H2A gene family by top down mass spectrometry. *J Proteome Res, 5(2),* 248-253.

[51] Jackson, J.D., Falciano, V.T., & Gorovsky, M.A. (1996). A likely histone H2A.F/Z variant in *Saccharomyces cerevisiae. Trends Biochem Sci, 21(12),* 466-467.

[52] Pehrson, J.R., & Fuji, R.N. (1998). Evolutionary conservation of histone macroH2A subtypes and domains. *Nucleic Acids Res, 26(12),* 2837-2842.

[53] Chadwick, B.P., & Willard, H.F. (2001). A novel chromatin protein, distantly related to histone H2A, is largely excluded from the inactive X chromosome. *J Cell Biol, 152(2),* 375-384.

[54] Kamakaka, R.T., & Biggins, S. (2005). Histone variants: Deviants? *Genes Dev, 19(3),* 295-316.

[55] Rogakou, E.P., Pilch, D.R., Orr, A.H., Ivanova, V.S., & Bonner, W.M. (1998). DNA double-stranded breaks induce histone H2AX phosphorylation on serine 139. *J Biol Chem, 273(10),* 5858-5868.

[56] Zhang, Y., Griffin, K., Mondal, N., & Parvin, J.D. (2004). Phosphorylation of histone H2A inhibits transcription on chromatin templates. *J Biol Chem, 279(21),* 21866-21872.

[57] Mah, L.J., El-Osta, A., & Karagiannis, T.C. (2010). γH2AX: A sensitive molecular marker of DNA damage and repair. *Leukemia, 24(4),* 679-686.

[58] Dickey, J.S., Redon, C.E., Nakamura, A.J., Baird, B.J., Sedelnikova, O.A., & Bonner, W.M. (2009). H2AX: Functional roles and potential applications. *Chromosoma, 118(6),* 683-692.

[59] McManus, K.J., & Hendzel, M.J. (2005). ATM-dependent DNA damage-independent mitotic phosphorylation of H2AX in normally growing mammalian cells. *Mol Biol Cell, 16(10),* 5013-5025.

[60] Stucki, M., Clapperton, J.A., Mohammad, D., Yaffe, M.B., Smerdon, S.J., & Jackson S.P. (2005). MDC1 directly binds phosphorylated histone H2AX to regulate cellular responses to DNA double-strand breaks. *Cell, 123(7),* 1213-1226.

[61] Huen, M.S., Grant, R., Manke, I., Minn, K., Yu, X., Yaffe, M.B., & Chen, J. (2007). RNF8 transduces the DNA-damage signal via histone ubiquitylation and checkpoint protein assembly. *Cell, 131(5),* 901-914.

[62] Keogh, M.C., Kim, J.A., Downey, M., Fillingham, J., Chowdhury, D., Harrison, J.C., Onishi, M., Datta, N., Galicia, S., Emili, A., Lieberman, J., Shen, X., Buratowski, S., Haber, J.E., Durocher, D., Greenblatt, J.F., & Krogan, N.J. (2006). A phosphatase complex that dephosphorylates γH2AX regulates DNA damage checkpoint recovery. *Nature, 439(7075),* 497-501.

[63] Chowdhury, D., Xu, X., Zhong, X., Ahmed, F., Zhong, J., Liao, J., Dykxhoorn, D.M., Weinstock, D.M., Pfeifer, G.P., & Lieberman, J. (2008). A PP4-phosphatase complex dephosphorylates γ-H2AX generated during DNA replication. *Mol Cell, 31(1),* 33-46.

[64] Heo, K., Kim, H., Choi, S.H., Choi, J., Kim, K., Gu, J., Lieber, M.R., Yang, A.S. & An, W. (2008). FACT-mediated exchange of histone variant H2AX regulated by phosphorylation of H2AX and ADP-ribosylation of Spt16. *Mol Cell, 30(1),* 86-97.

[65] Macůrek, L., Lindqvist, A., Voets, O., Kool, J., Vos, H.R., & Medema, R.H. (2010). Wip1 phosphatase is associated with chromatin and dephosphorylates γH2AX to promote checkpoint inhibition. *Oncogene, 29(15),* 2281-2291.

[66] van Attikum, H., Fritsch, O., Hohn, B., & Gasser, S.M. (2004). Recruitment of the INO80 complex by H2A phosphorylation links ATP-dependent chromatin remodeling with DNA double-strand break repair. *Cell, 119(6),* 777-788.

[67] Downs, J.A., Allard, S., Jobin-Robitaille, O., Javaheri, A., Auger, A., Bouchard, N., Kron, S.J., Jackson, S.P., & Coté, J. (2004). Binding of chromatin-modifying activities to phosphorylated histone H2A at DNA damage sites. *Mol Cell, 16(6),* 979-990.

[68] Morrison, A.J., Highland, J., Krogan, N.J., Arbel-Eden, A., Greenblatt, J.F., Haber, J.E., & Shen, X. (2004). INO80 and gamma-H2AX interaction links ATP-dependent chromatin remodeling to DNA damage repair. *Cell, 119(6),* 767-775.

[69] Kusch, T., Florens, L., MacDonald, W.H., Swanson, S.K., Glaser, R.L., Yates, J.R., III, Abmayr, S.M., Washburn, M.P., & Workman, J.L. (2004). Acetylation by Tip60 is required for selective histone variant exchange at DNA lesions. *Science, 306(5704),* 2084-2087.

[70] Artus, C., Boujrad, H., Bouharrour, A., Brunelle, M.N., Hoos, S., Yuste, V.J., Lenormand, P., Rousselle, J.C., Namane, A., England, P., Lorenzo, H.K., & Susin, S.A. (2010). AIF promotes chromatinolysis and caspase-independent programmed necrosis by interacting with histone H2AX. *EMBO J, 29(9),* 1585-1599.

[71] Xiao, A., Li, H., Shechter, D., Ahn, S.H., Fabrizio, L.A., Erdjument-Bromage, H., Ishibe-Murakami, S., Wang, B., Tempst, P.l., Hofmann, K., Patel, D.J., Elledge, S.J.,

Allis, C.D. (2009). WSTF regulates the H2A.X DNA damage response via a novel tyrosine kinase activity. *Nature, 457(7225),* 57-62.

[72] Cook, P.J., Ju, B.G., Telese, F., Wang, X., Glass, C.K., & Rosenfeld, M.G. (2009). Tyrosine dephosphorylation of H2AX modulates apoptosis and survival decisions. *Nature, 458(7238*), 591-596.

[73] Pehrson, J.R., & Fried, V.A. (1992). MacroH2A, a core histone containing a large nonhistone region. *Science, 257(5075),* 1398-1400.

[74] Chu, F., Nusinow, D.A., Chalkley, R.J., Plath, K., Panning, B., & Burlingame, A.L. (2006). Mapping post-translational modifications of the histone variant macroH2A1 using tandem mass spectrometry. *Mol Cell Proteomics, 5(1),* 194-203.

[75] Bernstein, E., Muratore-Schroeder, T.L., Diaz, R.L., Chow, J.C., Changolkar, L.N., Shabanowitz, J., Heard, E., Pehrson, J.R., Hunt, D.F., & Allis, C.D. (2008). A phosphorylated subpopulation of the histone variant macroH2A1 is excluded from the inactive X chromosome and enriched during mitosis. *Proc Natl Acad Sci USA, 105(5),* 1533-1538.

[76] Zhang, R., Poustovoitov, M.V., Ye, X., Santos, H.A., Chen, W., Daganzo, S.M., Erzberger, J.P., Serebriiskii, I.G., Canutescu, A.A., Dunbrack, R.L., Pehrson, J.R., Berger, J.M., Kaufman, P.D., & Adams, P.D. (2005). Formation of macroH2A-containing senescence-associated heterochromatin foci and senescence driven by ASF1a and HIRA. *Dev Cell, 8(1),* 19-30.

[77] Artificial Cells and Organs Research Centre: McGill University, Quebec, QC, Canada (2009). www.medicine.mcgill.ca/artcell

[78] Ajiro, K., Scoltock, A.B., Smith, L.K., Ashasima, M., & Cidlowski, J.A. (2010). Reciprocal epigenetic modification of histone H2B occurs in chromatin during apoptosis in vitro and in vivo. *Cell Death Differ, 17(6),* 984-993.

[79] Fernandez-Capetillo, O., Allis, C.D., & Nussenzweig, A. (2004). Phosphorylation of histone H2B at DNA double-strand breaks. *J Exp Med, 199(12),* 1671-1677.

[80] Yu, X., Caltagarone, J., Smith, M.A., & Bowser, R. (2005). DNA damage induces cdk 2 protein levels and histone H2B phosphorylation in SH-SY5Y neuroblastoma cells. *J Alzheimers Dis, 8(1),* 7-21.

[81] Mecklenbräuker, I., Kalled, S.L., Leitges, M., Mackay, F., & Tarakhovsky, A. (2004). Regulation of B-cell survival by BAFF-dependent PKC-δ-mediated nuclear signalling. *Nature, 431(7007),* 456-461.

[82] Solier, S., & Pommier, Y. (2009). The apoptotic ring: A novel entity with phosphorylated histones H2AX and H2B and activated DNA damage response kinases. *Cell Cycle, 8(12),* 1853-1859.

[83] Guo, A., Villén, J., Kornhauser, J., Lee, K.A., Stokes, M.P., Rikova, K., Possemato, A., Nardone, J., Innocenti, G., Wetzel, R., Wang, Y., MacNeill, J., Mitchell, J., Gygi, S.P., Rush, J., Polakiewicz, R.D., & Comb MJ. (2008). Signaling networks assembled by oncogenic EGFR and c-Met. *Proc Natl Acad Sci USA, 105(2),* 692-697.

[84] Vlaicu, S.I., Tegla, C.A., Cudrici, C.D., Fosbrink, M., Nguyen. V., Azimzadeh, P., Rus, V., Chen, H., Mircea, P.A., Shamsuddin, A., & Rus, H. (2010). Epigenetic modifications induced by RGC-32 in colon cancer. *Exp Mol Pathol, 88(1),* 67-76.

In: Histones: Class, Structure and Function
Editor: Chang-Hui Shen
ISBN: 978-1-62100-274-1

Chapter 8

BASIC PARAMETERS FOR PLANT NUCLEI AND 3D LOCALIZATION OF HISTONE MODIFICATION

***Toshiyuki Wako*[1] *and Kiichi Fukui*[2,*]**
[1]Division of Plant Sciences, National Institute of Agrobiological Sciences, Tsukuba, Japan
[2]Department of Biotechnology, Graduate School of Engineering, Osaka University, [2,1]Yamadaoka, Osaka, Japan

ABSTRACT

A nuclear genome physically exists as a nuclear DNA molecules or a set of chromosomes. The structural parameters of a nucleus or chromosome would determine the functional basis of a genome, i.e. regulation of gene expression and transmission of the genome to the next generation. In plants, there are distinct upper limits of structural parameters for maintaining genome functions. The genomic DNA content in a nucleus has an upper limit of 3% (v/v, DNA/nucleus) for construction and maintenance of a functional nucleus. The length of a chromosome arm is also restricted to an upper limit that is less than a half of the length between the two spindle poles. Moreover each chromosome functionally differentiates region from to region which is characterized with specific combinations of histone modifications. The three-dimensional distribution of the each chromosomal region with specific patterns of histone modification and the functional differentiation of the nuclear space are also closely related to the plants with large-sized genomes. This fact clearly indicates that the higher structure of the nucleus and the chromosomes with different histone modification set play essential roles in genome functions.

* Corresponding author e-mail: kfukui@bio.eng.osaka-u.ac.jp.

1. Introduction

The basic functions of a genome are the concerted regulation of gene expression through G1 to G2 phases and sound transmission of the genome to the next generation at M phase. Genomic DNAs are packed into chromosomes and the decondensed chromosomes form chromatins that fill the interphase nucleus. The interphase nucleus is an essential object in a cell for gene expression. Chromosomes comprising genomic DNAs are replicated and transferred to two daughter cells through mitosis by moving the inner space of a cell. Thus, the chromosomes play physical role in a cell. It is also important to clarify the physical properties of the chromosomes, although the relationships between genome functions and physical parameters of the nucleus and chromosomes are poorly understood. There have been two known types of nuclear structures, the Rabl orientation [1] and chromosome territories [reviewed by 2]. Here we show the physical parameters of the nucleus and the chromosome, and the spatial distribution of chromosomes within a plant nucleus with specific histone modifications.

2. Upper Limits for Nuclear DNA Contents and Chromosome Arm Length

The most important function of the nucleus is packaging of DNA molecules and concerted regulation of overall gene expression. The nucleus of eukaryotes contains not only whole genome DNAs, but also nucleosomal histones, other proteins, and RNAs. Thus nuclear volume is a sum of DNA and the other nuclear materials, and the volume ratio (VR) between the DNA and the nucleus comprises a physical parameter that reflects the overall nuclear structure. The nuclear volumes of twelve plant species whose DNA contents ranged from 0.3 to 100 Gbp/2C were examined [3]. Surprisingly, the VR of DNA to the nucleus is unexpectedly low. The VR does not exceed 3% through all the species examined (Table 1). These results clearly indicate that there is an upper limit of 3% for the VR of the plant nuclear DNA. When all the DNA molecules corresponds to 3% of nuclear volume form 30nm fibers, the VR of 30 nm chromatin fibers to the nucleus was estimated as 57 %. The remaining space in the nucleus might be essential for the other materials for maintaining nuclear function, such as transcription.

The other essential function of the genome is to transmit genomic DNA equally and safely to the daughter cells or the next generation through mitotic and meiotic cell divisions. Genomic DNA molecules are packaged into individual chromosomes and the chromosome sizes are determined by their DNA amounts. As a result, chromosome sizes differ between species and even within species. Chromosomes have similar shapes and structures even between plants and animals [6]. The typical morphology of a chromosome is a rod or a dot-like shape observed in mitotic metaphase when chromosomes are the most condensed and shortened. Among chromosomal functional structures, the centromere is the most important for cell division. Each chromosome is divided into two chromosome arms by the centromere. Spindle microtubules attach to the kinetochores of the metaphase chromosome to split the sister chromatids and to pull them toward the spindle poles.

Table 1. Relationship between nuclear DNA and nuclear volumes in twelve plant species [3]

Rabl orientations	-	-	-	-	+	+	+	+	+	-	-	-
Volume ratio (VR) DNA/Nucleus (%)	0.26	0.55	1.71	2.13	2.22	2.95	2.35	2.73	2.05	2.13	2.67	1.67
Nuclear volume (μm^3)[c] (Mean ± S.D.)	11.5 ± 41.0	15.2 ± 52.9	24.7 ± 115.4	35.3 ± 148.6	43.3 ± 168.6	75.7 ± 299.2	106.6 ± 471.3	87.9 ± 424.1	138.2 ± 655.6	157.4 ± 1148.0	237.8 ± 914.0	387.7 ± 2022.0
DNA volume (μm^3)[b]	0.107	0.289	1.972	3.162	3.740	8.840	11.1	11.6	13.5	24.4	24.4	33.8
2C Value (Mbp)	315[d]	850[d]	5,800[d]	9,300[d]	11,000[d]	26,000[d]	32,600[d]	34,000[d]	39,600	71,800	71,800	99,600
2*n*	10	24	20	48	14	28	16	42	18	24	24	36
Plant material (M, E)[a]	*Arabidopsis thaliana* (L.) Heynh. ecotype Columbia (E)	*Oryza sativa* L. cv. Nipponbare (M)	*Zea mays* L. var. saccbarata (M)	*Nicotiana tabacum* L. cv. SR1 (E)	*Hordeum vulgare* L. cv. Minorimugi (M)	*Triticum durum* Desf. cv. LD222 (M)	*Allium cepa* L. var. *cepa* (M)	*Triticum aestivum* L. cv. Shin Chunaga (M)	*Leucanthemella linearis* Matsum. (E)	*Lilium longiflorum* Thunb. cv. Nellie White (M)	*Fritillaria verticillata* Willd. var. thunbergii (M)	*Fritillaria camtschatcensis* (L.) Ker-gawl. (M)

(M, E)[a] *Arabidopsis thaliana* (L.) Heynh. ecotype Columbia (E) *Oryza sativa* L. cv. Nipponbare (M) *Zea mays* L. var. saccbarata (M) *Nicotiana tabacum* L. cv. SR1 (E) *Hordeum vulgare* L. cv. Minorimugi (M) *Triticum durum* Desf. cv. LD222 (M) *Allium cepa* L. var. *cepa* (M) *Triticum aestivum* L. cv. Shin Chunaga (M) *Leucanthemella linearis* Matsum. (E) *Lilium longiflorum* Thunb. cv. Nellie White (M) *Fritillaria verticillata* Willd. var. thunbergii (M)*Fritillaria camtschatcensis* (L.) Ker-gawl. (M)

The upper limit of chromosomal arm size in normal cell division was reported by Schubert and Oud, who developed several chromosome-reconstructed lines of *Vicia faba* where the lengths of the chromosome arms were increased by chromosome translocations [7]. The longest chromosome arm developed in the reconstructed lines was 1.57 times longer than that of the longest chromosome arm in wild-type. Lines with abnormally long chromosome arms are usually sterile, showing incomplete separation of sister chromatids and the formation of chromatin bridges between two daughter cells in mitotic telophase. From these observations, they proposed that the upper limit for the length of chromosome arm is the half of the length of the spindle axis at telophase, and that exceeding the limit length causes abnormal cell division, ultimately resulting in loss of the genomic DNA.

3. Morphology of Nucleus

Increase of genome size and/or VR (volume ratio) causes decrease of nuclear surface area ratio to genome DNA volume. Various nuclear materials such as RNAs and nuclear proteins pass through the nuclear pores at the nuclear surface. Relative decreasing the surface area to the nuclear genome size would be resulted in negative for transportation of the essential materials in and out of the nucleus. Some of plant nuclei deviate from ordinary round structures. Epidermal cells in onion and tobacco have stable grooves and invaginations that increase nuclear surface area [8]. The epidermal cells have often large amount of DNA caused by extensive DNA endoreduplication [9]. These findings suggest that increased surface area by grooves and invaginations might be necessary for mRNA export and protein import in the larger nuclei. However, the cell types with grooved nucleus are limited. Mitotic cells including root tip cells in plants with both small and large genome size have almost round-shaped nucleus. Thus it is suggested that increase of nuclear surface area with groove and invaginations is not necessary for efficient nuclear function even in plants with large genome size. Further investigation would be required to reveal roles of nuclear shape for nuclear function in relation with nuclear DNA content, nuclear volume and nuclear surface area.

The existence of non-globular shaped nuclei suggests that factor(s) for determining nuclear shape. Recent studies revealed that several proteins affect nuclear morphology. Depletion of human Lem2 that is the inner nuclear membrane leads to abnormally shaped nuclei with lobulations [10]. Mutation of *LMNA* gene that encoding nuclear lamin A causes lobulated nuclei [11-12]. These evidences indicate that rounded-shaped nuclei is maintained by functions of nuclear proteins rather than by surface tension of lipid bilayer of nuclear membrane. Nucleolar proteins also involve in nuclear shape even though they are localized in nucleolus through interphase. Depletion of SENP5, fibrillarin and nucleoplasmin in human cells resulted in abnormal nuclear morphology [13-15]. Detailed mechanism why depletion of nucleolar proteins cause abnormal nuclear shape is still unclear. These proteins are known to be localized at chromosome periphery during mitosis [14-15] and are categorized as the member of chromosome peripheral proteins [16].

4. Functional Differentiation of Nuclei with Rabl Orientation and Histone Modifications

It has been reported that different patterns of nuclear organization can be observed depending on the genome size in plant species [17]. One of the ordered chromatin organizations in the nucleus is the Rabl orientation, characterized by clustered centromeres in a narrow space near the nuclear membrane [1] and the another one is chromosome territories (CTs) that is characterized by distinct localization of each chromatin in a nucleus [2]. The Rabl orientation has been reported both in plants and other organisms, and in the case of plant it is known that the plants with large genomes of more than 10 Gbp/2C, such as barley, durum and common wheat, field bean and onion [17-19]. On the other hand, plants with a smaller genome size, less than 2 Gbp/2C, such as *A. thaliana*, rice and sorghum, have non-Rabl orientations [17, 20]. The existence of the Rabl orientation in plants with very large genomes (>60 Gbp/2C) is not yet clear, because specific probes for detecting centromeres in such species by FISH have not been identified [3].

It is suggested that the Rabl orientation in interphase nucleus is resulted in the polarized chromosomes at the mitotic anaphase [1]. If it is so, CTs would be formed by chromosome relocation in a nucleus and thus the Rabl orientation in interphase would be the result by inhibition of chromosome relocation. Rabl orientation is observed in plants with large genome size and with the VR figures close to 3%. When the figure of VR is 3% that is close to the upper limit of DNA density in the nucleus and all the total volume of chromatin fibers would occupy about 60 % of nuclear volume [3]. Thus it is reasonable to postulate that chromatins could not be relocating from Rabl orientation with clustered centromeres to CTs with dispersed centromeres. Recently, it was reported that a mutation of Lamin A, a protein of nuclear lamina causes Rabl-like centromere clustering and inhibits normal chromosome organization to CTs in human cell [12]. These results support that Rabl orientation would be originated from chromosome configuration at mitotic anaphase.

Barley shows a typical Rabl orientation, with the centromeres located in a very narrow region of the nuclear surface [Figure 2; 21-22]. Barley chromosomes are either metacentric or submetacentric; therefore, both chromosome arms have more or less similar lengths [23]. Barley chromosomes are also characterized by centromeric heterochromatin [23] and subtelomeric euchromatins with clusters of genes [24]. If the chromosome is considered as a linear object, the euchromatic regions on both chromosomal ends are the most distant to each other and thus the genes are very much separated from each other. However, Rabl orientation which makes clustering of the centromeres in barley results in gathering of the telomeres in the opposite hemisphere of the nucleus, and the nucleus can thus be divided into two hemispheres (centrosphere and telosphere) where centromeric heterochromatin and subtelomeric euchromatin are completely separated in their localization.

Eukaryotic chromosome is composed by several epigenetically defined subregions with different structure and function such as centromere, euchromatin and heterochromatin. Each region is characterized with specific post-translational modifications of nucleosomal histones and/or replacement of nucleosomal histones to histone variants [25-26]. Histone modifications include acetylation at lysines, methylation at lysines and arginines, phosphorylation at serines and threonines, sumoylation and ubiquitination at lysines. Recently, it is reported that nuclear histones are modified to β-*N*-acetyl-glucosamine [27].

Certain combinations of these histone modifications constitute a 'histone code' that can determine transcription status, replication, repair, recombination and chromosome structure [28-29]. In general, constitutive heterochromatin is characterized with H3 and H4 hypoacetylation and methylation at Lys 9 of H3, and euchromatin is marked with H3 and H4 acetylation and methylation at Lys 4 and Lys 36 of H3 [30]. Plant centromere is characterized with a histone H3 variant, CENH3 [31]. Mitotic chromosomes and interphase nucleus show specific pattern of histone modifications depending on plant species, modified residue and type of modification [32-33]. Euchromatin of metaphase chromosome in *Vicia faba* is marked with euchromatic histone modifications, and Giemsa-banded heterochromatin is marked with some of heterochromatic modifications [Figure 1a; 32].

Each barley chromosomes has euchromatin at subtelomeric regions [24], and the subtelomeric regions are localized in telosphere of interphase nucleus because of Rabl orientation (Figure 1b). The subtelomeric regions of the chromosomes and the telosphere of a nucleus of barley are marked with acetylated Lys 16 of H4 and di-methylated Lys 4 of H3, which are euchromatic histone modifications throughout mitosis (Figure 1b;21, 32]. The clustered centromeres are labeled with acetylated Lys 5, Lys 8 and Lys 12 of H4 in an interphase specific manner [21-22, 34]. These facts suggest that nuclear structures such as the Rabl orientation in plants with large genomes would divide the nucleus into functionally different hemispheres for efficient allocation of regulatory and transcriptionally machineries, and that centromeric acetylation contributes to the formation of the Rabl orientation.

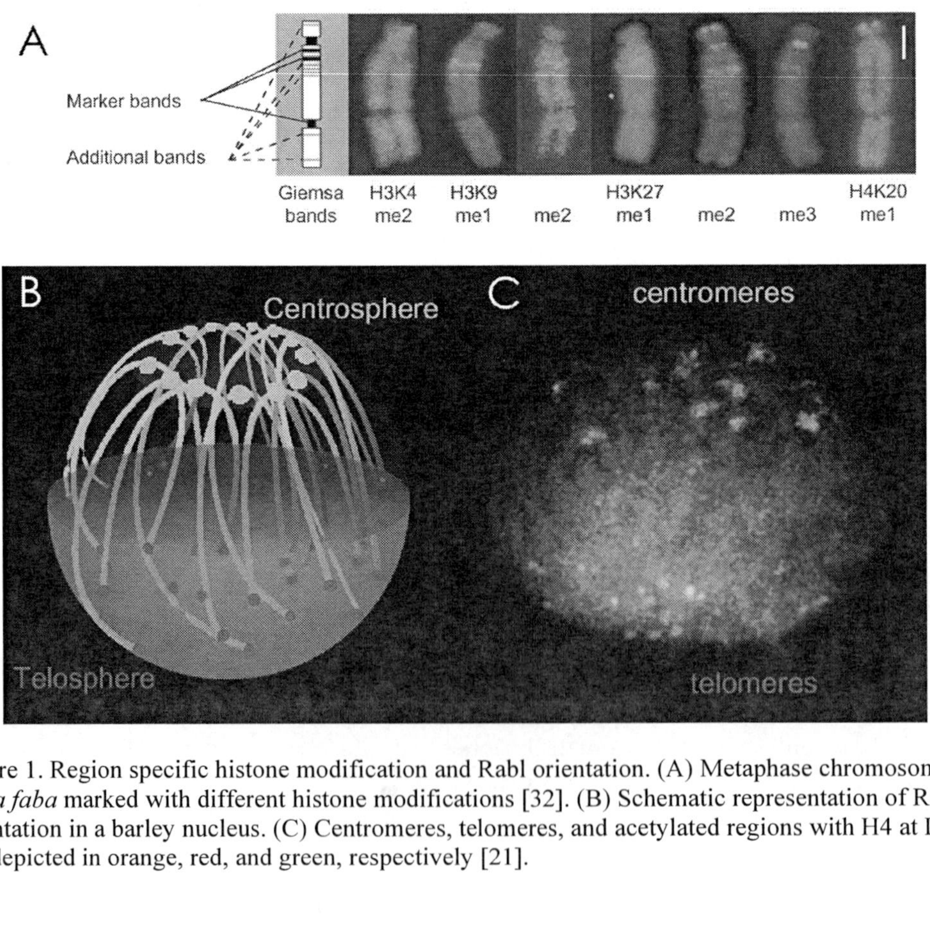

Figure 1. Region specific histone modification and Rabl orientation. (A) Metaphase chromosome of *Vicia faba* marked with different histone modifications [32]. (B) Schematic representation of Rabl orientation in a barley nucleus. (C) Centromeres, telomeres, and acetylated regions with H4 at Lys 16 are depicted in orange, red, and green, respectively [21].

Conclusions

The upper limits of the physical parameters for the nucleus and the chromosomes have been reported with reference to histone modification. The volume ratio (VR) of the genomic DNA volume to the nuclear volume should be less than 3% for maintaining adequate nuclear function. In addition, the lengths of the chromosome arms should be less than the distance between spindle axes at mitotic telophase for completion of normal mitosis. Plants with large genomes tend to show the higher DNA volume ratios and Rabl orientations at interphase. Rabl orientations and histone modification patterns in barley nuclei, a model system for plants with large genomes, unequivocally demonstrated that the nuclear structure characterized by different histone modifications is strongly related to the nuclear function. These findings also clearly indicate that the physical parameters, spatial distribution of the DNA within a nucleus and proper histone modifications are the three essential factors for the concerted genome functions.

References

[1] Rabl, C. (1885) *Uber Zelltheilung. Morphol. Jahrb., 10*, 214–330.

[2] Cremer, T., & Cremer, C. (2001) Chromosome territories, nuclear architecture and gene regulation in mammalian cells. *Nat. Rev. Genet., 2*, 292-301

[3] Fujimoto, S., Ito, M., Matsunaga, S. & Fukui, K. (2005). An upper limit of the ratio of DNA volume to nuclear volume exists in plants. *Genes Genet. Syst., 80*, 345-350.

[4] Bennett, M. D. & Leitch, I. J. (1995) Nuclear DNA amounts in angiosperms. *Ann. Bot., 76*, 113–176.

[5] Bennett, M. D. & Leitch, I. J. (2005) Plant DNA C-values database [online]. Available from http://data.kew.org/cvalues/ [release 4.0, October 2005].

[6] Fukui, K. & Uchiyama, S. (2007) Chromosome protein framework from proteome analysis of isolated human metaphase chromosomes. *Chem. Rec., 7*, 230-237.

[7] Schubert, I. & Oud, J. L. (1997). There is an upper limit of chromosome size for normal development of an organism. *Cell, 88*, 515-520.

[8] Collings, D. A., Carter, C. N., Rink, J. C., Scott, A. C., Wyatt, S. E. & Allen, N. S. (2000) Plant nuclei can contain extensive grooves and invaginations. *Plant Cell, 12*, 2425-2490.

[9] Traas J, Hülskamp M, Gendreau E, & Höfte H. (1998) Endoreduplication and development: rule without dividing? *Curr. Opin. Plant. Biol., 1*, 498-503.

[10] Ulbert, S., Antonin, W., Platani, M. & Mattaj, I.W. (2006) The inner nuclear membrane protein Lem2 is critical for normal nuclear envelope morphology. *FEBS Lett., 580*, 6435-6441.

[11] Wang, Y., Panteleyev, A. A., Owens, D. M., Djabali, K., Stewart, C.L. & Worman, H. J. (2008) Epidermal expression of the truncated prelamin A causing Hutchinson-Gilford progeria syndrome: effects on keratinocytes, hair and skin. *Hum. Mol. Genet., 17*, 2357-2369.

[12] Taimen, P., Pfleghaar, K., Shimi, T., Möller, D., Ben-Harush, K., Erdos, M. R., Adam, S. A., Herrmann, H., Medalia, O., Collins, F. S., Goldman, A. E., & Goldman, R. D.

(2009) A progeria mutation reveals functions for lamin A in nuclear assembly, architecture, and chromosome organization. *Proc. Natl. Acad. Sci. U. S. A.*, *106*, 20788-20793.

[13] Di Bacco, A., Ouyang, J., Lee, H. Y., Catic, A., Ploegh, H. & Gill, G. (2006) The SUMO-specific protease SENP5 is required for cell division. *Mol. Cell Biol.*, *26*, 4489-4498.

[14] Amin, M. A., Matsunaga, S., Ma, N., Takata, H., Yokoyama, M., Uchiyama, S. & Fukui, K. (2007) Fibrillarin, a nucleolar protein, is required for normal nuclear morphology and cellular growth in HeLa cells. *Biochem. Biophys. Res. Commun.*, *24*, 320-326.

[15] Amin, M. A., Matsunaga, S., Uchiyama, S. & Fukui, K. (2008) Depletion of nucleophosmin leads to distortion of nucleolar and nuclear structures in HeLa cells. *Biochem. J.*, *415*, 345-51.

[16] Uchiyama, S., Kobayashi, S., Takata, H., Ishihara, T., Hori, N., Higashi, T., Hayashihara, K., Sone, T., Higo, D., Nirasawa, T., Takao, T., Matsunaga, S., & Fukui, K. (2005) Proteome analysis of human metaphase chromosomes. *J. Biol. Chem.*, *280*, 16994-7004.

[17] Dong, F. G., & Jiang, J. M. (1998) Non-Rabl patterns of centromere and telomere distribution in the interphase nuclei of plant cells. *Chromosome Res.*, *6*, 551-558.

[18] Fussell, C. P. (1992) Rabl distribution of interphase and prophase telomeres in *Allium cepa* not altered by colchicine and or ultracentrifugation. *Am. J. Bot.*, *79*, 771-777.

[19] Houben, A., Guttenbach, M., Kress, W., Pich, U., Schubert, I., & Schmid, M. (1995) Immunostaining and interphase arrangement of field bean kinetochores. *Chromosome Res.*, *3*, 27-31.

[20] Talbert, P. B., Masuelli, R., Tyagi, A. P., Comai, L., & Henikoff, S. (2002) Centromeric localization and adaptive evolution of an *Arabidopsis* histone H3 variant. *Plant Cell*, *14*, 1053-1066.

[21] Wako, T., Fukuda, M., Furushima-Shimogawara, R., Belyaev, N. D. & Fukui, K. (2002) Cell cycle-dependent and lysine residue-specific dynamic changes of histone H4 acetylation in barley. *Plant Mol. Biol.*, *49*, 645-653.

[22] Wako, T., Houben, A., Furushima-Shimogawara, R., Belyaev, N. D. & Fukui, K. (2003) Centromere-specific acetylation of histone H4 in barley detected through three-dimensional microscopy. *Plant Mol. Biol.*, *51*, 533-541.

[23] Fukui, K. & Kakeda, K. (1990) Quantitative karyotyping of barley chromosomes by image-analysis methods. *Genome*, *33*, 450-458.

[24] Künzel, G., Korzun, L., & Meister, A. (2000) Cytologically integrated physical restriction fragment length polymorphism maps for the barley genome based on translocation breakpoints. *Genetics*, *154*, 397-412.

[25] Sims, R. J. 3rd. & Reinberg, D. (2008) Is there a code embedded in proteins that is based on post-translational modifications? *Nat. Rev. Mol. Cell Biol.* 9, 815–820.

[26] Talbert, P. B. & Henikoff, S. (2010) Histone variants - ancient wrap artists of the epigenome. Nat. Rev. Mol. Cell Biol., 11, 264-275.

[27] Sakabe, K., Wang, Z. & Hart, G. W. (2010) Beta-*N*-acetylglucosamine (O-GlcNAc) is part of the histone code. *Proc. Natl. Acad. Sci. U. S. A.*, 107, 19915-19920.

[28] Strahl, B. D. & Allis, C. D. (2000) The language of covalent histone modifications. *Nature*, 403, 41-45.

[29] Jenuwein, T. & Allis, C. D. (2001) Translating the histone code. *Science*, 293, 1074-1080.

[30] Wako, T. & Fukui, K. (2010) Higher organization and histone modification of the plant nucleus and chromosome. *Cytogenet. Genome Res.*, *129*, 55-63.

[31] Jiang, J., Birchler, J. A., Parrott, W. A., & Dawe, R. K. (2003) A molecular view of plant centromeres. *Trends Plant Sci., 8*, 570-575.

[32] Fuchs, J., Demidov, D., Houben, A. & Schubert, I. (2006) Chromosomal histone modification patterns - from conservation to diversity. *Trends Plant Sci.*, *11*, 212-212..

[33] Wako, T. & Fukui, K. (2008) Histone modifications in mitotic chromosomes of plant. Chromosome Sci., *11*, 47-52.

[34] Wako, T., Murakami, Y., & Fukui, K. (2005) Comprehensive analysis of dynamics of histone H4 acetylation in mitotic barley cells. *Genes Genet. Syst., 80*, 269-276.

In: Histones: Class, Structure and Function
Editor: Chang-Hui Shen
ISBN: 978-1-62100-274-1

Chapter 9

HISTONE MODIFICATION BY NON-ENZYMATIC GLYCATION AND PREVENTIVE STRATEGIES

Daniel Cervantes-Laurean[1] and Gilbert A. Boissonneault[2,*]

[1]Department of Chemistry and Biochemistry, South Dakota State University, Brookings, SD, US

[2]Division of Physician Assistant Studies, College of Health Sciences, University of Kentucky, Lexington, Kentucky, US

ABSTRACT

Oxidative stress causes DNA and protein damage, which may compromise genomic integrity. While oxidant-induced DNA damage has been extensively studied, less is known about nuclear protein damage, particularly of histones, which support chromatin structure and influence gene expression. Among other mechanisms, protein damage may be caused via non-enzymatic glycation initiated by reducing sugars (e.g., glucose and ADP-ribose) or reactive dicarbonyl compounds (e.g., glyoxal). Glycation is a non-enzymatic reaction between the reducing sugar carbonyl and protein amino groups, which undergo rearrangements to stable ketoamines and advanced glycation end products (AGEs) such as fluorescent (e.g., argpyrimidine) and non-fluorescent (e.g., N^{ε}-carboxymethyllysine, CML) protein adducts and crosslinks. Protein glycation, induced by hyperglycemia, has been implicated in diabetic complications, although the contribution of histone glycation to these processes is unknown. High local concentrations of ADP-ribose are produced in the chromatin microenvironment during DNA repair, potentially contributing to histone glycation.

This chapter describes our detection of non-enzymatic modifications occurring *in vivo* on histone H1 from calf thymus (Part I) and the basis for inhibition of these processes. In particular, we will discuss the roles of rutin metabolites as alternative, effective natural product AGE inhibitors for prevention of histone H1 modification and resulting disease complications (Part II).

* Corresponding author e-mail: gaboiss@email.uky.edu.

PART I: GLYCATION-INDUCED MODIFICATION OF NUCLEAR HISTONES

Genomic integrity requires that nuclear protein damage be minimized and, if it occurs, repaired. Alternately, damaged proteins must be degraded and replaced in order to maintain homeostatic regulation of gene expression. There are a variety of systems serving to repair DNA and histone damage [1,2]. Oxidatively modified [2] and glycoxidated [1] histones are removed by nuclear proteasomes, thereby preserving genomic integrity. Excessive protein damage due to non-enzymatic glycoxidation of histones may interfere with proteasomal degradation and affect normal regulation of gene expression. Non-enzymatic modifications of nuclear proteins may affect gene expression and contribute to the pathophysiology of diseases such as ethanol-induced liver toxicity [3] and diabetes [4], as well as contribute to aging. Proteins may be non-enzymatically glycated when reducing sugars react with amino groups of lysine residues, forming reactive α-oxoaldehydes such as glyoxal [5,6] and methylglyoxal [6]. These glycation products are up to 20,000 times more reactive than glucose. Protein glycation leads to the formation of stable advanced glycation end products (AGE) such as N^{ε}-carboxymethyllysine (CML) [7,8] and argpyrimidine [8]. CML is the most abundant AGE biomarker formed by non-enzymatic glycation of proteins. Thus, CML is a widely used reporter for glycation-induced protein modification. However, it should be noted that CML protein adducts can also be formed by reactions other than glycation, for example by lipid peroxidation via the glyoxal intermediate, and by deoxyribose DNA degradation [9]. Other glycation biomarkers have recently been characterized, such as argpyrimidine [8], which is formed by the reaction of methylglyoxal (formed by nonenzymatic degradation of glycolytic intermediates such glyceraldehyde 3-phosphate [10]) with guanidino groups of protein arginine residues. The enzyme glyoxalase utilizes glutathione to provide a cellular defense against protein glycation [11] by neutralizing glyoxal and methylglyoxal to non-reactive products, thereby protecting intracellular proteins from non-enzymatic glycation.

Investigations of protein glycation in biomedical systems initially focused on the effects of hyperglycemia on extra- and intracellular proteins such as collagen [7], laminin and hemoglobin [5]. Sugars other than glucose are also important physiological sources for inducing protein glycation [12]. In particular, pentoses were shown to be more reactive towards protein amino groups than hexoses, since pentoses dissociate more readily to the open carbonyl form than do hexoses [13]. Since the formation of high intracellular concentrations of reactive reducing sugars such as the pentose ADP-ribose is induced under conditions of oxidative stress resulting in DNA damage, we turned our attention to study nuclear proteins as targets for glycation.

In particular, acetylation and deacetylation have been correlated with active gene transcription and gene silencing, respectively, within localized regions of chromosomes [14]. If non-enzymatic modifications of histones interfere with acetylation and deacetylation patterns then it is likely that glycation may disrupt normal gene transcription and have implications in the pathophysiology of diseases that involve oxidative stress and DNA damage. Histone H1 binds to the linker DNA and is proposed to have a regulatory role in transcription via the modulation of chromatin structure [15].

When poly (ADP-ribose) polymerase (PARP) is activated after oxidative stress-induced DNA damage [16], polymers of ADP-ribose are formed and covalently react with nucleosomal histones as part of the DNA repair response. This process of introducing a high number of negative charges on histone molecules causes the separation of DNA from the nucleosomes and allows the DNA repair proteins access to initiate DNA repair [17]. On the other hand, histone-associated ADP-ribose polymers are quickly cleaved by the activity of poly (ADP-ribose) glycohydrolase (PARG), thereby restoring normal DNA and nucleosome interaction. High concentrations of ADP-ribose are then generated in the local microenvironment of histones, which contain high amounts of arginine and lysine residues [18]. This proximity makes the non-enzymatic condensation of sugar ADP-ribose and histones a likely event. This idea was tested as early as 1984 when Adamietz and Rudolph exposed hepatoma AH 7974 cells to DNA-damaging alkylating compounds, which resultedin activation of PARP and ADP-ribose polymer turnover when PARG activity was induced [19]. The high levels of ADP-ribose in the microenvironment of the chromatin resulted in the formation of histone H1-associated ADP-ribose monomers [20] (see figure 1). Chemical analyses of these monomers were consistent with ADP-ribose-induced glycation of histone H1. Because of a lack of evidence to suggest that nuclear protein glycation occurs *in vivo,* we looked for the presence of three biomarkers of protein AGE product formation, namely CML, argpyrimidine protein adducts and protein crosslinks, in histone H1 from calf thymus carefully collected and analyzed immediately after slaughter [21].

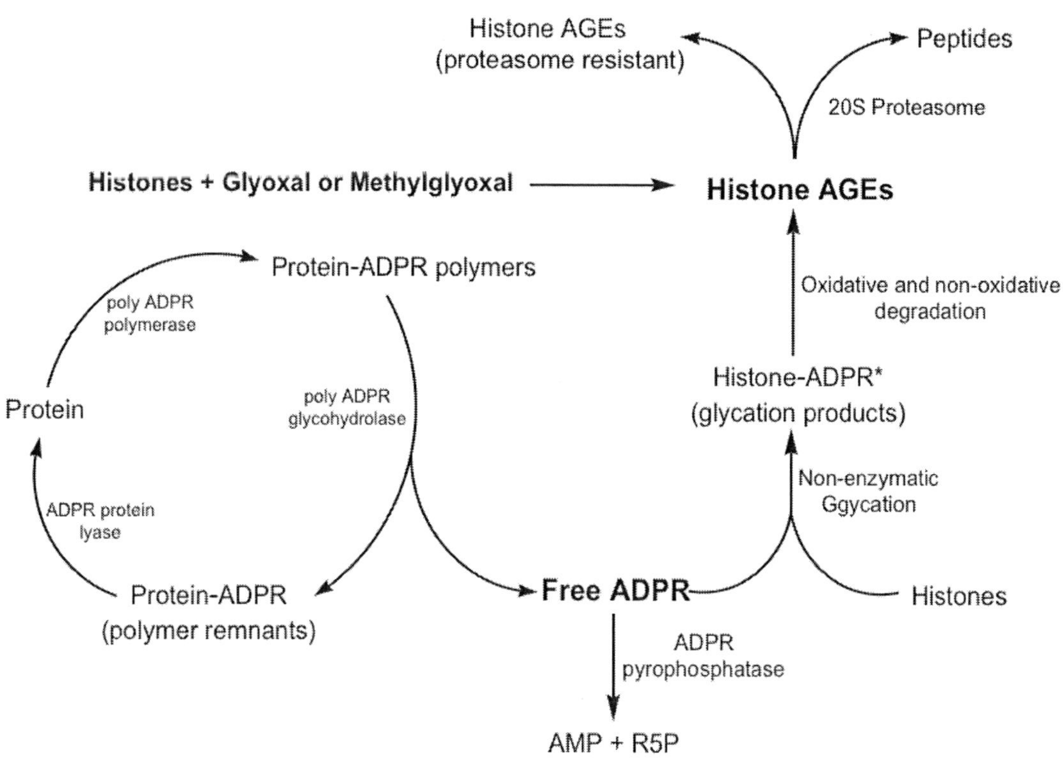

Figure 1. Schematic representation of possible intranuclear sources of ADP-ribose via polyADP-ribose turnover and extranuclear sources of glyoxal and methylglyoxal that may lead to formation of AGE histone adducts [21].

1. AGE Biomarkers in Calf Thymus Histone H1

It has been suggested that DNA damaging agents may induce nuclear protein modification through glycation. For example, it is widely known that alkylating agents such as MNNG, which cause DNA damage in cell culture, activate PARP as part of the DNA repair response [1,22-24]. PARP activation results in enhanced metabolism of ADP-ribose polymers, causing high local concentrations of ADP-ribose in the vicinity of lysine and arginine residue rich histones [17,20]. A number of reports demonstrated the presence of histone modification by monomers of ADP-ribose, particularly histone H1 [20, 22-25]. These monomers were presumed to be ketoamine adducts of ADP-ribosylated histones based on chemical stability studies. On the other hand, it is well established that chromatin structure and gene transcription is modulated through enzymatically controlled histone modifications such as acetylation, methylation, ubiquitination, and phosphorylation [26]. In particular, acetylation and deacetylation have been correlated with active gene transcription and gene silencing, respectively, within localized regions of chromosomes [14]. If non-enzymatic modifications of histones interfere with acetylation and deacetylation patterns then it is likely that glycation may disrupt normal gene transcription and have implications in the pathophysiology of diseases that involve oxidative stress and DNA damage.

Histone H1 binds to the linker DNA and is proposed to have a regulatory role in transcription via the modulation of chromatin structure [27]. Histone H1 has several isoforms. In mammals there are six somatic subtypes designated H1a-e and $H1_0$ [15]. A three-domain structure has been observed in the different histone H1 subtypes: a central globular domain, and the N and C terminal domains, which are highly basic and have little or no structure upon interaction with DNA.

Proteasome activity is not modulated by PARP activation under glyoxal-induced glycation condition of keratinocyte histones, including histone H1. Furthermore, some histone resistance to proteasome degradation was observed [1]. In contrast, histone modifications induced by hydrogen peroxide in cancer cells are degraded by proteasome activity, which is dependent on the activation of PARP [28].

2. Evidence for *In Vivo* AGE-Modified Histones

To look for evidence of *in vivo* AGE formation in nuclear proteins, histone H1 was chosen as the specific protein target since it was previously shown that this protein may be an important target of glycation-induced modification under conditions of oxidative stress [20]. AGE protein modifications include CML, argpyrimidine and crosslinks (see figure 2). The following experiments describe our recently published research investigating the presence of CML protein adducts in histone H1 isolated from calf thymus harvested immediately after slaughter [21]. CML protein adducts are the most abundant AGEs reported to occur in biological proteins [7]. To monitor for the presence of CML-histone H1 adducts, we employed reverse-phase high performance liquid chromatography (RP-HPLC), which took advantage of the dual acidic and basic properties of the CML-modified amino acids obtained from the hydrolyzed histone protein. CML contains two carboxyl groups and two amino groups, the charges of which can be modulated by pH [1]. These properties are unique, since no other amino acid has this dual occurrence of two carboxylic acids and two amino groups within the same molecule (see figure 2).

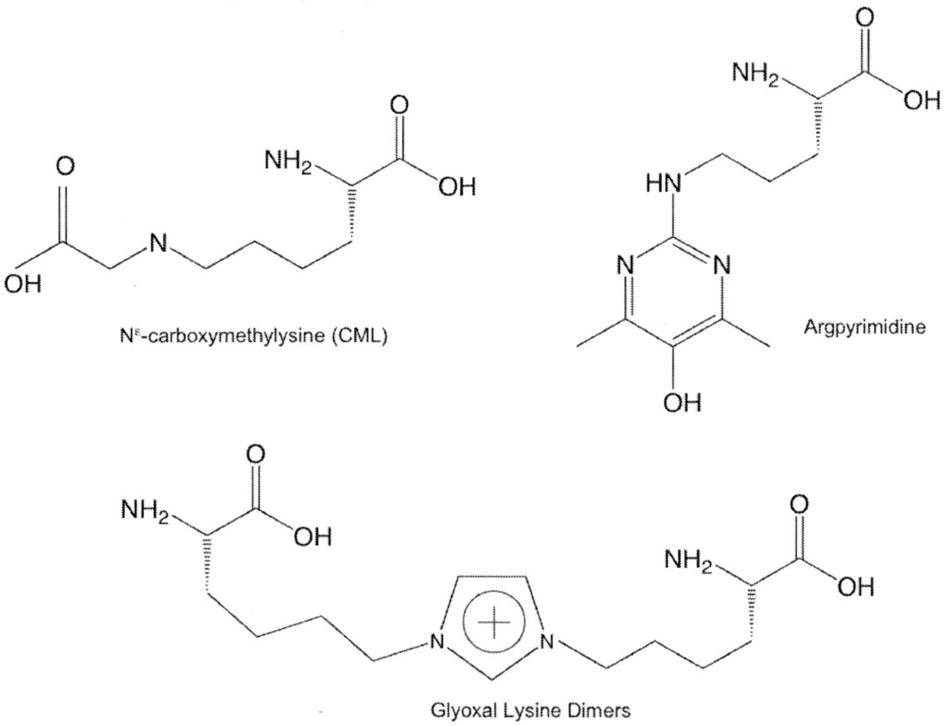

Figure 2. Structures of advanced glycation end product biomarkers [21].

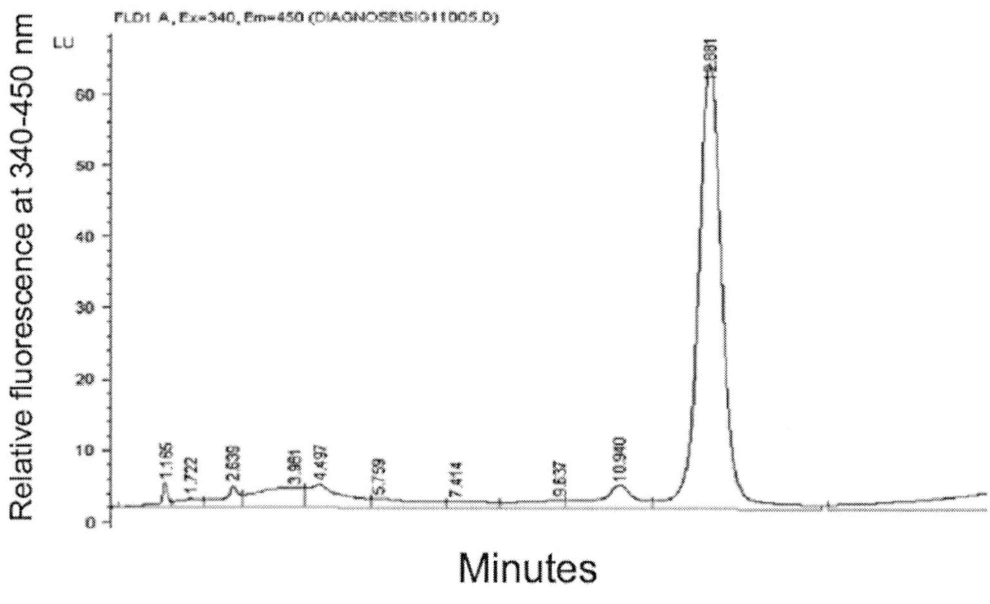

Figure 3. Reverse-phase HPLC analysis of CML after histone H1 hydrolysis [21].

When the CML-modified amino acids isolated by anion/cation chromatography from calf thymus histone H1 specimens were analyzed by RP-HPLC, a peak with a retention time of 12.9 minutes was observed, which co-migrated with an authentic CML standard, shown in figure 3. Under these conditions, 3.8 μmol CML was detected per mol of lysine, or 0.19 mmol/mol of histone H1.

3. Localization of the Lysine Histone H1 Residue Sites Modified with CML

Tandem mass spectroscopy was used to validate the presence of CML and also to locate the CML-modified lysine residues on histone H1. Freshly isolated calf thymus histone H1 purified by RP-HPLC was digested with trypsin and subjected to MS/MS analysis. The results shown in figure 4 demonstrated the presence of a peptide containing two modified lysine residues. This sequence is found in four different histone H1 subtypes (b-e [15]).

Since thymus contains a very heterogeneous population of cells we also studied human HaCaT keratinocyte cell cultures under stress conditions, which induce CML protein adduct formation, to investigate if CML modification also occurred in a homogenous population of viable, actively proliferating cells. Curiously, when human HaCaT keratinocytes were treated with glyoxal and histone H1 was isolated by HPLC and analyzed with MS/MS spectroscopy, the same peptide containing two CML-modified lysine residues was found, suggesting some sequence specificity for modification of lysine residues (results not shown). Thus, histone H1 appears to be selectively modified with CML on two lysine residues, whether the exposure to glycating compounds occurs *in vivo* or *in vitro*.

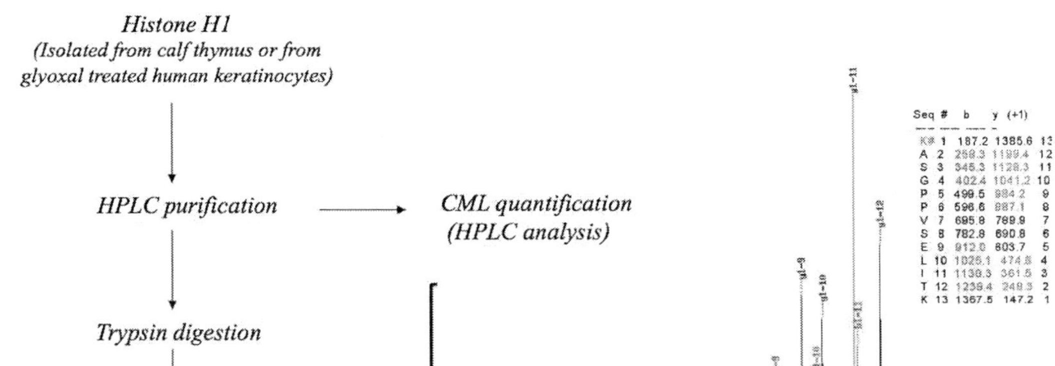


```
msetaplapt ipapaektpv kkkakkagat agkr**k**asgpp vselit**k**ava askersgvsl
aalkkalaaa gydveknnsr iklglkslvs kgtlvqtkgt gasgsfklnk kaasgegkpk
akkagaakpr kpagaakkpk kvagaatpkk sikktpkkvk kpataagtkk vaksakkvkt
pqpkkaaksp akakapkpka akpksgkpkv tkakkaapkk k
```

Figure 4. MS/MS analysis of the location of CML on histone H1 [21].

It should be noted that these two CML-modified lysine residues are located in the N-terminal region of the sequence of histone H1 close to the globular domain. This suggests that the modification of histone H1 with CML is restricted to areas of the protein that are exposed to high concentrations of the reducing sugar in the local microenvironment. While we do not know with certainty the identity of the sugar inducing CML formation and histone modification, ADP-ribose is a likely candidate (figure 1).

4. Modification of Histone H1 with Argpyrimidine Adducts

Another important AGE protein modification is the fluorescent adduct argpyrimidine, formed via the modification of the guanidino group from arginine residues. For this experiment histone H1 was loaded onto a SDS-PAGE gel and then transferred to a PVDF membrane for analysis with a monoclonal antibody against argpyrimidine. The results of the experiment are shown in figure 5. The signal for the presence of argpyrimidine on isolated histones from freshly collected calf thymus is shown in lane 1. Commercially obtained calf thymus histone H1 was also analyzed and is shown in lane 2. A positive signal for argpyrimidine in histone H1 was observed for the monomer and dimer versions of histone H1. A positive control was included using histone H1 modified *in vitro* with methylglyoxal (lane 3). Molecular weight markers used to assess antibody specificity were nonreactive.

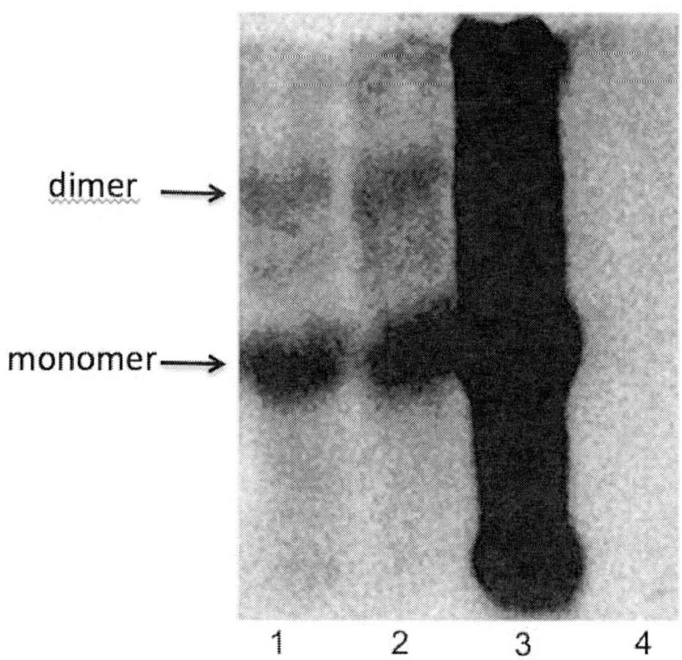

Figure 5. Immunodetection of argpyrimidine histone H1 protein adducts. Lane 1 represents histone H1 (isolated from freshly collected calf thymus); lane 2 represents commercial calf thymus histone H1; lane 3 is a positive control in which commercial calf thymus histone H1 was incubated with methylglyoxal; lane 4 is a negative control using molecular weight markers [21].

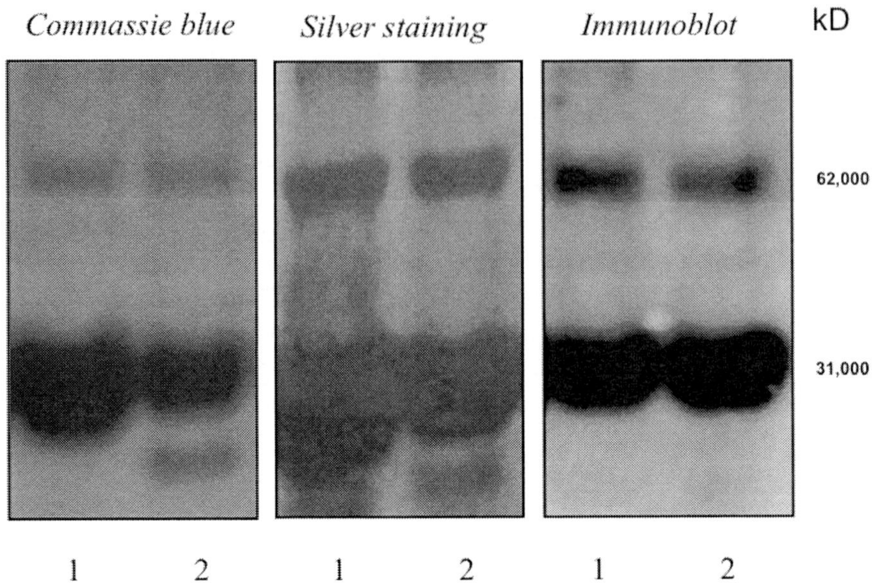

Figure 6. Electrophoretic and immunoanalysis of histone H1 crosslinks. Commercial obtained histone H1 (lane 1) or isolated from freshly collected calf thymus (lane 2) was analyzed by SDS-PAGE (12%) and Commassie blue staining (panel A), silver staining (panel B), and by immunodetection using an anti-histone H1 monoclonal antibody (panel C). The molecular weight of the protein bands observed is shown in the right lane [21].

5. Identification of Protein Cross-Links in Histone H1

To validate the presence of histone H1 crosslinks, histone H1 was further investigated using electrophoretic analysis. Figure 6 shows the SDS-PAGE analysis of freshly isolated calf thymus histone H1, purified by HPLC and stained with Commassie blue (panel A). A higher molecular weight band was observed at 62 KD, suggesting the presence of crosslinks. Because of the lower sensitivity associated with this assay it was necessary to use higher amounts of histone H1 in order to observe the presence of crosslinks. The visualization of histone H1 by SDS-PAGE was also performed using silver staining (panel B). The results observed in figure 6B demonstrate the increased intensity in both the monomer and the dimer histone H1 bands. To investigate whether the monomer and the dimer observed with SDS-PAGE and stained with Commassie blue and silver-staining analysis corresponded to histone H1, a western blot experiment was performed using a monoclonal antibody against histone H1. The results of these experiments are shown in figure 6C. The band at 62 KD was confirmed to be histone H1 in the form of a dimer.

6. Summary of *In Vivo* Glycative Histone H1 Modifications

Our findings that histone H1 isolated from freshly harvested calf thymus was modified with CML, argpyrimidine, and crosslinks upon isolation under stringent conditions strongly suggests that at least some of the tissue histones are modified *in vivo* with these, and perhaps other AGEs, under normal physiological conditions, and may accumulate with age. Previous studies in our laboratory using HaCaT keratinocytes treated with glyoxal showed that CML modified histones were degraded by proteasomal activity and that some modified histones were proteasome-resistant, resulting in their accumulation[1]. Other investigators also showed that histones that were extensively modified were resistant to proteasome degradation [28]. In addition, it was reported that total histones isolated from liver of chemically induced diabetic rats exhibited three-fold increased fluorescence compared to controls [29], a finding which was also seen with advancing age [29].

Recently it was shown that epigenetic changes induced by chronic hyperglycemia may influence the proinflammatory response, which is implicated in the development of vascular inflammation [30]. A combination of factors may be involved in these epigenetic modifications. One such factor is the activation of the enzymatic activity involved in regulating the activity of chromatin remodeling proteins, such acetyltransferases and methyltransferases, which modify histones. A second factor may be cellular oxidant stress-induced modification of histones, which may interfere with their normal enzymatic modification (e.g., acetylation) and ultimately affect chromatin structure and gene expression or repression. It will be important to better delineate the specific molecules and processes contributing to non-enzymatic histone modifications and to more fully understand their biological implications.

PART II: INHIBITION OF HISTONE H1 GLYCATION BY NATURAL PRODUCTS

Reactive dicarbonyl species (RCS) such as glyoxal and methylglyoxal, malondialdehyde, 4-hydroxynonenal, and acrolein are common intermediates in protein damage. Human plasma levels for glyoxal and methylglyoxal were estimated to be in the 0.3-1.5 μmol/L range [31-35] and their concentrations are increased in type I and II diabetes and chronic renal failure [31-35]. Protein damage by RCS is involved in a number of pathophysiological conditions including skin aging [36,37], diabetes [38], atherosclerosis [38], inflammation [39], Alzheimer's dementia, and is also important in "normal" physiologic processes such as ageing [40-43]. RCS are formed during reactions of normal cellular metabolism such as glycolysis, and are derived from non-enzymatic glycation [44], lipid peroxidation, and inflammation [45]. Glycation, a non-enzymatic reaction between reducing sugars and lysine residues, is an important source of RCS which leads to the formation of protein AGEs such as CML, N^{ε}-carboxyethyllysine (CEL), pentosidine, argpyrimidine, glyoxal lysine dimers (GOLD), and methylglyoxal lysine dimers (MOLD) [38]. As described above, CML is an important AGE that is formed during RCS-mediated glycoxidation processes. The CML content of proteins is indicative of overall protein glycation damage and it is a widely used tool for glycation research in *in vitro* and *in vivo* studies. Glycation-induced protein fluorescence at 335 and 385 nm (ex/em) is used to monitor the formation of pentosidine or pentosidine-like protein adducts. These adducts were AGE derivatives initially observed when pentoses caused protein glycation [46]. The most abundant non-fluorescent AGE protein-adduct is CML, hence this product is frequently used as a biomarker of protein glycation [40,46,47].

Administering the glycation inhibitor AG to rodents effectively suppresses secondary complications of experimental diabetes related to glycation and subsequent protein-AGE formation [48]. Unfortunately, these trials demonstrated that while AG was able to provide some benefit against development of diabetic complications, this hydrazine derivative resulted in toxicity upon long-term administration, presumably due to its potent inhibition of catalase [49,50] and inducible nitric oxide synthase [51]. Therefore, the toxicity profile of AG makes it unlikely that it will be used clinically, increasing the need for identifying and characterizing other safe and effective inhibitors of *in vivo* glycation.

Due to its abundance, glucose is assumed by many to be a major source of glycation and protein-AGE formation in extracellular proteins *in vivo*. However, glucose is only a weak glycation agent and the chemical reaction with proteins under physiological conditions only occurs under oxidative conditions over months to years. Moreover, glucose can form AGEs under oxidative [52,53] and non-oxidative conditions.

Unlike glucose, the highly reactive RCS glyoxal and methylglyoxal were implicated as instigators of non-oxidative protein modification, because they were much more efficient precursors for the formation of non-fluorescent AGEs such as CML and CEL and fluorescent AGEs such as argpyrimidine and peptide cross-links [13,18,54,55]. ADP-ribose, generated from NAD by multiple metabolic pathways, was shown to be an abundant intracellular pentose [56]. It was demonstrated that ADP-ribose was the most potent non-oxidative inducer of AGEs such as CML and protein fluorescent adducts, when histone H1 was the target protein [13,18,54]. ADP-ribose also induced protein cross-links of histone H1 under non-oxidative conditions [57].

The concentration of glyoxal in human plasma was reported to be 0.23 μmol/L in healthy human volunteers, 0.4 μmol/L in patients with mild/moderate uremia, and 0.76 μmol/L in patients with end-stage renal disease on hemodialysis [31]. Similar glyoxal levels in plasma samples were reported by Lapolla and colleagues [33]. Chronic renal failure resulted in accumulation of glyoxal, with a mean plasma level of 4 μmol/L, as reported by Odani *et al.* [34]. A potentially higher plasma concentration of glyoxal leading to local accumulation was suggested to occur in patients with diabetes-associated hyperglycemia [32]. In contrast to glyoxal, the physiological concentration of plasma methylglyoxal in healthy humans is 1.4 μmol/L [35]. The serum concentration of methylglyoxal was 5-6 fold higher in patients with type I diabetes mellitus and 2-3 fold greater in patients with type II diabetes than in healthy controls [58].

Rutin is a common dietary flavonoid that is consumed in fruits, vegetables and plant derived beverages such as tea, tomatoes and wine [59]. High doses of oral rutin protected diabetic rats against the formation of protein adducts in the skin [60]. Although the precise mechanism for this protection is not known, little or no dietary rutin is absorbed since gut microflora in the large intestine metabolize rutin to a variety of compounds that include quercetin and phenol derivatives such as 3,4-dihydroxyphenylacetic acid (DHPAA), 3,4-dihydroxy toluene (DHT), 3-hydroxyphenylacetic acid (HPAA) and 4-hydroxy-3-methoxyphenylacetic acid (homovanillic acid, HVA) [59, 61-65]. Quercetin plasma levels have been found to be 3.5 μmol/L when doses of 50 mg of either rutin or quercetin were supplemented in the diet of healthy human volunteers [66]. Moreover, up to 50% of an ingested dose of 75 mg rutin was recovered as microbial metabolites from urine of human volunteers [67], further supporting the appearance of μmol/L metabolite concentrations following rutin consumption.

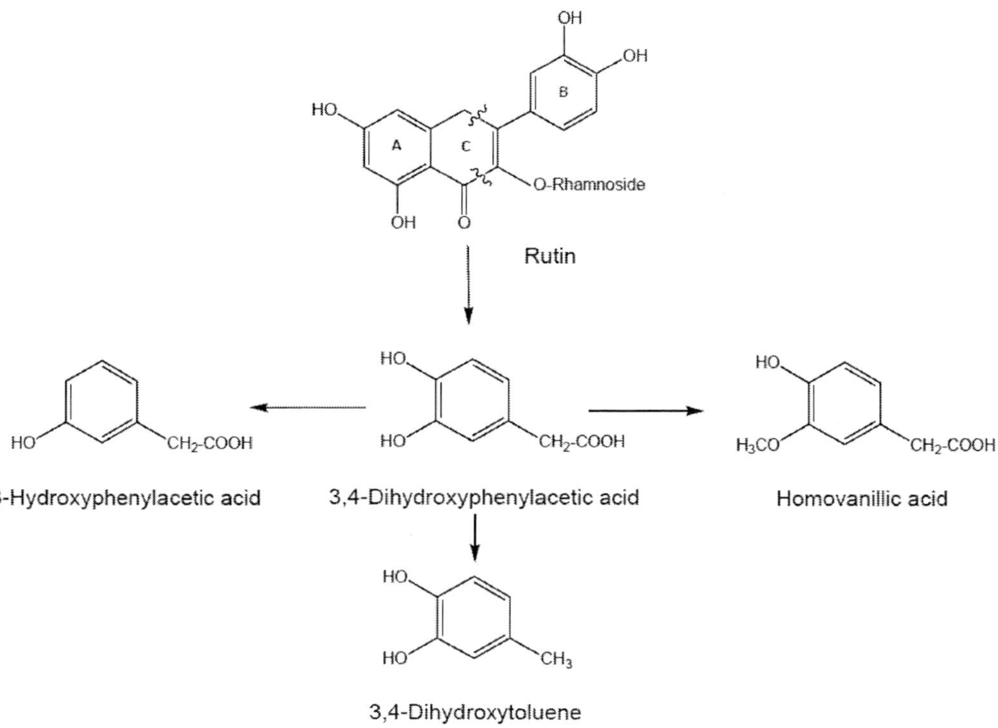

Figure 7. Structure of rutin and rutin metabolites [71].

Recent studies of the pharmacokinetics of rutin and its aglycone derivative quercetin in human volunteers reported blood concentrations of up to 3.5 μmol/L of quercetin and monophenolic rutin metabolites when doses of 50 mg of either rutin or its metabolite, quercetin, were given in the diet [66]. Most orally consumed rutin is likely broken down to low molecular weight phenolic acids by the colonic microflora [68], shown in figure 7. Furthermore, flavonoid intake has been estimated to be as high as 1 g/day [69], and plasma levels of such flavonoids and their metabolites are markedly increased following consumption [70]. Thus, the dietary intake of flavonoids may reasonably offer effective protection against glyoxal and methylglyoxal-induced protein damage inasmuch as our studies demonstrate that equimolar concentrations of rutin metabolites are effective inhibitors of RCS-induced protein damage.

We conducted a series of experiments to examine the potential effects of selected rutin-derived phenol derivatives on histone H1 glycation in an *in vitro* system [71]. Our purposes for selecting these metabolites were two-fold. First, as mentioned above, orally administered rutin has been shown in animal models of diabetes to have beneficial effects in complication reduction [60], making it likely that these effects were mediated by microbial metabolites [67]. Second, monophenols of rutin have poor antioxidant activity [68], which allowed us to directly assess their ability to inhibit non-oxidative AGE generation via glycation.

1. Inhibition of CML Formation on Histone H1 by Rutin Metabolites

Glucose-based protein glycation models have been used as *in vitro* models to search for potential inhibitors [72-74]. Since glucose mediated protein glycation may occur under oxidative and non-oxidative conditions to form AGE protein adducts [52], many potential protein glycation inhibitor candidates are difficult to distinguish from general antioxidants. In contrast to protein glycation by glucose, RCS (glyoxal and methylglyoxal) are both extracellular and intracellular glycating agents which are involved in *in vivo* non-oxidative glycation, with subsequent formation of AGE protein adducts [1,13,18,40,57,75,76]. Pentoses such as arabinose and ribose can also induce formation of AGEs via non-oxidative pathways [12]. Therefore, a non-oxidative model of protein glycation was chosen using histone H1, a protein target rich in lysine and arginine residues, and either glyoxal, methylglyoxal or ADP-ribose as reducing sugars. These models offered a very important advantage over glucose induced glycation models: they allowed us to test for pure glycation inhibitors, which act independently of antioxidant potential. We feel that this point is critical to underscore, since glycation inhibitors that use antioxidant properties to inhibit the formation of AGEs may not effectively inhibit non-oxidative protein glycation, an important contributor to intracellular and extracellular AGE formation.

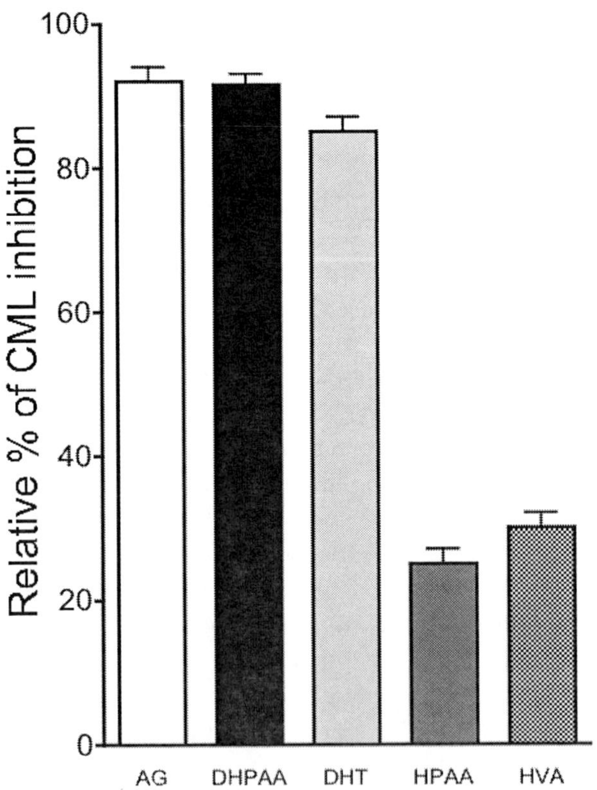

Figure 8. CML inhibition by rutin metabolites and AG during incubation of histone H1 and glyoxal [71].

Following consumption by humans, the polyphenol rutin derivatives quercetin-3-O-rhamnoglucoside and quercetin were not found in urine, however 3-hydroxyphenylacetic acid (HPAA), homovanillic acid (HVA), 3,4-dihydroxy-phenylacetic acid (DHPAA), and 3,4-dihydroxytoluene (DHT), metabolites resulting from colonic microflora degradation of these compounds, were detected [62]. Figure 7 shows the structures of the monophenol derivatives used in our study.

Histone H1 is particularly susceptible to non-enzymatic glycation because of its high percentage of arginine and lysine residues. To investigate the inhibitory capacity of various rutin metabolites on histone H1 glycation 1 g/L of histone H1 was incubated with 400-μmol/L glyoxal and production of CML was monitored as an AGE biomarker. As is seen in figure 8, 400 μmol/L of the known glycation inhibitor AG (control) effectively inhibited CML formation by 92% with a stoichiometry of 1:1. Similar results were obtained with rutin metabolites DHPAA and DHT (95 and 90%, respectively). In contrast, 400 μmol/L of the monohydroxyphenolic derivatives of rutin HPAA and HVA provided significantly less inhibition of CML formation, 25 and 30%, respectively. AG, DHT, and DHPAA were also used at ratios lower than 1:1 and a direct correlation with inhibition of CML formation was observed (data not shown). These results suggested that given reported plasma concentrations of glyoxal of 0.3-1.5 μmol/L [31-34] and plasma levels of quercetin and phenolic acid rutin metabolites of 3.5 μmol/L [62,66], diet-derived rutin metabolites such as DHPAA and DHT may be expected to inhibit and neutralize glyoxal under physiological conditions.

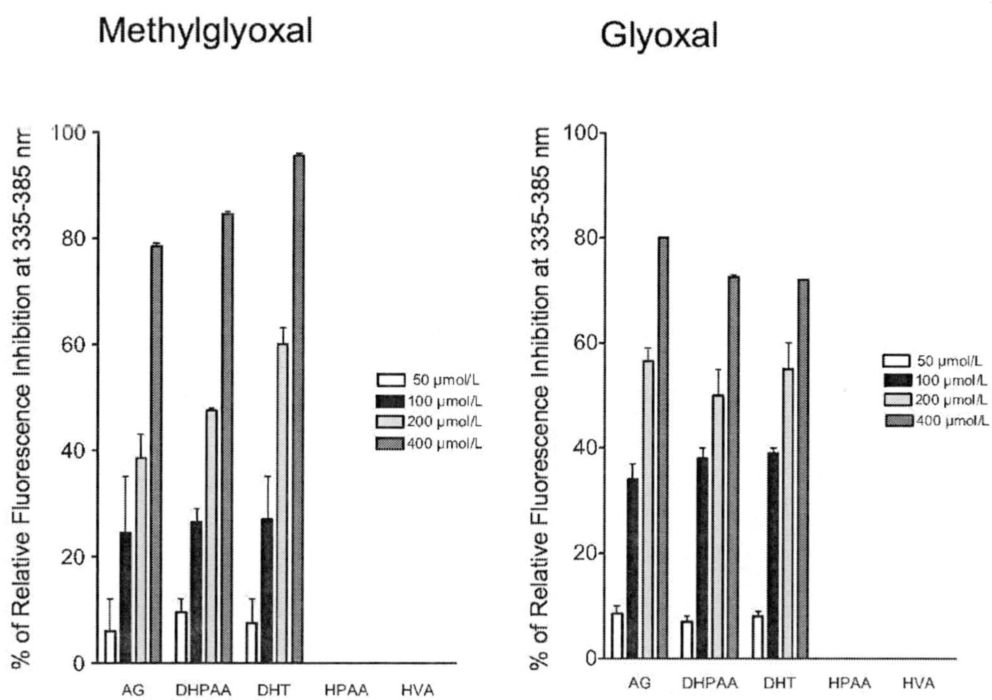

Figure 9. Fluorescence inhibition by rutin metabolites and AG during incubation of histone H1 and either glyoxal or methylglyoxal [71].

2. Inhibition of Fluorescent Adduct Formation on Histone H1

Certain AGE products such as pentosidine and argpyrimidine are produced by non-enzymatic glycation. These metabolites are known to fluoresce at 335-385 nm. Thus, we studied the effect of rutin metabolites on the formation of pentosidine-like fluorescent protein adducts by glycation of histone H1 with the physiologically relevant RCS glyoxal and methylglyoxal, monitoring fluorescence development at 335-385 nm. AG was used as a positive control. Figure 9 demonstrates that AG was effective in suppressing the relative glycation-induced histone H1 fluorescence induced by methylglyoxal (panel A) and glyoxal (panel B) in a dose-dependent manner. Similarly, DHPAA and DHT dose-dependently inhibited fluorescence resulting from glycation of histone H1 with 400-μmol/L methylglyoxal by up to 85 and 90%, respectively, or by 75% when using 400-μmol/L glyoxal as the glycating agent. Once again, the monohydroxy phenolic rutin derivatives HVA and HPAA were very poor inhibitors of either glyoxal or methylglyoxal glycation of histone H1. These results again demonstrated an approximate stoichiometry of 1:1 for neutralization of methylglyoxal and glyoxal-induced fluorescence development by AG and the rutin metabolites DHPAA and DHT. Thus, these results provided further reinforcement for physiological concentrations of rutin metabolites to neutralize reported plasma concentrations of glyoxal and methylglyoxal.

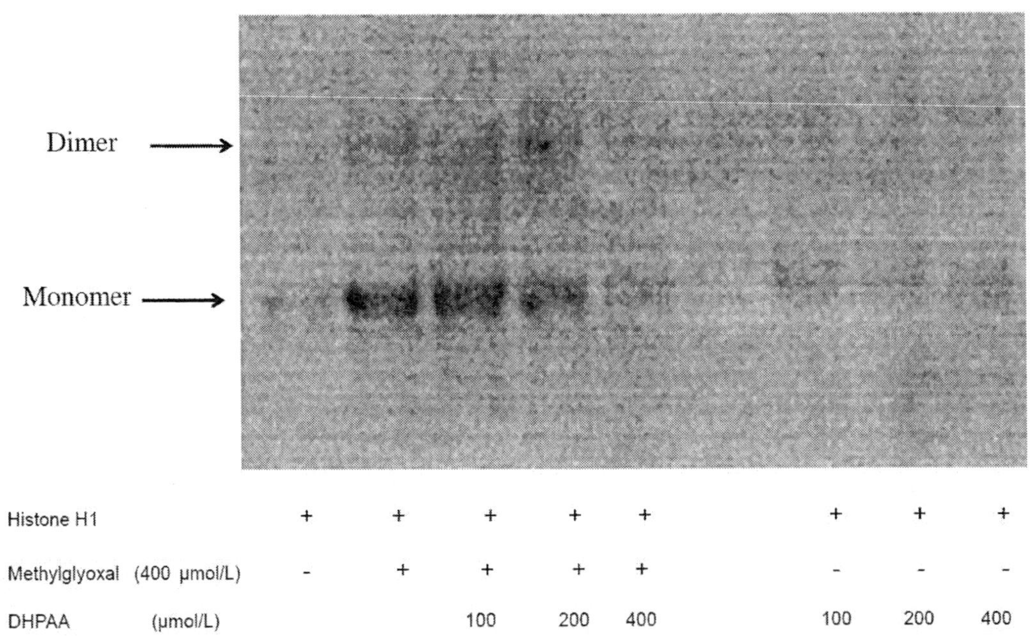

Figure 10. Argpyrimidine inhibition by DHPAA after glycation of histone H1 with methylglyoxal [71].

The formation of the fluorescent AGE product argpyrimidine, a protein adduct formed by methylglyoxal modification of arginine protein side chains, was monitored using monoclonal antibodies against argpyrimidine. Figure 10 shows the result of incubation of histone H1 (1mg/ml) and 400-μmol/L methylglyoxal in the presence of increasing concentrations of DHPAA, analyzed by immunochemistry analysis. Similar to the earlier experiments, DHPAA inhibited the formation of argpyrimidine induced by methylglyoxal in a 1:1 stoichiometric ratio. A control included in the right panel of figure 10 showed that when histone H1 was incubated with DHPAA in the absence of methylglyoxal there was no formation of the argpyrimidine protein adducts.

3. Inhibition of ADP-Ribose-Induced Fluorescence of Histone H1 by Rutin Metabolites

Figure 11 shows the relative fluorescence observed when rutin metabolites and AG were added in a dose-dependent manner to a glycation mixture consisting of 1 mg/ml histone H1 and 2.5 mmol/L ADP-ribose. Rutin metabolites and AG inhibited fluorescence in a dose-dependent manner at both 335-385 nm (specific for pentosidine and pentosidine-like fluorescent adducts) (figure 11A) and 370-440 nm (a general measure of fluorescence) (figure 11B). The IC_{50} of DHPAA was 260 and 320 µmol/L, for DHT the IC_{50} was 150 and 350 µmol/L, and for AG the IC_{50} was 275 and 320 µmol/L, at 335-385 nm and 370-440 nm, respectively. The flavonoids HPAA and HVA were significantly less potent (IC_{50}> 1600 µmol/L) at both 370-440 nm and 335-385 nm.

A control experiment was designed to test for the possibility of a false positive, which could result from fluorescence quenching by rutin metabolites. Following 7 days of reaction of ADP-ribose and histone H1 at 37°C, 1.6 mmol/L rutin metabolites or AG were added and the relative fluorescence at the corresponding wavelengths was measured before and after the addition. Under these conditions, neither rutin metabolites nor AG altered relative fluorescence values (data not shown). Our results showed that AG and rutin metabolites, particularly those metabolites that included vicinal hydroxyl groups in their structure (DHPAA and 3,4-DHT), were very potent in suppressing the formation of fluorescent derivatives (370-440 and 335-385 nm) induced by ADP-ribose glycation of histone H1 (figure 11). The stoichiometry of the inhibition by rutin metabolites was 0.7:1 when using ADP-ribose as a glycating agent and 1:1 when using RCS as glycating agents. In contrast, rutin metabolites containing monohydroxy groups in their structure, such as HPAA and HVA, were not as effective in suppressing fluorescence appearance. Thus, vicinal hydroxyl groups in the phenolic structure of rutin metabolites appear to contribute positively to the effectiveness of fluorescence inhibition during glyoxal, methylglyoxal and ADP-ribose glycation of histone H1. A possible explanation for this observation might involve the condensation of the adjacent phenolic hydroxyl moieties with dicarbonyl intermediates, forming a hemiacetal intermediate which can react further to form an acetal (see proposed structure in figure 12). We propose that these adducts more effectively trap and inactivate RCS such as glyoxal and methylglyoxal and thereby potently inhibit the glycation reaction.

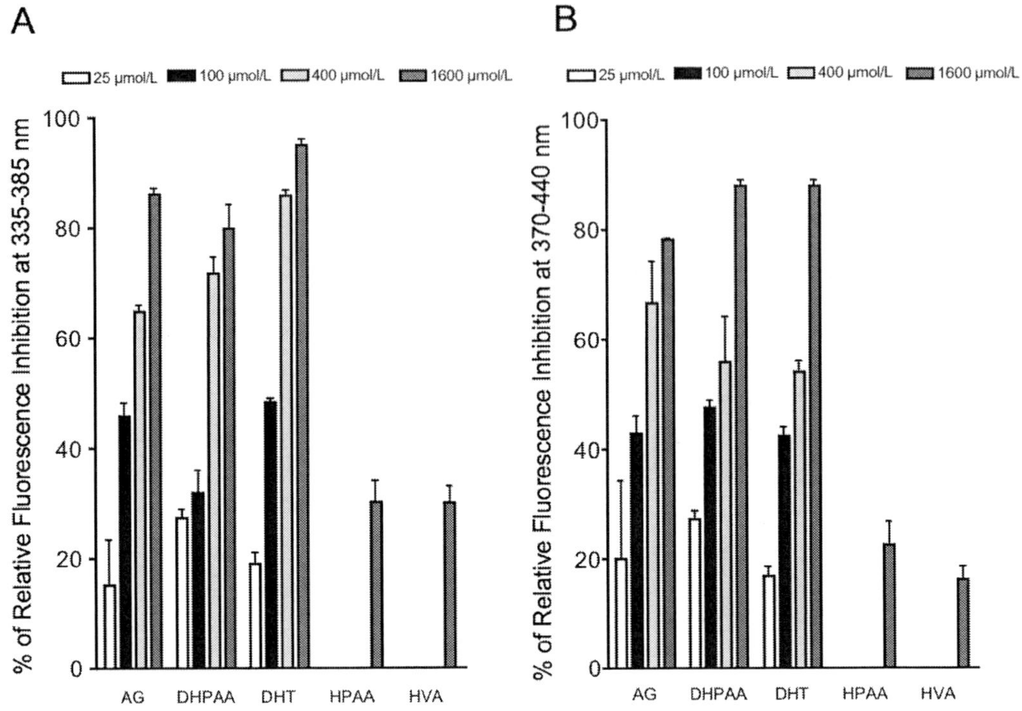

Figure 11. Fluorescence inhibition by rutin metabolites and AG during incubation of histone H1 and ADP-ribose [71].

CONCLUSIONS

In conclusion, our experimental data demonstrated that rutin metabolites, formed in the gut by microbial metabolism from dietary rutin following its consumption as a component of a variety of commonly consumed fruits and vegetables, are potent glycation inhibitors with antiglycative capability comparative to AG. In particular, it appears that those metabolites with vicinal hydroxyl groups are by far most effective. Plasma concentrations of these rutin metabolites can be expected to effectively neutralize the reported plasma concentrations of glyoxal and methylglyoxal. Furthermore, intracellular glycation of pentoses, which may result in the formation of AGEs via non-oxidative routes, is likely to be inhibited by these compounds at physiological concentrations.

Our results reveal very promising candidates for use as inhibitors of glycation-induced protein AGEs adduct formation. Because of their widespread availability within plant foods, their relative lack of toxicity is likely. Indeed, consumption of large quantities of fruit and vegetables, the traditional sources of dietary rutin, is associated with a healthy diet.

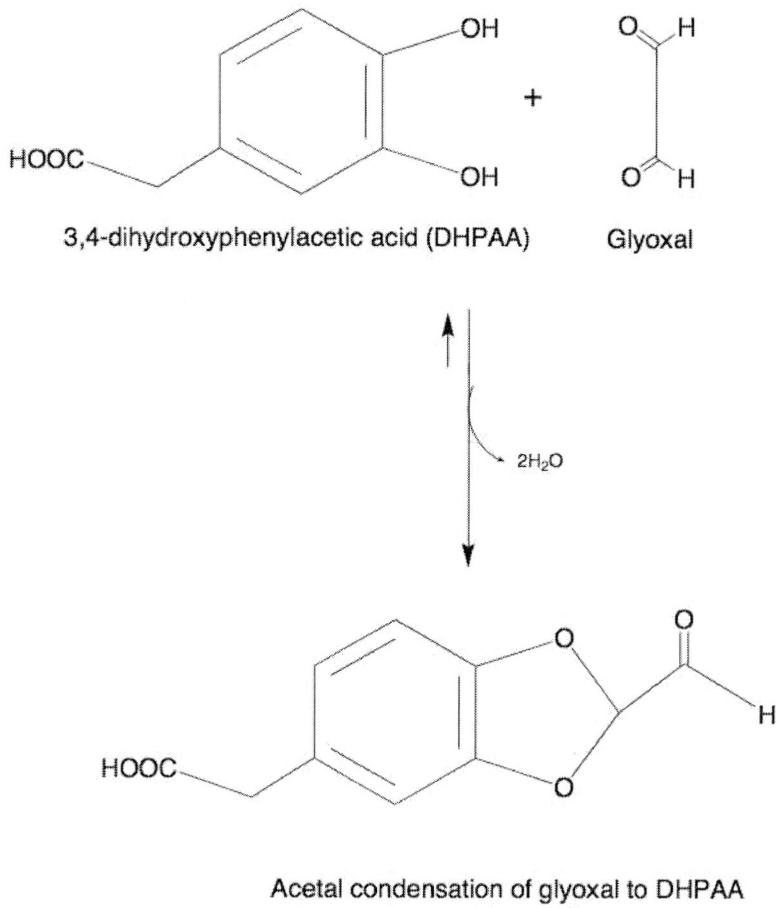

Figure 12. Proposed adducts formed by condensation dihydroxyphenol rutin metabolites and glyoxal [71].

REFERENCES

[1] Cervantes-Laurean, D., Roberts, M.J., Jacobson, E.L. & Jacobson, M.K. (2005). Nuclear proteasome activation and degradation of carboxymethylated histones in human keratinocytes following glyoxal treatment. *Free Radic Biol Med*, *38*, 786-95.

[2] Voss, P. & Grune, T. (2007). The nuclear proteasome and the degradation of oxidatively damaged proteins. *Amino Acids*, *32*, 527-34.

[3] Bardag-Gorce, F. (2009). Nuclear effects of ethanol-induced proteasome inhibition in liver cells. *World J Gastroenterol*, *15*, 1163-7.

[4] Rackova, L., Snirc, V., Jung, T., Stefek, M., Karasu, C. & Grune, T. (2009). Metabolism-induced oxidative stress is a mediator of glucose toxicity in HT22 neuronal cells. *Free Radic Res*, *43*, 876-86.

[5] Singh, R., Barden, A., Mori, T. & Beilin, L. (2001). Advanced glycation end-products: a review. *Diabetologia*, *44*, 129-46.

[6] Thornalley, P.J., Langborg, A. & Minhas, H.S. (1999). Formation of glyoxal, methylglyoxal and 3-deoxyglucosone in the glycation of proteins by glucose. *Biochem J, 344 Pt 1*, 109-16.

[7] Dunn, J.A., McCance, D.R., Thorpe, S.R., Lyons, T.J. & Baynes, J.W. (1991). Age-dependent accumulation of N epsilon-(carboxymethyl)lysine and N epsilon-(carboxymethyl)hydroxylysine in human skin collagen. *Biochemistry*, *30*, 1205-10.

[8] Shipanova, I.N., Glomb, M.A. & Nagaraj, R.H. (1997). Protein modification by methylglyoxal: chemical nature and synthetic mechanism of a major fluorescent adduct. *Arch Biochem Biophys*, *344*, 29-36.

[9] O'Brien, P.J., Siraki, A.G. & Shangari, N. (2005). Aldehyde sources, metabolism, molecular toxicity mechanisms, and possible effects on human health. *Crit Rev Toxicol*, *35*, 609-62.

[10] Richard, J.P. (1991). Kinetic parameters for the elimination reaction catalyzed by triosephosphate isomerase and an estimation of the reaction's physiological significance. *Biochemistry*, *30*, 4581-5.

[11] Thornalley, P.J. (1993). The glyoxalase system in health and disease. *Mol Aspects Med*, *14*, 287-371.

[12] Litchfield, J.E., Thorpe, S.R. & Baynes, J.W. (1999). Oxygen is not required for the browning and crosslinking of protein by pentoses: relevance to Maillard reactions in vivo. *Int J Biochem Cell Biol*, *31*, 1297-305.

[13] Cervantes-Laurean, D., Minter, D.E., Jacobson, E.L. & Jacobson, M.K. (1993). Protein glycation by ADP-ribose: studies of model conjugates. *Biochemistry*, *32*, 1528-34.

[14] Heintzman, N.D., Stuart, R.K., Hon, G., Fu, Y., Ching, C.W., Hawkins, R.D., Barrera, L.O., Van Calcar, S., Qu, C., Ching, K.A., Wang, W., Weng, Z., Green, R.D., Crawford, G.E. & Ren, B. (2007). Distinct and predictive chromatin signatures of transcriptional promoters and enhancers in the human genome. *Nat Genet*, *39*, 311-8.

[15] Ponte, I., Vila, R. & Suau, P. (2003). Sequence complexity of histone H1 subtypes. *Mol Biol Evol*, *20*, 371-80.

[16] Hassa, P.O., Haenni, S.S., Elser, M. & Hottiger, M.O. (2006). Nuclear ADP-ribosylation reactions in mammalian cells: where are we today and where are we going? *Microbiol Mol Biol Rev*, *70*, 789-829.

[17] Althaus, F.R., Kleczkowska, H.E., Malanga, M., Müntener, C.R., Pleschke, J.M., Ebner, M. & Auer, B. (1999). Poly ADP-ribosylation: a DNA break signal mechanism. *Mol Cell Biochem*, *193*, 5-11.

[18] Cervantes-Laurean, D., Jacobson, E.L. & Jacobson, M.K. (1996). Glycation and glycoxidation of histones by ADP-ribose. *J Biol Chem*, *271*, 10461-9.

[19] Alvarez-Gonzalez, R. & Althaus, F.R. (1989). Poly(ADP-ribose) catabolism in mammalian cells exposed to DNA-damaging agents. *Mutat Res*, *218*, 67-74.

[20] Adamietz, P. & Rudolph, A. (1984). ADP-ribosylation of nuclear proteins in vivo. Identification of histone H2B as a major acceptor for mono- and poly(ADP-ribose) in dimethyl sulfate-treated hepatoma AH 7974 cells. *J Biol Chem*, *259*, 6841-6.

[21] Pashikanti, S., Boissonneault, G.A. & Cervantes-Laurean, D. (2011). *Ex vivo* detection of histone H1 modified with advanced glycation end products. *Free Rad Biol Med*, doi:10.1016/j.freeradbiomed.2011.01.034.

[22] Hilz, H., Adamietz, P., Bredehorst, R. & Wielckens, K. (1978). ADP-ribosylation of nuclear proteins. *Adv Enzyme Regul*, *17*, 195-211.

[23] Kreimeyer, A., Adamietz, P. & Hilz, H. (1985). Alkylation-induced mono(ADP-ribosyl)-histones H1 and H2B. Hydroxylamine-resistant linkage in hepatoma cells. *Biol Chem Hoppe Seyler*, *366*, 537-44.

[24] Kreimeyer, A., Wielckens, K., Adamietz, P. & Hilz, H. (1984). DNA repair-associated ADP-ribosylation in vivo. Modification of histone H1 differs from that of the principal acceptor proteins. *J Biol Chem*, *259*, 890-6.

[25] Hilz, H., Wielckens, K., Adamietz, P., Bredehorst, R. & Kreymeier, A. (1983). Functional aspects of mono- and poly(ADP-ribosyl)ation: subcellular distribution and ADP-ribosyl turnover under conditions of repair and 'starvation'. *Princess Takamatsu Symp*, *13*, 155-63.

[26] Cosgrove, M.S., Boeke, J.D. & Wolberger, C. (2004). Regulated nucleosome mobility and the histone code. *Nat Struct Mol Biol*, *11*, 1037-43.

[27] Ponte, I., Vidal-Taboada, J.M. & Suau, P. (1998). Evolution of the vertebrate H1 histone class: evidence for the functional differentiation of the subtypes. *Mol Biol Evol*, *15*, 702-8.

[28] Merker, K., Ullrich, O., Schmidt, H., Sitte, N. & Grune, T. (2003). Stability of the nuclear protein turnover during cellular senescence of human fibroblasts. *FASEB J*, *17*, 1963-5.

[29] Gugliucci, A. & Bendayan, M. (1995). Histones from diabetic rats contain increased levels of advanced glycation end products. *Biochem Biophys Res Commun*, *212*, 56-62.

[30] Ling, C. & Groop, L. (2009). Epigenetics: a molecular link between environmental factors and type 2 diabetes. *Diabetes*, *58*, 2718-25.

[31] Agalou, S., Ahmed, N., Dawnay, A. & Thornalley, P.J. (2003). Removal of advanced glycation end products in clinical renal failure by peritoneal dialysis and haemodialysis. *Biochem Soc Tran*, *31*, 1394-6.

[32] Akhand, A.A., Hossain, K., Kato, M., Miyata, T., Du, J., Suzuki, H., Kurokawa, K. & Nakashima, I. (2001). Glyoxal and methylglyoxal induce lyoxal and methyglyoxal induce aggregation and inactivation of ERK in human endothelial cells. *Free Radic Biol Med*, *31*, 1228-35.

[33] Lapolla, A., Flamini, R., Tonus, T., Fedele, D., Senesi, A., Reitano, R., Marotta, E., Pace, G., Seraglia, R. & Traldi, P. (2003). An effective derivatization method for quantitative determination of glyoxal and methylglyoxal in plasma samples by gas chromatography/mass spectrometry. *Rapid Commun Mass Spectrom*, *17*, 876-8.

[34] Odani, H., Shinzato, T., Matsumoto, Y., Usami, J. & Maeda, K. (1999). Increase in three alpha,beta-dicarbonyl compound levels in human uremic plasma: specific in vivo determination of intermediates in advanced Maillard reaction. *Biochem Biophys Res Commun*, *256*, 89-93.

[35] Thornalley, P.J., Hooper, N.I., Jennings, P.E., Florkowski, C.M., Jones, A.F., Lunec, J. & Barnett, A.H. (1989). The human red blood cell glyoxalase system in diabetes mellitus. *Diabetes Res Clin Pract*, *7*, 115-20.

[36] Lyons, T.J., Bailie, K.E., Dyer, D.G., Dunn, J.A. & Baynes, J.W. (1991). Decrease in skin collagen glycation with improved glycemic control in patients with insulin-dependent diabetes mellitus. *J Clin Invest*, *87*, 1910-5.

[37] Mizutari, K., Ono, T., Ikeda, K., Kayashima, K. & Horiuchi, S. (1997). Photo-enhanced modification of human skin elastin in actinic elastosis by N(epsilon)-

(carboxymethyl)lysine, one of the glycoxidation products of the Maillard reaction. *J Invest Dermatol*, *108*, 797-802.

[38] Baynes, J.W. & Thorpe, S.R. (2000). Glycoxidation and lipoxidation in atherogenesis. *Free Radic Biol Med*, *28*, 1708-16.

[39] Anderson, M.M., Hazen, S.L., Hsu, F.F. & Heinecke, J.W. (1997). Human neutrophils employ the myeloperoxidase-hydrogen peroxide-chloride system to convert hydroxy-amino acids into glycolaldehyde, 2- hydroxypropanal, and acrolein. A mechanism for the generation of highly reactive alpha-hydroxy and alpha,beta-unsaturated aldehydes by phagocytes at sites of inflammation. *J Clin Invest*, *99*, 424-32.

[40] Baynes, J.W. (1991). Role of oxidative stress in development of complications in diabetes. *Diabetes*, *40*, 405-12.

[41] Brownlee, M. (1995). Advanced protein glycosylation in diabetes and aging. *Annu Rev Med*, *46*, 223-34.

[42] Manske, C.L. (1993). Coronary artery disease in diabetic patients with nephropathy. *Am J Hypertens*, *6*, 367S-74S.

[43] Yan, S.D., Yan, S.F., Chen, X., Fu, J., Chen, M., Kuppusamy, P., Smith, M.A., Perry, G., Godman, G.C., Nawroth, P., *et al.* (1995). Non-enzymatically glycated tau in Alzheimer's disease induces neuronal oxidant stress resulting in cytokine gene expression and release of amyloid beta-peptide. *Nat Med*, *1*, 693-9.

[44] Glomb, M.A. & Monnier, V.M. (1995). Mechanism of protein modification by glyoxal and glycolaldehyde, reactive intermediates of the Maillard reaction. *J Biol Chem*, *270*, 10017-26.

[45] Anderson, M.M., Requena, J.R., Crowley, J.R., Thorpe, S.R. & Heinecke, J.W. (1999). The myeloperoxidase system of human phagocytes generates Nepsilon-(carboxymethyl)lysine on proteins: a mechanism for producing advanced glycation end products at sites of inflammation. *J Clin Invest*, *104*, 103-13.

[46] Monnier, V.M., Sell, D.R., Nagaraj, R.H., Miyata, S., Grandhee, S., Odetti, P. & Ibrahim, S.A. (1992). Maillard reaction-mediated molecular damage to extracellular matrix and other tissue proteins in diabetes, aging, and uremia. *Diabete*, *41*, *Suppl 2*, 36-41.

[47] Baynes, J.W., Watkins, N.G., Fisher, C.I., Hull, C.J., Patrick, J.S., Ahmed, M.U., Dunn, J.A. & Thorpe, S.R. (1989). The Amadori product on protein: structure and reactions. *Prog Clin Biol Res*, *304*, 43-67.

[48] Edelstein, D. & Brownlee, M. (1992). Aminoguanidine ameliorates albuminuria in diabetic hypertensive rats. *Diabetologia*, *35*, 96-7.

[49] Freedman, B.I., Wuerth, J.P., Cartwright, K., Bain, R.P., Dippe, S., Hershon, K., Mooradian, A.D. & Spinowitz, B.S. (1999). Design and baseline characteristics for the aminoguanidine Clinical Trial in Overt Type 2 Diabetic Nephropathy (ACTION II). *Control Clin Trials*, *20*, 493-510.

[50] Ou, P. & Wolff, S.P. (1993). Aminoguanidine: a drug proposed for prophylaxis in diabetes inhibits catalase and generates hydrogen peroxide in vitro. *Biochem Pharmacol*, *46*, 1139-44.

[51] Okuda, Y., Sakoda, S., Fujimura, H. & Yanagihara, T. (1998). Aminoguanidine, a selective inhibitor of the inducible nitric oxide synthase, has different effects on experimental allergic encephalomyelitis in the induction and progression phase. *J Neuroimmunol*, *81*, 201-10.

[52] Fu, M.X., Wells-Knecht, K.J., Blackledge, J.A., Lyons, T.J., Thorpe, S.R. & Baynes, J.W. (1994). Glycation, glycoxidation, and cross-linking of collagen by glucose. Kinetics, mechanisms, and inhibition of late stages of the Maillard reaction. *Diabetes*, *43*, 676-83.

[53] Sell, D.R., Biemel, K.M., Reihl, O., Lederer, M.O., Strauch, C.M. & Monnier, V.M. (2005). Glucosepane is a major protein cross-link of the senescent human extracellular matrix. Relationship with diabetes. *J Biol Chem*, *280*, 12310-5.

[54] Jacobson, E.L., Cervantes-Laurean, D. & Jacobson, M.K. (1994). Glycation of proteins by ADP-ribose. *Mol Cell Biochem*, *138*, 207-12.

[55] Jacobson, E.L., Cervantes-Laurean, D. & Jacobson, M.K. (1997). ADP-ribose in glycation and glycoxidation reactions. *Adv Exp Med Biol*, *419*, 371-9.

[56] Cardenas-Corona, M.E., Jacobson, E.L. & Jacobson, M.K. (1987). Endogenous polymers of ADP-ribose are associated with the nuclear matrix. *J Biol Chem*, *262*, 14863-6.

[57] Wondrak, G.T., Cervantes-Laurean, D., Jacobson, E.L. & Jacobson, M.K. (2000). Histone carbonylation in vivo and in vitro. *Biochem J*, *351 Pt 3*, 769-77.

[58] McLellan, A.C., Thornalley, P.J., Benn, J. & Sonksen, P.H. (1994). Glyoxalase system in clinical diabetes mellitus and correlation with diabetic complications. *Clin Sci (Lond)*, *87*, 21-9.

[59] Winter, J., Moore, L.H., Dowell, V.R., Jr. & Bokkenheuser, V.D. (1989). C-ring cleavage of flavonoids by human intestinal bacteria. *Appl Environ Microbiol*, *55*, 1203-8.

[60] Odetti, P.R., Borgoglio, A., De Pascale, A., Rolandi, R. & Adezati, L. (1990). Prevention of diabetes-increased aging effect on rat collagen-linked fluorescence by aminoguanidine and rutin. *Diabetes*, *39*, 796-801.

[61] Baba, S., Furuta, T., Fujioka, M. & Goromaru, T. (1983). Studies on drug metabolism by use of isotopes XXVII: urinary metabolites of rutin in rats and the role of intestinal microflora in the metabolism of rutin. *J Pharm Sci*, *72*, 1155-8.

[62] Baba, S., Furuta, T., Horie, M. & Nakagawa, H. (1981). Studies on drug metabolism by use of isotopes XXVI: Determination of urinary metabolites of rutin in humans. *J Pharm Sci*, *70*, 780-2.

[63] Braune, A., Gutschow, M., Engst, W. & Blaut, M. (2001). Degradation of quercetin and luteolin by Eubacterium ramulus. *Appl Environ Microbiol*, *67*, 5558-67.

[64] Griffiths, L.A. & Barrow, A. (1972). Metabolism of flavonoid compounds in germ-free rats. *Biochem J*, *130*, 1161-2.

[65] Schneider, H., Simmering, R., Hartmann, L., Pforte, H. & Blaut, M. (2000). Degradation of quercetin-3-glucoside in gnotobiotic rats associated with human intestinal bacteria. *J Appl Microbiol*, *89*, 1027-37.

[66] Erlund, I., Kosonen, T., Alfthan, G., Mäenpää, J., Perttunen, K., Kenraali, J., Parantainen, J. & Aro, A. (2000). Pharmacokinetics of quercetin from quercetin aglycone and rutin in healthy volunteers. *Eur J Clin Pharmacol*, *56*, 545-53.

[67] Sawai, Y., Kohsaka, K., Nishiyama, Y. & Ando, K. (1987). Serum concentrations of rutoside metabolites after oral administration of a rutoside formulation to humans. *Arzneimittelforschung*, *37*, 729-32.

[68] Olthof, M.R., Hollman, P.C., Buijsman, M.N., van Amelsvoort, J.M. & Katan, M.B. (2003). Chlorogenic acid, quercetin-3-rutinoside and black tea phenols are extensively metabolized in humans. *J Nutr*, *133*, 1806-14.

[69] Kuhnau, J. (1976). The flavonoids. A class of semi-essential food components: their role in human nutrition. *World Rev Nutr Diet*, *24*, 117-91.

[70] Conquer, J.A., Maiani, G., Azzini, E., Raguzzini, A. & Holub, B.J. (1998). Supplementation with quercetin markedly increases plasma quercetin concentration without effect on selected risk factors for heart disease in healthy subjects. *J Nutr*, *128*, 593-7.

[71] Pashikanti, S., de Alba, D.R., Boissonneault, G.A. & Cervantes-Laurean, D. (2010). Rutin metabolites: novel inhibitors of nonoxidative advanced glycation end products. *Free Radic Biol Med*, *48*, 656-63.

[72] Cervantes-Laurean, D., Schramm, D.D., Jacobson, E.L., Halaweish, I., Bruckner, G.G. & Boissonneault, G.A. (2006). Inhibition of advanced glycation end product formation on collagen by rutin and its metabolites. *J Nutr Biochem*, *17*, 531-40.

[73] Nagasawa, T., Tabata, N., Ito, Y., Aiba, Y., Nishizawa, N. & Kitts, D.D. (2003). Dietary G-rutin suppresses glycation in tissue proteins of streptozotocin-induced diabetic rats. *Mol Cell Biochem*, *252*, 141-7.

[74] Nagasawa, T., Tabata, N., Ito, Y., Nishizawa, N., Aiba, Y. & Kitts, D.D. (2003). Inhibition of glycation reaction in tissue protein incubations by water soluble rutin derivative. *Mol Cell Biochem*, *249*, 3-10.

[75] Brownlee M. The pathological implications of protein glycation. *Clin Invest Med* 1995;18:275-81.

[76] Wondrak GT, Cervantes-Laurean D, Roberts MJ, *et al*. Identification of alpha-dicarbonyl scavengers for cellular protection against carbonyl stress. *Biochem Pharmacol* 2002;63:361-73.

In: Histones: Class, Structure and Function
Editor: Chang-Hui Shen
ISBN: 978-1-62100-274-1

Chapter 10

THE ROLE OF CHROMATIN IN INDUCIBLE GENE REGULATION IN THE IMMUNE SYSTEM

Lucy D. Apps and Adele F. Holloway*
Menzies Research Institute Tasmania, University of Tasmania, Australia

ABSTRACT

The organisation of DNA and histone proteins into chromatin plays a central role in the regulation of eukaryotic gene expression. The coordinated regulation of gene expression programs is critical in directing developmental programs and in enabling cells to respond to extracellular signals. The combination of transcription factor binding sites found in gene promoters and enhancers, and the availability of these transcription factors, is important in co-ordinating these gene expression programs. However transcription factors must operate within the chromatin landscape, which regulates their ability to access their binding sites and drive transcription. The chromatin architecture of a gene can vary dramatically in different cell types, with differences in histone density, composition, modification and the presence of chromatin remodelers. While there would appear to be almost unlimited permutations of chromatin architectures, recent data generated using genome-wide technologies has begun to define chromatin signatures that are associated with particular gene states, and provide insight into how these architectures interact with extracellular signals. In this chapter we review current information on the role of chromatin in regulating inducible gene expression, using the immune system as a model.

1. INTRODUCTION

The initiation, duration and resolution of an appropriate immune response are vital for defending the host against invading pathogens, toxins and cancerous cells. Such immune responses are reliant on the coordinated regulation of gene expression programs, which direct immune function. These programs are driven by networks of transcription factors which bind

* Corresponding author e-mail : A.F.Holloway@utas.edu.au.

to their cognate binding sites in the regulatory regions of genes encoding transcription factors, cytokines and other proteins required for an immune response [1]. Control of gene transcription is therefore essential for proper immune cell function, but rather than being reliant on transcription factor binding alone, an intricate and multi-layered system of transcriptional regulation is in place in eukaryotes, as reviewed in [2]. The action of transcription factors is in part, dictated by the chromatin structure established at the regulatory regions of a gene, which controls the ability of transcription factors to access DNA binding sites and drive transcription [1]. The chromatin structure may be subject to extensive modification and reorganisation and as such the chromatin architecture of a gene can vary dramatically in different cell types, with differences in histone density, composition, modification and the presence of chromatin remodelling complexes [1].

In the immune system, changes in chromatin architecture are important on two levels: developmentally and acutely. Chromatin changes can occur in response to developmental signals, with different chromatin architectures established across individual genes during the differentiation of immune cell lineages [3, 4]. Once established these chromatin architectures must be maintained in the differentiated cell. Acute modifications to chromatin architectures also occur in response to an immune stimuli, facilitating changes in gene expression [5, 6]. Investigation of individual gene loci important in immune cell activation, and more recently genome-wide studies of immune cells have provided considerable insight into the role of the basal chromatin architecture and acute changes to this architecture in regulating inducible gene expression. This is the focus of this chapter, in which we review current information on the role of chromatin in regulating inducible gene expression, using the immune system as a model.

2. Chromatin Architecture

Within each eukaryotic cell DNA is packaged into a highly dynamic chromatin structure. Chromatin is a nucleoprotein complex in which the DNA double helix is integrated into regularly spaced nucleosomes, the basic structural unit of chromatin. Each nucleosome consists of an octamer of core histone proteins, containing two each of H2A, H2B, H3 and H4 [7]. Histone chaperone proteins deposit H3/H4 tetramers on newly replicated DNA. A second chaperone protein then adds H2A/H2B dimers and modulates the completion of nucleosome assembly [8]. This nucleosome structure is then further compacted into a higher order 30nm chromatin fibre and ultimately the chromosome [9]. Histone 1 (H1) proteins are linker histones structurally unrelated to the core histones, which occupy approximately 20bp within the linker regions of DNA, located in between nucleosomes. H1 proteins have an essential role in mediating chromatin condensation and stability [10]. While the assembly of DNA into chromatin is essential for the ordered storage of large amounts of genetic information within the cell nucleus, it is clear that it also provides an additional layer of regulatory information superimposed upon the DNA sequence [11]. This information may be generated by the precise positioning of nucleosomes, chemical modification and physical remodelling of the histones, and also by the incorporation of variant histone proteins.

Two distinct regions of chromatin have been identified within the cell nucleus and these have been termed heterochromatin and euchromatin, as reviewed in [12]. Heterochromatic regions are typically regions where the chromatin fibre is highly condensed; such regions are usually associated with transcriptionally silent genes. Euchromatin on the other hand is less condensed and hence allows greater accessibility of the underlying DNA to gene transcription. The maintenance of euchromatin and heterochromatin regions is highly complex and rather than being maintained as a static structure allows plasticity between the two regions [12].

2.1. Histone Modification

Each of the core histone proteins consists of an inner globular domain into which the DNA is embedded and an N terminal tail region of between 14 and 28 amino acids, which extends beyond the nucleosome complex [10]. The histone proteins are subject to a range of modifications, occurring primarily on the histone tails. There have now been more than 10 forms of histone modification described, the most extensively studied being acetylation, methylation and phosphorylation [11, 13].

Acetylation of histones on lysine residues is strongly correlated with active gene transcription [14], although the particular mechanisms by which histone acetylation facilitates gene expression have yet to be fully elucidated. Initially the correlation between histone acetylation and increased gene transcription was attributed to acetylation neutralising the positive charge of lysine residues within the histone. This is thought to reduce the affinity between the histone tails and the negatively charged DNA on surrounding nucleosomes, thus loosening the overall chromatin structure and enabling transcription from the underlying DNA [15, 16]. Acetylation marks may also serve as recognition sites for bromodomain containing protein complexes. Bromodomains recognise acetylated lysine residues and are contained within a number of chromatin modifying complexes such as histone acetyltrasferases, histone methyltransferases and chromatin remodelling complexes. Thus histone acetylation may mark regions of chromatin for downstream modification and subsequent chromatin remodelling events [17].

Histone methylation can occur on either arginine residues, which may be mono- or di-methylated, or lysine residues which may be mono-, di- or tri- methylated [18]. Methylation is associated with both activation and repression of gene expression depending on the precise location of the amino acid residue on which the methylation occurs, and the number of methylation groups added. This observation led to the proposal of the so-called "histone code" hypothesis, in which histone tail modifications act as receptors to be recognised and bound by other proteins which further modify chromatin structure [11, 14]. For example, histone 3 lysine 27 (H3K27) di- and tri-methylation is well documented as being a repressive histone modification required for mediating gene silencing. This repressive histone modification is recognised and bound by polycomb repressor complex 2 (PCRC2), a methyltransferase which further catalyses the addition of repressive histone modifications, thus maintaining gene silencing throughout processes such as DNA replication [19].

The functional role of many of the histone modifications remains elusive, although it is clear that such modifications can have a profound effect on the overall chromatin structure and function. However, recent data, particularly that generated using genome-wide technologies, has begun to define chromatin signatures that are associated with particular gene states, providing insight into how different chromatin architectures interact with extracellular signals and regulate gene expression.

2.2. Chromatin Remodelling Complexes

In addition to the proteins that modify histones, a group of chromatin remodelling complexes (CRCs) have been described, which utilise the energy released by ATP hydrolysis to physically alter chromatin structure through changes in the association between histone proteins and the DNA double helix. This may result in changes in nucleosome conformation, position, composition and occupancy [20]. These mechanisms provide another layer of complexion to gene regulation by dictating the accessibility of regulatory regions of DNA to transcription factors. The sliding and partial or full disassembly of the core nucleosome proteins by chromatin remodelling complexes may disrupt the DNA/ histone protein interaction, allowing the transcription factors to access and bind DNA regulatory regions and initiate transcription. Likewise re-assembly of the nucleosome complex may conclude gene transcription [21]. It has been proposed that remodelling enzymes may be targeted by interaction with transcription factors, non-coding RNA, histone modifications and DNA damage in addition to functioning independently of targeting [22]. CRCs may be classified into subfamilies by their ATPase subunit including SWI/SNF, ISWI and Mi2, as reviewed in [23]. These families have both complementary and apposing, but distinct roles in regulating gene expression.

2.3. Histone Variants

Chromatin structure can also be modified by the incorporation of variants of the core histone proteins. These variants are so named as they are variants of the major histone genes. They may vary in only a single amino acid or exhibit more significant alterations in their composition. The histone variant H2A.Z is a replacement histone for H2A, the sequence of which varies significantly from that of canonical H2A [24]. H2A.Z appears to have a key role in chromatin structure determination [25]. H3.3 is another frequently occurring histone variant, but in contrast to H2A.Z, H3.3 is highly structurally similar to H3 [26]. The integration of H3.3 into nucleosomes is highly associated with gene transcription [26, 27]. Variant histone proteins have been shown to be enriched at target genes in embryonic stem cells and appear to be necessary for lineage commitment to occur. The incorporation of histone variants has also been implicated as having a role in X inactivation, DNA repair and regulating gene expression [25].

3. Regulation of Immune Gene Expression by Chromatin

As immune responses are transient and involve the coordinated activity of a number of immune cell types, the majority of immune response genes are inducible and many are cell-type specific. Inducible genes are usually maintained in a repressed state with minimal or no basal gene transcription; with gene expression increased when cells receive the appropriate activation signals. Post initiation, a transient period of high level transcription is seen before returning to basal levels as the immune response progresses. The role of chromatin in mediating these responses is now being unravelled through the study of chromatin structure at single gene loci in response to activation signals, as well as through genome-wide analysis of chromatin structure. These studies are revealing how nucleosome positioning and composition, as well as the incorporation of histone variants contribute to cell-type and signal-specific gene expression patterns.

3.1. Nucleosome Positioning and Changes in Nucleosome Occupancy

The chromatin structure established at the regulatory regions of immune genes partially dictates the response of the gene to stimuli and hence contributes to the effectiveness of an immune response. During the process of immune cell differentiation lineage specific chromatin structure is established along with lineage specific gene expression patterns. Variation in chromatin structure between different cell lineages may be visualised as the appearance of lineage specific DNase I hypersensitive (DH) sites [28]. These sites represent regions of DNA that are more sensitive to DNase I and have long been appreciated to coincide with regulatory elements of a gene. Further, the appearance of DH sites has been well documented to coincide with the activation of inducible genes. Inducible DH sites typically represent regions of chromatin which have been remodelled so as to be more accessible to transcription factors and regulatory molecules [29, 30]. Changes in gene transcription are commonly associated with changes in chromatin structure and inducible DH sites have been described at the regulatory regions of a number of different cytokine genes in immune cells in response to appropriate activation signals, as depicted in Figure 1 [30, 31].

Data from both single gene and genome-wide studies suggest that the promoter region of many inducible genes is encompassed within a nucleosome, and that nucleosome rearrangement or disassembly is required to enable the transcription machinery to access the gene promoter and facilitate gene activation [32]. At a single gene level, such chromatin changes have been well described during activation of many inducible genes. The granulocyte macrophage – colony stimulating factor (GM-CSF) gene and closely linked interleukin 3 (IL-3) gene have been well characterised by DH mapping in T cells in order to identify key regulatory elements [30]. Both genes appear to be regulated by proximal promoters located immediately upstream of the transcriptional start site in addition to three enhancers that are identifiable by the induction of DH sites in response to stimulation [33]. Similarly at the interleukin 2 (IL-2) gene, DH sites appear in both the proximal and distal regulatory regions upon stimulation [31]. Further studies revealed that these inducible DH sites at the GM-CSF and IL-2 promoters result from an increase in accessibility of the promoters, due the depletion of histones from regions of the promoter corresponding to one (GM-CSF) or two (IL-2) nucleosomes, in response to activation signals [34-37]. Likewise, micrococcal nuclease analysis of the IL-12 p40 promoter has demonstrated that the selective remodelling of a single positioned nucleosome covering the promoter accompanies its activation in macrophages [38].

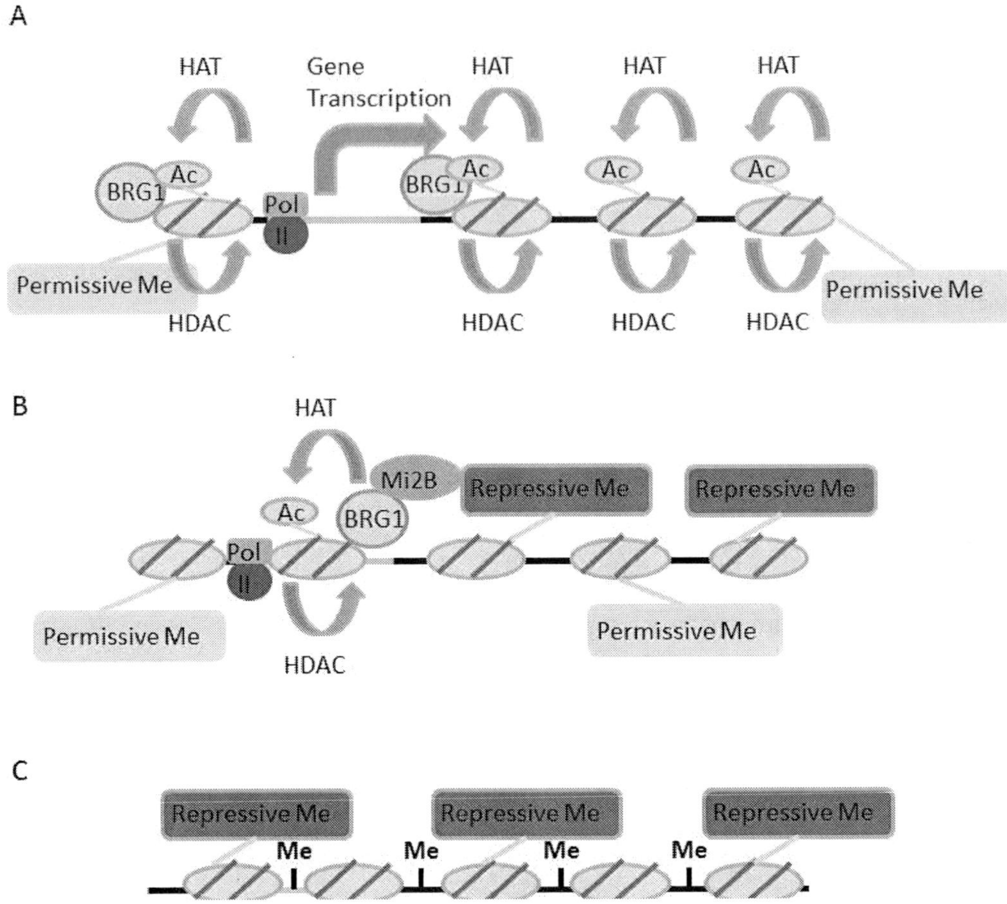

Figure 1. Diagrammatic representation of the chromatin features commonly associated with different states of gene activity. (A) Active genes are characterised by elongating PolII, remodelled nucleosomes, permissive histone modifications, and high activity of HATs and HDACs. (B) Repressed but inducible genes are typically characterised by the presence of stalled RNA Pol II, histone acetylation, a combination of permissive and repressive methylation marks, highly phased nucleosomes and in some cases the presence of chromation remodelling complexes. (C) Silenced genes are characterised by the absence of RNA PolII, lack of nucleosome phasing, absence of acetylation and presence of repressive histone methylation and DNA methylation marks.

Genome-wide analysis of histone occupancy and positioning suggest that the remodelling events observed at these individual cytokine promoters are a relatively common feature of inducible genes. Genome-wide maps of nucleosome positioning in both resting and activated CD4+ T cells have recently been generated using direct sequencing [39], and these data suggest that nucleosomes are highly phased relative to the transcriptional start site (TSS) of expressed genes, but that this phasing is not seen at unexpressed genes [39]. Nucleosome phasing appears to be influenced by surrounding histone features such as the incorporation of histone variants and the presence of poised or elongating RNA Pol II at the TSS [39]. Further, data from this study suggest that many inducible genes are in a permissive basal state in terms of nucleosome positioning but that significant nucleosome reorganisation occurs following TCR stimulation, with a decrease in histone density preferentially occurring at the promoters of activated genes. This remodelling appears to alter the chromatin surrounding the promoters of inducible genes so that they are more accessible to the transcription machinery [39].

3.2. Histone Modification Signatures

While changes in nucleosome occupancy are clearly important in regulating gene expression, the modification state of these nucleosomes also influences gene activity, see Figure 1. Early studies investigating histone acetylation patterns associated with the IL-4 and IFNγ genes during T cell differentiation, indicated that specific histone acetylation patterns where generated at the gene loci during differentiation, establishing permissive chromatin environments [40]. Differentiation of naive T cells down the Th1 or Th2 lineage results in cells which express either IFNγ or IL-4 respectively, correlating with the accumulation of histone acetylation at the active locus. More recently, genome-wide studies utilising high–throughput technologies such as ChIP-chip and ChIP-sequencing, many of which have been carried out in CD4+ T cells, have shed new light on the distribution and role of permissive and repressive histone modifications in regulating gene expression [39, 41-44].

Early genome-wide analysis of histone acetylation in T cells described an association of H3K9ac/K14ac with regulatory regions, including both promoters and enhancers, with the acetylation levels correlating well with gene expression. This discovery that promoter regions are often occupied by highly acetylated histones in comparison to the rest of the genome and the finding that promoters containing unmethylated CpG islands are often marked by highly acetylated histones, led to the proposition of "acetylation islands'. It was hypothesised that histone hyperacetylation may function as a mechanism to prevent DNA methylation and hence is correlated with genes being maintained in an inducible state [43]. Subsequent ChIP-chip experiments in T cells revealed that inducible genes exhibit higher levels of acetylation as well as permissive methylation marks when compared to non-expressive genes, suggesting that permissive histone modifications may act to prime genes for transcriptional response upon activation [44]. Further this study found that early response genes have significantly greater acetylation than delayed response genes, suggesting that the histone modifications may determine not only the ability of a gene to respond to activating signals, but also the kinetics of the gene response. While histone acetylation is generally associated with gene transcription, highlighted by a further genome-wide study that profiled 18 different acetylated lysine residues on the four core histones and demonstrated that all were associated with gene transcription [42], there are also reports of particular histone acetylation marks being associated with gene silencing. For example H4K12ac has been demonstrated to have a role in transcriptional silencing in heterochromatin in yeast [45].

In contrast to histone acetylation, which is largely associated with gene transcription, histone methylation is associated with either activation or repression depending on the particular modification. Genome-wide analysis of the distribution of H3K27me3 and H3K9 methylation suggest these are repressive histone modifications as they are more prevalent at silenced genes compared to active promoters [41]. The association of these marks with a repressive role was confirmed in a further study, along with H3K27me2 and H4K20me3 [42]. However, in addition to marking promoters of inactive genes these modifications are also present in intergenic regions [41, 46]. In contrast, H3K4 methylation appears to be a permissive methylation mark, with H3K4Me3 enriched at the TSS of active genes, and H3K4me1/2 enriched across the transcribed regions of active genes [41]. Similarly, H3K36me3 and H3K79 methylation are associated with actively transcribed genes, although H3K36me3 is enriched at the 3' end, while H3K79 methylation is enriched at the 5' end of these genes. Furthermore, while di- and tri methylation of H3K9, H3K27 and H4K20 are associated with gene repression the monomethylation of these residues is enriched at active genes.

While specific modifications have therefore now become clearly associated with gene activation or gene repression, genome wide studies, initially in embryonic stem cells, have identified genes which carry so-called 'bivalent' marks. These genes are marked by both active (H3K4me3) and repressive (H3K27me3) histone modifications [47-49]. Bivalent marking is suggested to enable a gene to be held in a non-transcribed state by the repressive markers, while it is also poised to respond to appropriate signals by the permissive marks. These bivalent histone modifications were initially thought to be a unique feature of stem cells and to contribute to the pluripotency of these cells, but have subsequently been observed in differentiated cells. It has therefore been suggested that this bivalent chromatin state may be involved in regulating genes that respond to developmental and environmental signals in many cell types [41, 44, 50-52].

The level of the various histone modifications is maintained by the activity of groups of modifying enzymes. Acetylation levels are determined by the opposing activity of histone acetyltransferases (HATs) and histone deacetylases (HDACs), whilst methylation modifications are mediated by histone methyltransferases (HMTs) and demethylases as reviewed in [53, 54]. Thus the activity of these enzymes determines the level of permissive and repressive markers at each gene. Recent ChIP-sequencing studies in T cells have profiled the association of HATs with gene promoters and analysis with gene expression data reveals that, as may be expected, a high level of HAT activity correlates with higher levels of acetylation, in addition to elevated gene transcription. In addition HAT binding is positively correlated with the binding of RNA Pol II at active genes [55]. Surprisingly, ChIP-sequencing revealed that HDACs are also correlated with active gene transcription, with only a small fraction associated with silent genes. These results suggest that rather than having a primary role in maintaining a silent state, the main role of HDACs may be the removal of acetyl groups added by HATs at active genes. HDACs are recruited to actively transcribed genes by interaction with elongating RNA Pol II and may therefore play an important role in resetting chromatin following remodelling [55]. Examining the binding of HATs and HDACs to transcriptionally silent genes reveals that these genes are also subject to constant acetylation and deacetylation cycles mediated by the transient and unstable binding of HATs and HDACs. Thus like 'bivalent' chromatin marks, the activity of HATs and HDACs at repressed genes poises these genes for future activation [55], see Figure 1. Furthermore, activity of HATs and HDACs at these 'poised' genes appears to be further modulated by the presence of other chromatin modifying factors including permissive H3K4 methylation and the histone variant protein H2A.Z [55].

Ten years ago the histone code hypothesis was proposed to explain how specific histone modifications may control gene activity [56], but at that time no-one could have imagined the enormous complexity of histone modifications that exists. For example, genome-wide analysis of 38 histone modifications in T cells has identified a number of common combinatorial patterns, including a 'backbone' consisting of 17 modifications, commonly associated with gene promoters [42]. Furthermore, mass spectrometric analysis has detected more than 150 modification combinations associated with histone H3.2 in HeLa cells [57]. Such analyses have revealed the incredible complexity of histone modifications, but further work is required to elucidate the functional consequences of specific modifications and particularly the combinatorial effects. While specific modifications are associated with particular gene states, it is also apparent that particular histone modification patterns can be associated with quite diverse gene states. Further, the nature of the 'histone code' remains a matter of conjecture, with considerable on-going debate regarding whether histone modifications actually regulate gene expression or are instead 'markers' associated with particular transcriptional states generated by sequence-specific transcription factors.

3.3. Variant Histone Proteins in Inducible Gene Expression

The incorporation of variant histone proteins further contributes to the diversity of overall chromatin architecture, as reviewed in [58]. Of the many different histone variants, H2A.Z and H3.3 are the most extensively studied. H2A.Z associates with functional regulatory elements and thus is proposed to contribute to the regulation of gene expression. Studies in yeast have found that the H2A.Z homologue Hzt1 is preferentially associated with inactive promoters [59], but depleted from active promoters, with H2A.Z containing and modified nucleosomes preferentially lost from the promoter regions in response to stimulation [39, 41, 60]. Stimulation of T cells via the T cell receptor is accompanied by nucleosome reorganisation at regulatory enhancer and promoter regions, and it is thought that H2A.Z incorporation may aid in the nucleosome repositioning or eviction of these histones from promoter regions. In support of this, H2A.Z containing and modified histones appear to be preferentially lost from around the TSS [39]. Importantly, H2A.Z has been reported to interact extensively with other permissive factors including RNA Pol II and histone acetylation [39, 41, 55].

Histone variant H3.3 is also reported to be enriched at active genes in yeast where it is documented to replace H3 whilst active gene transcription is taking place [61]. Furthermore the localisation of both H2A.Z and H3.3 variants into the same nucleosome results in nucleosome instability which may facilitate transcription machinery assembly at promoter regions occupied by H3.3/H2A.Z double variant nucleosomes [62]. Genome-wide analysis of H3.3/H2A.Z nucleosome distribution in epithelial derived HeLa cells revealed that such double variant nucleosomes are relatively common throughout the genome, often enriched at regions such as promoters and enhancers which have previously been described as 'nucleosome-free', as under normal experimental conditions these nucleosomes are displaced from the DNA [62]. Further, a recent study of the highly inducible CD69 and heparanase genes in T cells found that chromatin remodelling events occurring at these promoters upon gene activation do not involve nucleosome depletion, but rather exchange of histone variants, with depletion of H2A.Z and deposition of H3.3 [63].

3.4. Chromatin Remodelling Complexes (CRCs) in Inducible Gene Expression

Chromatin remodelling complexes (CRCs) are commonly required to mediate the changes in chromatin accessibility required for transcriptional activation of inducible genes. CRCs utilise ATP-hydrolysis to reposition, assemble or dismantle nucleosome complexes from the regulatory regions of genes [23, 64]. However chromatin remodelling is not required for activation of all genes. Genes that respond rapidly to immune stimuli often exhibit high sensitivity to nucleases at regulatory regions prior to stimulation. Further analysis revealed that this is due to low nucleosome occupancy at these promoters [65]. These genes therefore are able to increase gene expression independent of chromatin remodelling as the promoter is subject to less nucleosome interference with transcription factor association. Genome- wide analysis of BRG1 (an ATPase component of the SWI/SNF complex) binding in non-stimulated and stimulated T cells in conjunction with mRNA gene expression analysis reveals that BRG1 is important for the response to cell activating signals and that BRG1 binding is positively correlated with gene transcription [66]. Knockdown of the BRG1/BRM complex in murine macrophages has shown that a subset of inducible genes require SWI/SNF for activation. Analysis of these genes revealed a strong reliance of inducible genes on SWI/SNF for activation [65]. The recruitment of BRG1 to these genes is reported to be mediated by the presence of permissive histone acetylation marks [66]. It is therefore apparent that for chromatin remodelling to occur via the recruitment of BRG1 the chromatin environment established at the gene must be appropriate.

SNF2H is an ISWI chromatin remodelling complex which is abundantly expressed in immune T cells. Knockdown of SNF2H revealed that SNF2H has a regulatory role at a number of cytokine genes, typically repressing gene transcription, but at some cytokine genes such as IL-3, increasing transcription [67]. SNF2H appears to mediate changes in chromatin structure by directly binding in association with its partners including ACF1 to regulatory sites; however cell stimulation seems to have minimal influence on SNF2H binding, which remains comparatively uniform across cytokine loci. Knockdown experiments have demonstrated that SNF2H is crucial in mediating development and as there is little evidence of dynamic recruitment of the remodeller, it appears that SNF2H association with chromatin at cytokine genes may be initiated during differentiation [67].

Mi2/NuRD complexes interact with DNA-binding proteins such as Ikaros which are implicated in repression as well as HDACs and have been shown to contribute to transcriptional repression, however the complex has also been implicated in transcriptional activation [68]. In macrophage cells the Mi2β/NuRD complex associates with pro-inflammatory cytokine genes in a similar way to the BRG1 SWI/SNF remodeller. Knockdown of Mi2β revealed that Mi2β acts as an antagonist of gene transcription, moderating the response of late primary and early secondary response genes. When BRG1 is depleted the association of Mi2β with cytokine genes is also depleted, suggesting that Mi2β is dependent on BRG1 for recruitment to regulatory regions to occur [68].

3.5. Interferon β: A Model Inducible Gene

While genome-wide studies have provided considerable insight into the role of chromatin in regulating inducible gene expression, a mechanistic understanding of the regulatory processes has primarily come from the analysis of a small number of inducible genes in immune cell models and several inducible genes in yeast. One of the most extensively studied cytokines in terms of understanding the contribution of chromatin to its regulation is interferon beta (IFNβ). While the IFNβ enhancer is contained in nucleosome-free linker region, a nucleosome is positioned over the IFNβ promoter. In response to activation signals transcription factors bind to their binding sites in the enhancer, forming an enhanceosome complex which recruits cofactors with histone acetyltransferase activity to acetylate a nucleosome positioned over the TATA box at the IFNβ promoter [69]. This increased acetylation mediates binding of BRG1, the ATPase component of the SWI/SNF chromatin remodelling complex [70]. Recruitment of the SWI/SNF complex results in the repositioning of the promoter nucleosome downstream of the TATA box, permitting assembly of the transcription machinery at the promoter [70, 71]. Whilst the recruitment of BRG1 is required for nucleosome repositioning to occur, the nucleosome retains the remodelled position for a prolonged period despite BRG1 being present for a short time [70].

CONCLUSIONS

Analysis of the IFNβ gene has provided considerable insight into the chromatin remodelling events required for activation of an inducible gene, and the basic concepts are likely to be applicable to other genes. Genome-wide studies, discussed above have demonstrated that enrichment of permissive histone modifications, and recruitment of chromatin remodelling proteins to increase accessibility for the transcription machinery is a common theme in inducible gene transcription. However, activation of many other inducible genes is likely to be mechanistically different to that of the IFNβ gene. In contrast to the IFNβ enhancer where the transcription factor binding sites are in a nucleosome-free region prior to gene activation, for most of the other inducible genes that have been studied in detail, the regulatory elements are encompassed within nucleosomes. Activation of the IL-2 and GM-CSF genes, as well as the IL-12 p40 gene requires remodelling of nucleosomes encompassing the gene promoters, as discussed above. Activation of these genes appears similar to the activation of the well-described yeast Pho5 and Pho8 genes, where stimulus-driven increases in histone acetylation across the promoter facilitates binding of SWI/SNF complexes, resulting in depletion of histones from the promoter region [72-74]. It therefore seems apparent that while genome-wide studies are beginning to generate rules which govern regulation of gene expression by chromatin, the mechanistic details are likely to be at least subtly different from gene to gene.

The advent of genome-wide technologies to study chromatin architecture has provided remarkable insight into the multitude of chromatin factors which mediate gene activity, including histone modifications, activity of modifying enzymes, the presence and state of RNA polymerase II and the incorporation of histone variants. Such analysis has provided general rules which govern the regulation of gene expression, and are deciphering chromatin signatures associated with particular gene states. These genome-wide studies continue to be complimented by gene-centric studies which are uncovering the key mechanisms that regulate gene expression, and how particular chromatin features impact gene activity.

References

[1] Wilson, C. B., Rowell, E. & Sekimata, M. (2009) Epigenetic control of T-helper-cell differentiation. *Nat Rev Immunol*, *9*, 91-105.

[2] Venters, B. J. & Pugh, B. F. (2009) How eukaryotic genes are transcribed. *Crit Rev Biochem Mol Biol*, *44*, 117-41.

[3] Tagoh, H., Melnik, S., Lefevre, P., Chong, S., Riggs, A. D. & Bonifer, C. (2004) Dynamic reorganization of chromatin structure and selective DNA demethylation prior to stable enhancer complex formation during differentiation of primary hematopoietic cells in vitro. *Blood*, *103*, 2950-5.

[4] Wurster, A. L. & Pazin, M. J. (2008) BRG1-Mediated Chromatin Remodeling Regulates Differentiation and Gene Expression of T Helper Cells. *Mol Cell Biol*, *28*, 7274-85.

[5] Holloway, A. F., Rao, S. & Shannon, M. F. (2002) Regulation of cytokine gene transcription in the immune system. *Mol Immunol*, *38*, 567-80.

[6] Smale, S. T. & Fisher, A. G. (2002) Chromatin structure and gene regulation in the immune system. *Annual Review of Immunology*, *20*, 427-62.

[7] Kornberg, R. D. & Lorch, Y. L. (1999) Twenty-five years of the nucleosome, fundamental particle of the eukaryote chromosome. *Cell*, *98*, 285-94.

[8] Lorch, Y., Maier-Davis, B. & Kornberg, R. D. (2006) Chromatin remodeling by nucleosome disassembly in vitro. *Proc Natl Acad Sci U S A*, *103*, 3090-3.

[9] Tremethick, D. J. (2007) Higher-order structures of chromatin: The elusive 30 nm fiber. *Cell*, *128*, 651-4.

[10] McBryant, S. J., Adams, V. H. & Hansen, J. C. (2006) Chromatin architectural proteins. *Chromosome Res*, *14*, 39-51.

[11] Gardner, K. E., Allis, C. D. & Strahl, B. D. (2011) OPERating ON Chromatin, a Colorful Language where Context Matters. *J Mol Biol*.

[12] Tamaru, H. (2010) Confining euchromatin/heterochromatin territory: jumonji crosses the line. *Genes & Development*, *24*, 1465-78.

[13] Kouzarides, T. (2007) Chromatin modifications and their function. *Cell*, *128*, 693-705.

[14] Strahl, B. D. & Allis, C. D. (2000) The language of covalent histone modifications. *Nature*, *403*, 41-5.

[15] Grunstein, M. (1997) Histone acetylation in chromatin structure and transcription. *Nature*, *389*, 349-52.

[16] Shogren-Knaak, M., Ishii, H., Sun, J. M., Pazin, M. J., Davie, J. R. & Peterson, C. L. (2006) Histone H4-K16 acetylation controls chromatin structure and protein interactions. *Science*, *311*, 844-7.

[17] de la Cruz, X., Lois, S., Sanchez-Molina, S. & Martinez-Balbas, M. A. (2005) Do protein motifs read the histone code? *Bioessays*, *27*, 164-75.

[18] Rice, J. C., Briggs, S. D., Ueberheide, B., Barber, C. M., Shabanowitz, J., Hunt, D. F., *et al.* (2003) Histone methyltransferases direct different degrees of methylation to define distinct chromatin domains. *Molecular Cell, 12*, 1591-8.

[19] Xu, C., Bian, C. B., Yang, W., Galka, M., Hui, O. Y., Chen, C., *et al.* (2010) Binding of different histone marks differentially regulates the activity and specificity of polycomb repressive complex 2 (PRC2). *Proc Natl Acad Sci U S A, 107*, 19266-71.

[20] Li, B., Carey, M. & Workman, J. L. (2007) The role of chromatin during transcription. *Cell, 128*, 707-19.

[21] Kornberg, R. D. & Lorch, Y. (2007) Chromatin rules. *Nat Struct Mol Biol, 14*, 986-8.

[22] de la Serna, I. L., Ohkawa, Y. & Imbalzano, A. N. (2006) Chromatin remodelling in mammalian differentiation: lessons from ATP-dependent remodellers. *Nature Reviews Genetics*, *7*, 461-73.

[23] Lusser, A. & Kadonaga, J. T. (2003) Chromatin remodeling by ATP-dependent molecular machines. *Bioessays*, *25*, 1192-200.

[24] Bonenfant, D., Coulot, M., Towbin, H., Schindler, P. & van Oostrum, J. (2006) Characterization of histone H2A and H2B variants and their post-translational modifications by mass spectrometry. *Mol Cell Proteomics*, *5*, 541-52.

[25] Creyghton, M. P., Markoulaki, S., Levine, S. S., Hanna, J., Lodato, M. A., Sha, K., *et al.* (2008) H2AZ Is Enriched at Polycomb Complex Target Genes in ES Cells and Is Necessary for Lineage Commitment. *Cell, 135*, 649-61.

[26] Szenker, E., Ray-Gallet, D. & Almouzni, G. (2011) The double face of the histone variant H3.3. *Cell Res*, *21*, 421-34.

[27] Kafer, G. R., Lehnert, S. A., Pantaleon, M., Kaye, P. L. & Moser, R. J. (2010) Expression of genes coding for histone variants and histone-associated proteins in pluripotent stem cells and mouse preimplantation embryos. *Gene Expression Patterns*, *10*, 299-305.

[28] Hawwari, A., Burrows, J., Vadas, M. A. & Cockerill, P. N. (2002) The human IL-3 locus is regulated cooperatively by two NFAT-dependent enhancers that have distinct tissue-specific activities. *Journal of Immunology*, *169*, 1876-86.

[29] Takemoto, N., Kamogawa, Y., Lee, H. J., Kurata, H., Arai, K., O'Garra, A., *et al.* (2000) Cutting edge: Chromatin remodeling at the IL-4/IL-13 intergenic regulatory region for Th2-specific cytokine gene cluster. *Journal of Immunology*, *165*, 6687-91.

[30] Cockerill, P. N. (2004) Mechanisms of transcriptional regulation of the human IL-3/GM-CSF locus by inducible tissue-specific promoters and enhancers. *Critical Reviews in Immunology*, *24*, 385-408.

[31] Ward, S. B., Hernandez-Hoyos, G., Chen, F., Waterman, M., Reeves, R. & Rothenberg, E. V. (1998) Chromatin remodeling of the interleukin-2 gene: distinct alterations in the proximal versus distal enhancer regions. *Nucleic Acids Res*, *26*, 2923-34.

[32] Workman, J. L. (2006) Nucleosome displacement in transcription. *Genes & Development*, *20*, 2009-17.

[33] Cockerill, P. N., Shannon, M. F., Bert, A. G., Ryan, G. R. & Vadas, M. A. (1993) The Granulocyte Macrophage Colony-Stimulating Factor Interleukin-3 Locus Is Regulated by an Inducible Cyclosporin A-Sensitive Enhancer. *Proc Natl Acad Sci U S A*, *90*, 2466-70.

[34] Rao, S. H., Procko, E. & Shannon, M. F. (2001) Chromatin remodeling, measured by a novel real-time polymerase chain reaction assay, across the proximal promoter region of the IL-2 gene. *Journal of Immunology, 167*, 4494-503.

[35] Holloway, A. F., Rao, S., Chen, X. X. & Shannon, M. F. (2003) Changes in chromatin accessibility across the GM-CSF promoter upon T cell activation are dependent on nuclear factor kappa B proteins. *Journal of Experimental Medicine, 197*, 413-23.

[36] Chen, X. X., Wang, J., Woltring, D., Gerondakis, S. & Shannon, M. F. (2005) Histone dynamics on the interleukin-2 gene in response to T-Cell activation. *Mol Cell Biol, 25*, 3209-19.

[37] Brettingham-Moore, K. H., Sprod, O. R., Chen, X., Oakford, P., Shannon, M. F. & Holloway, A. F. (2008) Determinants of a transcriptionally competent environment at the GM-CSF promoter. *Nucleic Acids Res, 36*, 2639-53.

[38] Weinmann, A. S., Plevy, S. E. & Smale, S. T. (1999) Rapid and selective remodeling of a positioned nucleosome during the induction of IL-12 p40 transcription. *Immunity, 11*, 665-75.

[39] Schones, D. E., Cui, K. R., Cuddapah, S., Roh, T. Y., Barski, A., Wang, Z. B., *et al.* (2008) Dynamic regulation of nucleosome positioning in the human genome. *Cell, 132*, 887-98.

[40] Avni, O., Lee, D., Macian, F., Szabo, S. J., Glimcher, L. H. & Rao, A. (2002) T-H cell differentiation is accompanied by dynamic changes in histone acetylation of cytokine genes. *Nature Immunology, 3*, 643-51.

[41] Barski, A., Cuddapah, S., Cui, K. R., Roh, T. Y., Schones, D. E., Wang, Z. B., *et al.* (2007) High-resolution profiling of histone methylations in the human genome. *Cell, 129*, 823-37.

[42] Wang, Z. B., Zang, C. Z., Rosenfeld, J. A., Schones, D. E., Barski, A., Cuddapah, S., *et al.* (2008) Combinatorial patterns of histone acetylations and methylations in the human genome. *Nature Genetics, 40*, 897-903.

[43] Roh, T. Y., Cuddapah, S. & Zhao, K. (2005) Active chromatin domains are defined by acetylation islands revealed by genome-wide mapping. *Genes & Development, 19*, 542-52.

[44] Lim, P. S., Hardy, K., Bunting, K. L., Ma, L., Peng, K. M., Chen, X. X., *et al.* (2009) Defining the chromatin signature of inducible genes in T cells. *Genome Biology, 10.*

[45] Braunstein, M., Sobel, R. E., Allis, C. D., Turner, B. M. & Broach, J. R. (1996) Efficient transcriptional silencing in Saccharomyces cerevisiae requires a heterochromatin histone acetylation pattern. *Mol Cell Biol, 16*, 4349-56.

[46] Roh, T. Y., Cuddapah, S., Cui, K. R. & Zhao, K. J. (2006) The genomic landscape of histone modifications in human T cells. *Proc Natl Acad Sci U S A, 103*, 15782-7.

[47] Azuara, V., Perry, P., Sauer, S., Spivakov, M., Jorgensen, H. F., John, R. M., *et al.* (2006) Chromatin signatures of pluripotent cell lines. *Nature Cell Biology, 8*, 532-U189.

[48] Bernstein, B. E., Mikkelsen, T. S., Xie, X. H., Kamal, M., Huebert, D. J., Cuff, J., *et al.* (2006) A bivalent chromatin structure marks key developmental genes in embryonic stem cells. *Cell, 125*, 315-26.

[49] Mikkelsen, T. S., Ku, M. C., Jaffe, D. B., Issac, B., Lieberman, E., Giannoukos, G., *et al.* (2007) Genome-wide maps of chromatin state in pluripotent and lineage-committed cells. *Nature, 448*, 553-U2.

[50] Araki, Y., Wang, Z. B., Zang, C. Z., Wood, W. H., Schones, D., Cui, K. R., *et al.* (2009) Genome-wide Analysis of Histone Methylation Reveals Chromatin State-Based Regulation of Gene Transcription and Function of Memory CD8(+) T Cells. *Immunity, 30*, 912-25.

[51] Cui, K. R., Zang, C. Z., Roh, T. Y., Schones, D. E., Childs, R. W., Peng, W. Q., *et al.* (2009) Chromatin Signatures in Multipotent Human Hematopoietic Stem Cells Indicate the Fate of Bivalent Genes during Differentiation. *Cell Stem Cell, 4*, 80-93.

[52] Rodriguez, J., Munoz, M., Vives, L., Frangou, C. G., Groudine, M. & Peinado, M. A. (2008) Bivalent domains enforce transcriptional memory of DNA methylated genes in cancer cells. *Proc Natl Acad Sci U S A, 105*, 19809-14.

[53] Couture, J. F. & Trievel, R. C. (2006) Histone-modifying enzymes: encrypting an enigmatic epigenetic code. *Current Opinion in Structural Biology, 16*, 753-60.

[54] Smith, B. C. & Denu, J. M. (2009) Chemical mechanisms of histone lysine and arginine modifications. *Biochim Biophys Acta-Gene Regul Mech, 1789*, 45-57.

[55] Wang, Z. B., Zang, C. Z., Cui, K. R., Schones, D. E., Barski, A., Peng, W. Q., *et al.* (2009) Genome-wide Mapping of HATs and HDACs Reveals Distinct Functions in Active and Inactive Genes. *Cell, 138*, 1019-31.

[56] Jenuwein, T. & Allis, C. D. (2001) Translating the histone code. *Science, 293*, 1074-80.

[57] Garcia, B. A., Pesavento, J. J., Mizzen, C. A. & Kelleher, N. L. (2007) Pervasive combinatorial modification of histone H3 in human cells. *Nature Methods, 4*, 487-9.

[58] Henikoff, S. & Ahmad, K. (2005) Assembly of variant histones into chromatin. *Annu Rev Cell Dev Biol, 21*, 133-53.

[59] Guillemette, B., Bataille, A. R., Gevry, N., Adam, M., Blanchette, M., Robert, F., *et al.* (2005) Variant histone H2A.Z is globally localized to the promoters of inactive yeast genes and regulates nucleosome positioning. *Plos Biology, 3*, 2100-10.

[60] Zhang, H. Y., Roberts, D. N. & Cairns, B. R. (2005) Genome-wide dynamics of Htz1, a histone H2A variant that poises repressed/basal promoters for activation through histone loss. *Cell, 123*, 219-31.

[61] Ahmad, K. & Henikoff, S. (2002) The histone variant H3.3 marks active chromatin by replication-independent nucleosome assembly. *Molecular Cell, 9*, 1191-200.

[62] Jin, C. Y., Zang, C. Z., Wei, G., Cui, K. R., Peng, W. Q., Zhao, K. J., *et al.* (2009) H3.3/H2A.Z double variant-containing nucleosomes mark 'nucleosome-free regions' of active promoters and other regulatory regions. *Nature Genetics, 41*, 941-U112.

[63] Sutcliffe, E. L., Parish, I. A., He, Y. Q., Juelich, T., Tierney, M. L., Rangasamy, D., *et al.* (2009) Dynamic Histone Variant Exchange Accompanies Gene Induction in T Cells. *Mol Cell Biol, 29*, 1972-86.

[64] Saha, A., Wittmeyer, J. & Cairns, B. R. (2006) Chromatin remodelling: the industrial revolution of DNA around histones. *Nature Reviews Molecular Cell Biology, 7*, 437-47.

[65] Ramirez-Carrozzi, V. R., Braas, D., Bhatt, D. M., Cheng, C. S., Hong, C., Doty, K. R., *et al.* (2009) A Unifying Model for the Selective Regulation of Inducible Transcription by CpG Islands and Nucleosome Remodeling. *Cell, 138*, 114-28.

[66] De, S., Wurster, A. L., Precht, P., Wood, W. H., Becker, K. G. & Pazin, M. J. (2011) Dynamic BRG1 recruitment during T helper differentiation and activation reveals distal regulatory elements. *Mol Cell Biol.*

[67] Precht, P., Wurster, A. L. & Pazin, M. J. (2010) The SNF2H chromatin remodeling enzyme has opposing effects on cytokine gene expression. *Mol Immunol, 47*, 2038-46.

[68] Ramirez-Carrozzi, V. R., Nazarian, A. A., Li, C. C., Gore, S. L., Sridharan, R., Imbalzano, A. N., *et al.* (2006) Selective and antagonistic functions of SWI/SNF and Mi-2 beta nucleosome remodeling complexes during an inflammatory response. *Genes & Development*, *20*, 282-96.

[69] Merika, M., Williams, A. J., Chen, G. Y., Collins, T. & Thanos, D. (1998) Recruitment of CBP/p300 by the IFN beta enhanceosome is required for synergistic activation of transcription. *Molecular Cell*, *1*, 277-87.

[70] Agalioti, T., Lomvardas, S., Parekh, B., Yie, J. M., Maniatis, T. & Thanos, D. (2000) Ordered recruitment of chromatin modifying and general transcription factors to the IFN-beta promoter. *Cell*, *103*, 667-78.

[71] Lomvardas, S. & Thanos, D. (2002) Modifying gene expression programs by altering core promoter chromatin architecture. *Cell*, *110*, 261-71.

[72] Barbaric, S., Gregory, P. D., Munsterkotter, H., Korica, T. & Horz, W. (1999) Regulation of the yeast PHO5 and PHO8 genes: A network of regulatory proteins, transcription factors and chromatin. *Food Technology and Biotechnology*, *37*, 1-8.

[73] Barbaric, S. & Horz, W. (1998) Transcriptional regulation of the yeast PHO5 gene: a complex interplay of protein-protein and protein-DNA interactions. *Periodicum Biologorum*, *100*, 345-52.

[74] Gregory, P. D., Barbaric, S. & Horz, W. (2000) Transcriptional control of phosphate-regulated genes in yeast: the role of specific transcription factors and chromatin remodeling complexes in vivo. *Food Technology and Biotechnology*, *38*, 295-303.

In: Histones: Class, Structure and Function
Editor: Chang-Hui Shen
ISBN: 978-1-62100-274-1

Chapter 11

THE CHROMATIN STRUCTURE OF RIBOSOMAL DNA

Benoît Leblanc*
Département de Biologie, Faculté des Sciences, Université de Sherbrooke
Sherbrooke, QC, Canada

ABSTRACT

The production of ribosomal RNA (rRNA) accounts for more than half the transcriptional activity of an actively growing eukaryotic cell. Not surprisingly given its importance to the cell's well-being, the entire system of rRNA production is subject to several levels of regulation that often distinguish it from those of the protein-coding genes. The regulation of rRNA gene expression occurs at many levels, including that of chromatin structure. Our current understanding of rRNA gene chromatin (r-chromatin) in higher eukaryotes is that it exists in several forms ranging from a nucleosomal and silenced chromatin in which promoter DNA is methylated at several positions to a non-nucleosomal, methylation-poor chromatin associated with the ribosomal transcription factor UBF. The roughly equal ratio of silenced to active rDNA copies usually changes very little between resting and actively growing cells, and most of the expression regulation is due to the rate of transcription initiation and elongation at the active rDNA copies. This peculiar observation suggests that the organization of r-chromatin plays an important role in more than gene expression regulation, and is likely to involve the stability of the recombination-prone ribosomal RNA genes. In this chapter, we describe different structural aspects of the r-chromatin and their possible relation to the regulation of its expression.

1. INTRODUCTION: THE NUCLEOLUS, A RIBOSOME FACTORY

Most people who have attended a biology class will remember having had to look at a cell through a microscope and being tasked to picture what they saw. In the majority of cases, their first representation of the most basic unit of life would have looked somewhat like three

* Email: benoit.leblanc@usherbrooke.ca.

concentric circles: a large one representing the cell membrane, a smaller one representing the nucleus and, within this, a third, probably shaded in black, representing a visible but quickly dismissed structure... the nucleolus. Most budding biologists could be excused for overlooking the importance of this nuclear particularity. After all, their attention was diverted by the many other mysteries elsewhere in the field, and only a few would later have the privilege to learn more about this diminutive black spot despite its incontestable importance to the existence of the cell.

A lot can be said about this nuclear feature, whose unassuming aspect belies its frantic activity. To begin with, the nucleolus is not necessarily alone in the nucleus and it is not infrequent to observe multiple nucleoli in a growing cell; indeed, several hundred can be seen in a single amphibian oocyte. The nucleolus is also remarkably protean and dynamic, as the most artistically prone of students will doubtless have noticed; mutliple nucleoli can even be seen to fuse to form larger ones. This is doubtless made easier by the fact that nucleoli are not constrained by a membrane but are formed by a local assemblage of factors that include parts of certain chromosomes, RNA molecules and numerous proteins. All these agents meet to form the scaffold of a molecular factory that has the responsibility of producing ribosomes, the machines that manage the not inconsequential feat of turning genetic information into actual protein. Considering the importance of this task, it comes as no wonder that formation of the nucleolus entails the use of highly regulated mechanisms. Among these mechanisms, we must count the dynamic chromatin structure of the genes responsible for nucleolus formation. These genes form a locus distiguishable as a band in the small arm of the human acrocentric chromosomes. And because these genes nucleate the formation of the nucleolus, their locus is appropriately referred to as the nucleolar organizer region or NOR. [1].

In human cells, NORs can be found on the short arm of the five acrocentric chromosomes (chromosomes 13, 14, 15, 21 and 23) [2]. In other evolutionarily very distant model organism like the yeast *S. cerevisiae* or the amphibian *Xenopus laevis*, there is however only one NOR lying on a single chromosome. Each NOR consists of a repeated cluster of a single gene that encodes, within a single transcript, three of the four structural RNAs that, upon maturation and association with over 80 proteins, will form the ribosome.

Considering their connection to the very purpose of the nucleolus, let us devote a little time to these RNAs of the ribosome (rRNAs). There are four of them altogether, and they are designated by their sedimentation coefficients as 5S, 5.8S, 18S and 28S [3]. All but the 5S RNA are transcribed together as a precursor molecule, called the pre-rRNA, in which the sequences of the 18S, 5.8S and 28S RNAs follow each other in tandem. Upon maturation, the pre-rRNA (now part of a large pre-ribosome) is cleaved into the three rRNA molecules, releasing the large and small subunits of the ribosome. The pre-rRNA, which in mammals is refered to as the 45S transcript, is transribed by a dedicated polymerase, RNA polymerase I (POL I). This dedication is well suited to such a hard worker, as more than half of the total RNA production in a rapidly growing cell can be attributed to POL I [4-6]. For simplicity's sake, the 45S pre-rRNA genes are often referred to as simply "rDNA", and they are the ones found in the NORs.

The remaining rRNA molecule required for the ribosome, designated 5S, is transcribed by another RNA polymerase (RNA Polymerase III), which is also responsible for the production of tRNAs and several other low molecular weight RNAs. 5S rRNA genes are not found in the same chromosomal position as the 45S gene in vertebrate cells; for example, *in situ* hybridization localized the multiple copies of the human 5S rRNA gene to sites on chromosome 1 [7]. Unlike the repeated 45S genes, the 5S gene repeats do not form a NOR on their own.

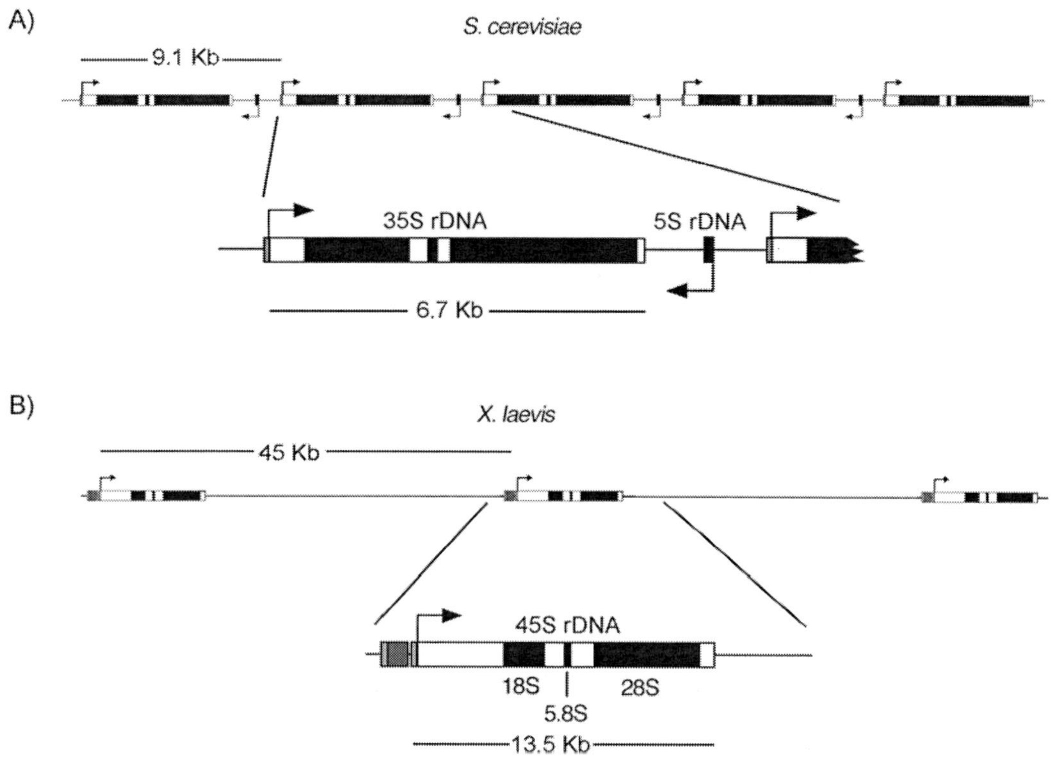

Figure 1. The repeated structure of the rDNA gene loci is conserved in eukaryotes. Solid boxes represent the sequence of the mature rRNAs while open boxes represent transcribed sequences that are lost upon maturation. A) *S. cerevisiae* rDNA locus. B) *X. laevis* rDNA locus, typical of that of most multicellular eukaryotes.

2. The Ribosomal RNA Genes Are Repeated

Saul may have had his thousands and David his tens of thousands, but when it comes to ribosomes the cell may well have nothing less than its millions. No wonder either, since actively growing and proliferating cells require a constant production of proteins and a proportionate number of ribosomes. To meet demand, one rDNA gene would probably not be sufficient; it is therefore one of the few heavily repeated genes in the genome. In the haploid human genome, for example, we can find roughly 200 copies of the rDNA gene. These copies are organized as repeated arrays mostly oriented in tandem and divided between the five chromosomal locations mentioned above. Each gene in a repeat array is separated from its neighbour by an intergenic spacer [8,9]. Transcription at any of these loci is accompanied by a secondary constriction of the acrocentric arm and by a ten-fold decondensation of chromatin relative to neighboring heterochromatic sequences [10]. It is interesting to note that any active rDNA gene can act as a NOR, even from an ectopic position. This is made very clear in the amphibian oocyte, in which hundreds of nucleoli can be found. These multiple nucleoli form not only on the chromosomal rDNA copies, but also on hundreds of free-floating episomes that originate from the selective amplification of the chromosomal rDNA genes [11] . Figure 1 depicts the arrangement of rDNA genes in yeast and in vertebrates.

Within a particular rDNA repeat, most genes are organized in a head-to-tail manner, although some have been shown to adopt a palindromic orientation [9]. In between the genes, the size and organization of the rDNA intergenic spacers can vary from species to species and even from gene to gene. Interestingly, in some organisms such as *S. cerevisiae* but also in more complex eukaryotes like the copepod *C. finmarchicus*, the intergenic spacer contains a copy of the 5S gene, in reverse orientation, depite its being transcribed by POL III [8,12]. The intergenic spacer contains several regulatory elements such as transcription enhancers and terminators, as well as origins of replication and replication fork barriers. Intriguingly, it is frequent to encounter one or many functional duplications of the rDNA promoter the intergenic region. The role of these promoters, which are not connected to an actual gene, has been the object of several interpretations; however, they definitely seem to play a role in enhancing rDNA gene expression [13-15]. Figure 2 depicts the intergenic spacers of yeast and mouse.

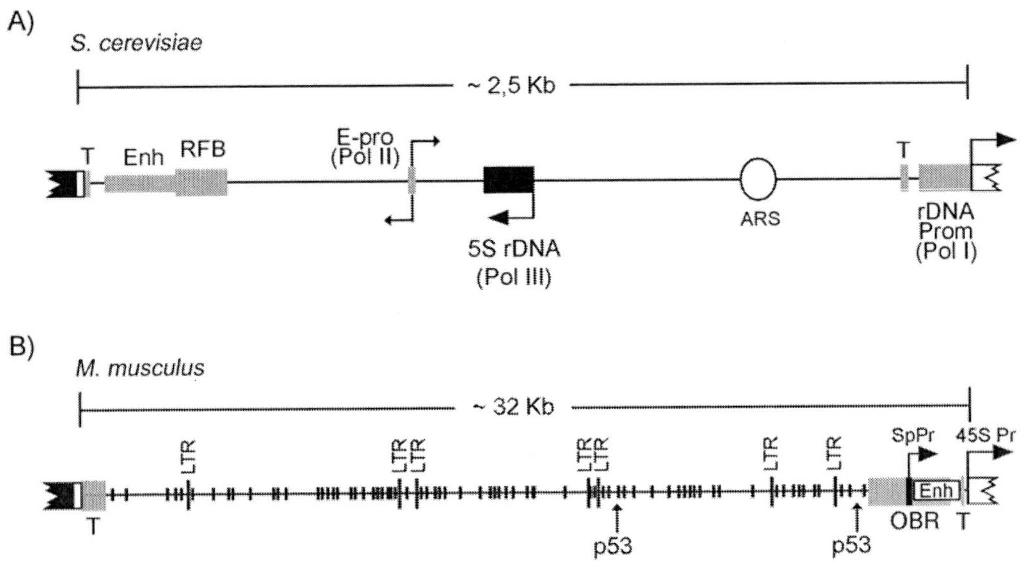

Figure 2. Structure of the rDNA intergenic spacer in yeast and in mouse. (A) In yeast, the relatively short intergenic spacer between each repetition of the rDNA gene contains a copy of the 5S rRNA gene, which is transcribed by POL III. (B) In mouse, the intergenic spacer does not contain the 5S gene but does contain several Alu repeats that might act as POL III promoters. The spacer also contains duplications of the promoter and enhancer regions (represented as gray boxes immediately upstream of the 45S rRNA gene). The position of the different elements was obtained from the Saccharomyces cerevisiae genome database [108] and Genbank accession number BK000964. T: transcription terminator. Enh: enhancer sequence. RFB: replication fork barrier. E-Pro (Pol II): bidirectional, non-coding Pol II promoter. 5S rDNA: transcription unit for the 5S rRNA gene. ARS: autonomous replication sequence. Prom: Promoter. LTR: Long Terminal repeat (retrotransposon integration mark). P53: binding site for p53. SpPr: Spacer Promoter. OBR: origin of bidirectional replication. 45S Pr: 45S rRNA promoter.

3. ACTIVE AND SILENT rDNA COPIES

The structure of the rDNA chromatin and its dynamic regulation make for an interesting puzzle. Each new major advance in their analysis over the past twenty years has increased our awareness of their complexity, even though on the face of it the regulation of something as basic as rRNA production would be expected to be pretty straightforward. Considering the need for a massive production of rRNA during cell growth, the large number of rDNA genes seems logical enough: more RNA needed, more genes to supply it. However it seems counterintuitive, to say the least, that only half of these genes would be used at any given time; if nothing else, it would appear rather wasteful. And yet, that is precisely what we generally observe. There may be variations this way or that, depending on the growth or differentiation status of the cell, but in most cases roughly half the rDNA genes are expressed while the other half are silent.

Electron microscopy studies performed in yeast cells allow us to freeze a moment in time and to visualize the rDNA as it is being transcribed. Under the microscope, stretched out ribosomal gene repeats show structures evoking Christmas trees, where the DNA being transcribed by POL I represents the trunk and the transcribed rRNA molecules, stretched on either side but still attached by one end to the DNA-bound polymerases, form the branches. These elegant so-called "Miller spreads" clearly show that while some Christmas trees are very bushy indeed, a roughly equal number is devoid of transcription [16,17]. Since a typical rDNA Christmas tree counts a great many branches, we know that each active promoter allows multiple, shortly-spaced initiations. Successive initiations on particular promoters are therefore favoured over initiation from an increasing number of rDNA copies as rRNA demand grows, suggesting that some promoters are "open for business" while others are kept silenced; a state which, in higher eukaryotes at least, is associated with typical sliencing marks [18]. These marks seem to follow the same type of code that is observed for POL II genes. For example, using *Arabidopsis* hybrids and the phenomenon of nucleolar dominance, in which the rDNA set of one of the parent species is silenced, it was demonstrated that active rDNA genes are associated with trimethylation of lysine 4 on histone H3 (H3K4me3) and hypomethylated promoters, while inactive genes are associated with H3K9me2 with methylated promoter DNA [19]. Likewise in mouse, H3K9- and rDNA promoter methylation, as well as recruitment of Heterochromatin Protein 1 (HP1) are associated with inactive genes [20].

As we will see below, the use of several model systems (yeast, plant, amphibian and mammalian) have provided different insights into these marks, which may overlap or differ according to circumstances. For example, while DNA methylation clearly plays an important role in higher eukaryotes when it comes to silencing, this cannot be the case in yeast since yeast DNA is not methylated. This said, even in human cells DNA methylation is not indispensable for the silencing of 50% of the genes, exactly paralleling the yeast situation [21]. However, to what extent observations made in one system are relevant to another remains a little uncertain at the moment, and the reader is invited to remain aware of it as we refer to the properties of different organisms.

Since the production of rRNA parallels protein demand, it is expected to rise with the rate of growth or proliferation, as occurs when cells undergo changes in culture conditions [22]. One could then assume that even if a large number of rDNA copies are silent in standard growth conditions, the inactive copies could eventually be activated in conditions requiring a large increase in ribosome production. Accordingly, a moderate increase in the number of active copies has indeed been reported in yeast cells in response to more favorable growth conditions [23] and in higher eukaryotes during development and differentiation [24,25]. Conversely, and logically enough, growth of yeast cells to stationary phase can lead to a reduction in the number of open copies, an effect that depends on the histone deacetylase (HDAC) Rpd3 but not on the HDAC Sir2 [26]. This requirement will be discussed later on. However, even if the number of open copies can vary somewhat depending on growth conditions, this is probably not the mechanism by which rRNA transcription levels are regulated; rather, it satisfies other biological constraints. Instead, most of the silent rDNA copies remain silent, and the cell modulates rRNA production from its already active rDNA copies so that transcription can be increased or decreased while the open copy numbers remain mostly unchanged [17,21,27,28]. This surprising turn of events can be demonstrated in several ways.

An elegant approach that reduces threefold the number of rDNA repeats in a strain of *S. cerevisiae* makes possible the comparison of these two regulation strategies. In a strain developped in the laboratory of M. Nomura (NOY886) the rDNA locus has had its number of copies reduced from ~143 to about ~42, and even as low as 25 [17,29]). Considering that roughly half the genes of a normally growing yeast cell are active, this reduction of the total number of rDNA copies to less than the normal number of active copies should logically lead to a reduced level of rRNA production. This is however not the case, and the production of rRNA remains unimpaired. In these rDNA-poor cells, electron microscopy confirms that most if not all of the lower number of rDNA copies are expressed. This stands to reason if a normal complement of ribosomes is to be maintained, as the total available number of copies is now made artificially smaller than that of the normally active copies. Another level of response is observed here: on each rDNA copy the density of transcribing POL I is doubled [17]. This demonstrates that the cell is sensitive to its level of rRNA synthesis and is capable of reacting to an inappropriately low level by increasing production from each active gene. However, in mammalian cells, it was demonstrated that neither the change in active copies nor POL I recruitment is the most important response to an increased demand in rRNA. In response to growth factors, which stimulate cell proliferation and increase rRNA demand, the rate of rRNA elongation increases five-fold (a rate that precisely parallels the increase in 45S production), suggesting that most of the growth condition-dependent regulation occurs at the elongation level on active rDNA copies [30]. These experiments explain the previous counter-intuitive observation that the ratio of active to inactive copies in Friend cells remains constant throughout the cell cycle even though the rate of rDNA run-on transcription varies significantly: the cells do not silence active rDNA copies when the total output of rRNA production must be reduced, but reduce their individual output [31]. Naturally, as with other systems in the cell, rDNA transcription initiation will also depend on growth conditions and both POL I and the rDNA transcription factors can respond to growth-related signaling leading to certain reactions of phosphoprylation and acetylation [32].

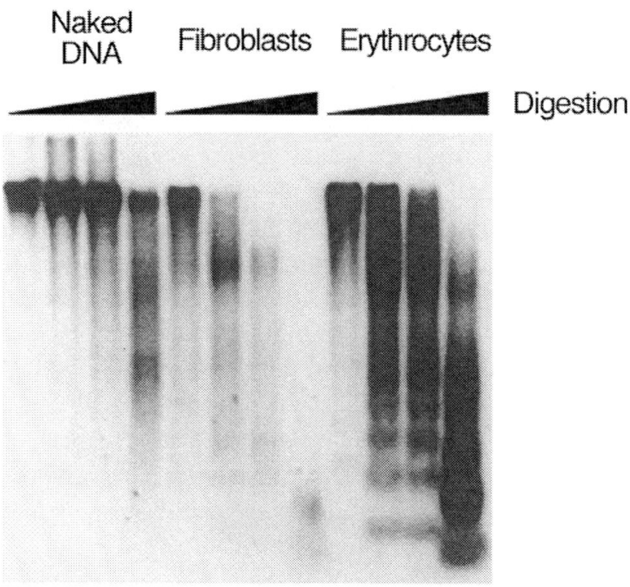

Figure 3. Chromatin structure of the X. laevis 28S rRNA gene. Purified DNA ("Naked DNA") or purified nuclei from fibroblasts or erythrocytes were submitted to limited micrococcal nuclease digestion. The recovered DNA in each case was analyzed by Southern blotting using a 28S rDNA probe to reveal local phasing of the nucleosomes. The naked DNA control shows a smear typical of cuts spread over the entire region. Erythrocytes, which are transcriptionally inactive, show a 28S gene chromatin organized as a series of regularly spaced nucleosomes. Actively growing fibroblasts show a weak regular pattern (caused by inactive gene copies) superimposed to a smear (caused by a chromatin with missing or unphased nucleosomes).

4. The Structure of Ribosomal Chromatin

While electron microscopy can show whether individual rDNA copies are associated with transcription complexes or not, the bulk of rDNA chromatin analyzed by biochemical means reveals two populations differing by their sensitivity to the cross-linking activity of psoralen. One part of the ribosomal chromatin is nucleosomal (or "closed"), making it resistant to the reagent; the other part is more open and available to cross-linking. As a result, the two forms can be distinguished by gel electrophoresis since cross-linked, open genes migrate more slowly than uncrosslinked and closed ones. These two forms are in roughly equal proportions in most cells and most conditions considered, and it is tempting to conclude that they correspond to active and inactive rDNA copies respectively. After all, it seems logical for chromatin that is available to psoralen to be likewise available to transcription factors, and other experiments show that the open conformation is actually associated with nascent rRNA [23,31,33]. Furthermore, experiments performed in mammalian cells show that DNA methylation marks associated with silent chromatin are mostly found in the enhancers and promoter of the closed population [34].

Agreeing with observations from the electron microscopy studies, the psoralen cross-linking approach shows that the active (open) and inactive (closed) copies are not segregated in discrete groups but are interspersed, at least in *S. cerevisiae*. The regulation of opening up or tightly packing the rDNA chromatin is therefore contingent on a mechanism capable of acting at the level of the individual gene and not only at the level of an entire locus [31,35].

While two states of ribosomal chromatin have long been a given, the nature of their association with histones and of their packaging in nucleosomes is still a controversial subject. Figure 3 shows a typical micrococcal nuclease digestion of *X. laevis* DNA probed with 28S rDNA sequence. A naked, purified DNA control shows that in the absence of nucleosomes the nuclease degrades the rDNA randomly and generates a smear upon electrophoresis. Erythrocytes provide a control for resolutely inactive rDNA genes, and we can see that in this case rDNA chromatin is arranged as a succession of very regular nucleosomes. (As is the case with all non-mammalian vertebrates, *Xenopus* red blood cells retain their nucleus, which makes this control possible). Finally, actively growing *Xenopus* cultured fibroblasts show a weak nucleosomal ladder (caused by the inactive copies) superimposed to a smear (arguably coming from the active copies) caused either by the digestion of totally unprotected DNA or by that of irregularly spaced nucleosomes. But which is it? Are the active, open rDNA copies organized as a series of unphased or perturbed nucleosomes, or are the nucleosomes missing altogether?

A big technical hurdle in trying to analyze the active and inactive rDNA copies is that they have the same sequence. To assess whether a common histone modification such as histone H4 acetylation is enriched on the active rDNA copies, for example, it is not possible to perform a straightforward chromatin immunoprecipitation (ChIP) experiment because both active and inactive copies would be detected during the DNA detection step.

One early way around this limitation was to compare the results of two parallel ChIP assays using an antibody against hyperacetylated histone H4 (H4ac). One would compare the level of H4ac associated with the 28S gene (active and inactive copies being indistinguishable) to the level of H4ac associated with the 28S gene found in purified mononucleosomes. This approach showed a greater level of H4ac associated with rDNA in general than with nucleosomal rDNA, suggesting that H4ac is enriched on non-nucleosomal ("active") rDNA copies [36]. That active rDNA copies are not assembled as regularly-spaced canonical nucleosomes has long been observed in several organisms [31,37,38]. In fact, recent work goes so far as to suggest a dramatic reduction in the number of histone molecules over the active rDNA copies.

More precise results were recently obtained thanks to the development of the Chromatin Endogenous Cleavage (ChEC) technique. This approach fuses a micrococcal nuclease moiety to a DNA-binding protein of interest, which leads to the targeted recruitment of the nuclease to the protein's binding sites and to their eventual identification [39]. Using this approach in combination with psoralen crosslinking, it was possible to show that a nuclease-fused POL I is enriched on the open copies of *S. cerevisiae* rDNA (confirming the expressed nature of the open copies). It was also possible to reveal that the actively transcribed rDNA repeats are not only lacking well-phased nucleosomes, but are also largely devoid of histone molecules [10]. Such a paucity in histones over the transcribed genes has been alluded to (even if not reaching such an important level) in other studies in *S. cerevisiae* [40] and in human cells. In the latter case, a reduction of ~45% in H3 and H4 was observed over the transcribed rDNA sequence of a colorectal carcinoma cell line where the double knock-out of DNA methytransferases DNMT-1 and 3b leads to an increase in open rDNA copies. The histones still present carry marks associated with transcription: in this case, increased penta-acetylation of H4 and reduced trimethylation of H3K9 [21].

Other characteristics of r-chromatin may be defined. ChIP-on-chip experiments show that the histone variant H2A.Z is enriched in the rDNA of *S. cerevisiae* [41]. This histone variant is frequently found at the 5' end of POL II-transcribed genes, where it helps maintain transcriptional competence if not outright activation. H2A.Z depletion in *S. cerevisiae* seems to lead to a moderate decrease in rDNA expression (pers. observation); this suggests that it may preferentially associate with active copies or conditionally active ones. This would agree with the mutually antagonistic presence of H2A.Z and DNA methylation reported in higher eukaryotes by Zilberman *et al* [42].

Together, these diverse observations show that the two populations of rDNA chromatin initially identified can be defined as follows: one that is not expressed and is associated with histones that form a series of phased nucleosomes; and another that is histone-poor (or at least nucleosome-poor) and is actively transcribed. This, however, is still not the end of the story. As we will now see, the ribosomal chromatin is more dynamic than at first suspected, and these two populations can be further distinguished into different types.

Heavy DNA methylation of CpGs found in promoter sequences is very frequently associated with gene silencing. This was reported for example in rat rDNA, with methylated promoter sequences mostly found in the inactive rDNA copies [34]. In mouse cells, at one site in particular (-133 relative to the transcription start site), CpG methylation correlates with gene silencing [43]. In human cells no such singular methylation switch is apparent, although the level of DNA methylation can vary according to activation status. Silencing of rDNA copies does not however systematicaly require methylation, since in a DNMT-1, DNMT3b double knock-out human cell line half the rDNA copies are still psoralen-insensitive [21]. Methylation-dependent and methylation-independent silencing thus seem to co-exist on the rDNA genes, which probably reflects different requirements in the regulation of r-chromatin structure, either regarding rDNA expression or chromosome stability.

5. The Upstream Binding Factor UBF Helps Define R-chromatin

The chromatin protein UBF (the Upstream Binding Factor), found in multicellular organisms as widely separated as *Arabidopsis* and vertebrates, has little DNA sequence specificity and relies on its multiple HMG box domains to bind the double helix as a dimer [44]. It was first identified as a rDNA promoter-binding and as an enhancer-binding transcription factor [45,46] but was later shown to also associate with the entire length of the transcribed gene to various levels [47,48].

The role of UBF is manyfold. At the chromatin level, it can trigger a large-scale, Swi/Snf- and acetylated H3K9- independent decondensation, even from an unusual chromosomal position to which it is artificially recruited [49]. That ability would be useful in the transition between a repressive chromatin structure and a permissive one, and in agreement with this UBF has been reported to inhibit the binding of histone H1 [50,51]. When it comes to transcription, UBF facilitates the formation of the pre-initiation complex by helping in the recruitment of POL I and of the TBP-containing rDNA transcription factor SL-1 to NORs [52]. UBF also plays a vital role in the growth factor-dependent regulation of elongation [53]. The association of a UBF homodimer with rDNA allows its multiple HMG boxes to induce a pronounced looping in the double helix, forming a nucleosome-like structure initially called enhancesome [54-56]. Phosphorylation of the protein has an effect both on its ability to form this structure and on the regulation of elongation. Dephosphorylated UBF has been shown to arrest POL I elongation *in vitro*, and in this dephosphorylated state it can readily form the enhancesome. However, phosphorylation of ERK sites in the first two HMG boxes (or mutations that mimick such phosphorylations) impede the formation of the looped structure and facilitate rRNA elongation [30,56-58]. In tetrapods, alternative splicing from a single transcript leads to the roughly equal production of two forms of the UBF protein. In mammals, this produces UBF1 and UBF2, which differ by a 37-residue deletion in the second HMG box that makes UBF2 less active as a transcription factor [59]. Functionally, UBF2 may behave as a partially phosphorylated UBF1, since in *in vitro* transcription reactions it presents less of an obstacle to elongation, and one that is easier to alleviate through ERK-dependent phosphorylation [60].

UBF affects all stages of transcription to a certain level and apparently plays an important role in the structure of active ribosomal chromatin. The induced depletion of UBF1 and 2 in human cells leads to an increase in closed rDNA copies that are stably maintained for many cell generations, as revealed by psoralen analysis, suggesting that one of the roles of UBF is to help maintain the open status of the rDNA chromatin. This dearth of UBF is however not accompanied by a higher methylation level of the promoter, which prompts a distinction between two forms of silenced rDNA chromatin: one that has a heavily-methylated promoter and is likely difficult to reactivate, and a second one that does not involve promoter methylation. This form, one supposes, could be derepressed quickly in response to the appropriate signaling, and has been dubbed "pseudo-silenced" r-chromatin [25].

UBF and the POL I machinery can be maintained on mitotic chromosomes. It is possible that this unexpectedly continued presence (even in the absence of gene expression) is due to the need for a rapid reactivation of transcription following cell division; the concerned genes would be kept in a stand-by state rather than be properly silenced. And indeed, renewed mitotic rDNA transcription is possible if okadaic acid is used to reverse the effects of cdc2/cyclin B-dependent silencing of the POL I machinery [47,61].

This would bring our different possible types of ribosomal chromatin to four: methylation-dependent, silenced chromatin; methylation-independent, pseudo-silenced chromatin; open but inactive chromatin, associated with UBF and the POL I machinery but not producing mature rRNA (as in mitotic chromosomes [62] or other reported inactive genes [63-65]); and finally open and active chromatin, with an elongation-permissive form of UBF (see Figure 4).

Unlike higher eukaryotes, yeast cells do not contain UBF. Nevertheless, it is possible that some other factor fills its shoes since ChEC analysis suggests DNA looping at the rDNA promoter in *S. cerevisiae,* a UBF-induced feature of higher eukaryotes [39]. The role of UBF at the promoter may be partly assumed by the Upstream Activating Factor (UAF) complex, which contains histones H3 and H4 as well as several other subunits and which could be the basis for some looping mechanism. Another UBF functional equivalent in yeast could be Hmo1, a factor that contains two HMG-homology domains and that plays a role in optimal rDNA expression as well as in plasmid maintenance and general chromatin structure [66,67]. Hmo1, like UBF, associates with the entire length of the open rDNA copies; furthermore, it binds to the promoters of a large number of ribosomal protein genes, thus probably playing a role in the concerted expression of several ribosomal components [68,69]. It should however be noted that Hmo1 is not essential for the formation of the open rDNA copies [10], while depletion of UBF severely reduces their number in higher eukaryotes [25]. Given the evolutionary distance between yeast and plants or vertebrate, however, it may be that the respective roles of Hmo1 and UBF are no longer exactly the same.

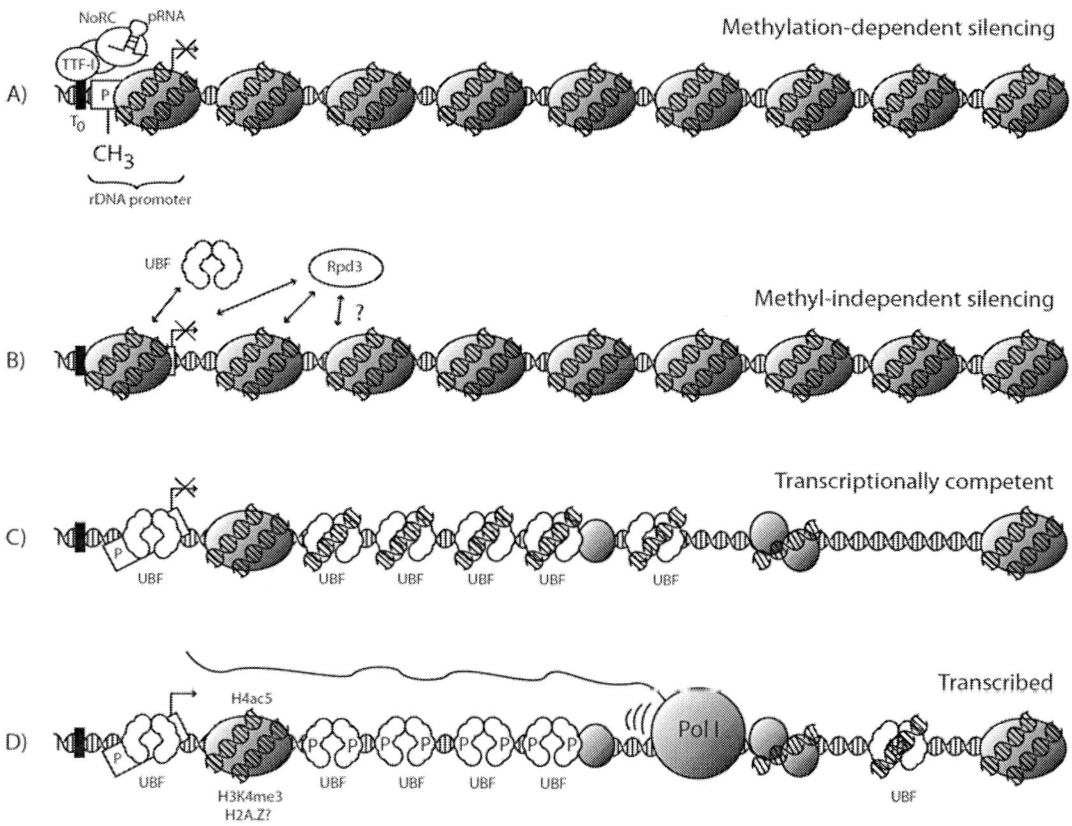

Figure 4: Four possible forms of ribosomal chromatin. The promoter-proximal terminator is labelled T_0; the rDNA promoter is labelled P and the transcription start site is marked by an arrow. Only a few of all the proteins involved are depicted. A) Nucleosomal, psoralen-resistant and silenced chromatin, with enriched promoter methylation. The chromatin remodeling complex NoRC, recruited to the promoter by the terminator binding factor TTF-I, moves a nucleosome into a position that prevents initiation. H3K9me2 represents one of the possible epigenetic marks associated with silencing. This type of chromatin would be the "permanently" silenced chromatin inherited from generation to generation. B) Nucleosomal, methylation-independent pseudo-silenced chromatin as observed in a UBF knock-down experiment [25]. Although there is no UBF in yeast, this form could correspond to the Hmo-1 poor and Rpd3-dependent silenced chromatin associated with growth saturation of *S. cerevisiae* [26]. C) Psoralen-sensitive and transcriptionally competent competent. UBF is shown as looping DNA. Perturbed, unphased or incomplete nucleosomes are depicted although their density (assumed to be low) and position are both uncertain. This form might the one preserved in mitotic chromosomes. D) Open, psoralen-sensitive chromatin, associated with elongation-permissive phosphorylated UBF and actively transcribed. Some epigenetic marks associated with activation are indicated (H4ac5 for penta-acetylated histone H4, for example).

6. Ribosomal Chromatin Remodeling

We have mentioned active and inactive rDNA copies, nucleosomes that come and go, and ribosomal chromatin factors that compete with some histones. Clearly, there is a lot going on with the architecture of the r-chromatin, begging the question of whether chromatin remodelers are involved in its dynamic nature.

Ribosomal chromatin is associated with at least three chromatin remodelers. Two of them contain the ATPase Snf2h: they are called WICH (WSTF-ISWI chromatin remodeling complex) and NoRC (Nucleolar Remodeling Complex). A third remodeler, the Cockayne Syndrome protein B (CSB), is a member of the Swi/Snf family of ATPases [70-73].

NoRC and CSB can both associate with a protein involved in POL I transcription termination, called TTF-I in mammals and Reb1 in *S. cerevisiae* [74,75]. In mammals, TTF-I binds to several terminator sequences located downstream of the transcription unit; it also binds to a terminator located immediately upstream of the promoter, called T_0. It is from this particular position that TTF-I recruits remodelers to the promoter (see figure 2 for the location of the terminators in the mouse intergenic spacer).

CSB is better known as an important ATPase involved in transcription-coupled nucleotide excision repair [76]. It has been linked to rDNA transcription and co-immunoprecipitates with several POL I transcription factors; its overexpression leads to increased rDNA expression while its depletion negatively affects it. Furthermore, since its ATPase activity is required in POL I transcription, it appears that it is indeed its chromatin remodeling activity is necessary for this effect [70,77].

The second remodeler that TTF-I can recruit to the promoter, NoRC, is linked to silencing. The NoRC complex consists of the protein Tip5, which binds TTF-I, and of the ATPase Snf2h, alongside the co-repressor Sin3 [78-80]. At the mouse rDNA promoter, NoRC will displace a positioned nucleosome that covers bases -157 to -2 in active genes to a new position located 25 base pairs further downstream [20]. This new location happens to lie just downstream of the CpG at -133, the methylation of which has been linked to silencing (although whether this is significant or not is not yet known) [81,82].

One of the requirements of the NoRC complex, which remodels ribosomal chromatin into a silenced form, may help with a long-standing puzzle concerning the rDNA locus of higher eukaryotes: the presence of functional rDNA promoters in its intergenic spacers (with no gene following them) [83]. These spacer promoters are known to recruit POL I and to allow the transcription of downstream sequences, generating RNAs that are complementary to the sequence of the actual gene promoter and its neighbouring terminator. A small RNA generated by the processing of these transcripts, called pRNA (for promoter-associated RNA), binds to the Tip5 subunit of NorC and targets the remodeling complex to the nucleolus and to the rDNA promoter [84]. Not unexpectedly, ectopic overexpression or pRNA leads to a lower expression of rDNA and to a local increase in heterochromatic marks. Conversely, its antisense-mediated depletion leads to an increase in rDNA expression [15,84,85]. A mutant Tip5 deficient in RNA binding can no longer recruit NoRC to the promoter, underscoring the importance of pRNA in its targeting.

Interestingly, this mutant Tip5 can still recruit a DNA methyltransferase activity to the promoter region and induce gene silencing, suggesting that it is the methylation of the promoter, rather than the NoRC-dependent displacement of a nucleosome, that is the ultimate cause of Tip5-dependent silencing [15]. Meanwhile, an ectopically-overexpressed mutant form of pRNA that does not interact with Tip5 has been shown to bind to the promoter directly, likely forming a DNA-RNA triplex. This triplex can recruit the methyltransferase DNMT3b to the promoter and trigger its methylation [85]. This emphasizes the dynamic nature of rDNA chromatin once again, since it suggests that methylation-driven silencing may be achieved by different mechanisms: one that is Tip5-dependent, and the other strictly pRNA-dependent.

Interestingly, TTF-I and pRNA binding to T_0 are mutually exclusive [85]. While this is surprising since the pRNA-binding Tip5 was first described (and even named!) as a TTF-I binding protein, one may speculate that pRNA occupation of the sequence leads only to methylation-dependent silencing of the neighbouring promoter, while TTF-I binding could possibly allow both NoRC-dependent *silencing* (via the Tip5-recruited DNMT3b and displacement of the nucleosome) and CSB-dependent *activation* (since CSB also binds to TTF-I). In that regard, it is interesting to remember that a promoter-proximal terminator has long since been shown to positively regulate rDNA transcription from the neighbouring promoter [86]. At his point, however, this is still speculation and more work will be required to understand the many roles of TTF-I in both silencing and activation.

7. Looping

Long-range DNA looping that brings the terminator in proximity to the promoter has been demonstrated in mouse cells by the technique of chromatin conformation capture (3C). A decreased level of CpG methylation in the terminator and promoter regions is involved in this looping, and the loss of looping upon serum starvation strongly suggest that it is the active copies that are looped [87]. This long range looping has prompted the development of new models for the higher-order three dimensional structure of r-chromatin, and illustrates the recent realization that genomic organization does not rely solely on activation and deactivation switches, but also on the actual position of chromosomal domains in the nucleus [48]. This field of study, made possible by powerful large-scale analysis tools, will doubtless reveal a whole new world of long-distance interactions and higher-order chromosomal arrangements in the regulation of all nuclear functions.

8. Wherefore Silencing of the Ribosomal Chromatin?

There are two forms of transcriptional silencing occuring in the rDNA loci. One targets rDNA transcription by POL I. The other, often referred to as nucleolar silencing, targets transcription by POL II and causes Pol II-dependent genes ectopically introduced in a rDNA locus to be transcriptionally repressed [88-90]. In yeast, nucleolar silencing is dependent on the histone deacetylase Sir2 which is recruited to the rDNA locus as part of a complex called RENT [91]. Both forms of silencing, however, could ultimately share a common goal: that of ensuring local genomic stability. Demonstrating the importance of gene silencing to this stability, studies performed in *Drosophila* show that interfering with H3K9 methylation and with the RNA interference pathway leads to disorganized nucleoli and to recombination in the rDNA loci [92], while loss of CpG methylation has a similar effect in human cells [21].

In yeast, an effect of Sir2 activity is to reduce by one order of magnitude the level of unequal recombination between the rDNA genes, which are choice targets for such an event because of their repetitive nature. This recombination normally increases with age and causes the appearance of extrachromosomal rDNA minicircles that are hallmarks of yeast senescence. Sir2 therefore increases the yeast cell's longevity, a discovery that prompted high hopes regarding the extension of life in other eukaryotes [93-95].

The intergenic spacer in yeast is not intrinsically silent, even if we do not consider its 5S rRNA gene. A cryptic, noncoding and bidirectional promoter can drive POL II to transcribe part of its sequence. This activity, which can be silenced by Sir2, causes the dissociation of the cohesin protein complex that inhibits unequal recombination events; this is apparently how the loss of Sir2 can lead to a change in the number of rDNA repeats [96-98]. Thus, it would appear that the role of nucleolar silencing at the yeast rDNA locus is not to help maintain a heterochromatic structure that prevents POL I from transcribing rDNA so much as to prevent the POL II-mediated displacement of cohesin in the intergenic spacer. In human cells, where silencing depends in part on the methylation of DNA, interfering with methylation (using 5-aza-deoxycyttidine or a DNMT1 / DNMT3b double knock-out) likewise leads to the induction of cryptic POL II promoters. In this case, spurious transcription is associated with a lower rRNA synthesis and processing, despite the increase in the level of open r-chromatin that accompanies lower levels of methylation (and also episomal rDNA) [21]. In both lower and higher eukaryotes, then, it seems very advantageous to maintain conditions that prevent POL II transcription of r-chromatin.

Sir2 has something of an ambiguous relation to rDNA: while the deletion of yeast Sir2 has no effect on rDNA accumulation [99], its multiple homologs in human cells, called sirtuins, will act either as POL I activators or as POL I repressors [88,90,100,101]. The sirtuin SirT1, closest homolog to yeast Sir2, can repress POL I transcription *in vitro* by deacetylating a subunit of the essential POL I transcription factor SL1, TAF_I68 [102]. Meanwhile, the sirtuin SirT7 is an essential nucleolar protein that associates with the rDNA genes and with POL I. Unlike SirT1, SirT7's expression positively correlates with POL I activity, and its knock-down by siRNA or the suppression of its enzymatic activity by nicotinamide treatment both reduce pre-rRNA synthesis. SirT1 is enriched in the nucleolus only at the onset of mitosis (when there is no rDNA transcription) while SirT7 is present throughout interphase. It has been suggested that an as-yet unidentified POL I elongation factor could be SirT7's target, and that in this way it would be possible for the cell to link rRNA elongation rate with NAD^+ levels [101].

Regardless of the mechanisms involved, it appears quite certain that rDNA silencing is not a simple system to keep back some "reserve ammunition", as it were, in case of an increased need in rRNA. Many investigations hint instead at a major role in genome stability. Homologous recombination normally tends to reduce the number of repetitive genes in a cell, but the rDNA loci manage to retain a fairly constant number of copies; when copies are lost in yeast, for example, unequal sister-chromatid recombination manages to increase the copy number anew [96]. This maintenance of a large number of copies that do not contribute to the production of more rRNA must be justified by some other important role in the continued survival of the cell. Studies performed in *S. cerevisiae* showed that active rDNA copies see their association with condensin greatly reduced relative to inactive copies. This would mean that in inactive copies, there would be a greater sister-chromatid cohesion, which would in turn increase the efficiency of replication-coupled recombination (see figure 5). Yeast strains with a five-fold reduction in rDNA copies did not see their rRNA levels affected, but this deletion did make the cells far more sensitive to DNA damaging agents [103]. Maintaining a large number of inactive copies could therefore arguably be a system allowing the cell to maintain the integrity of its rDNA locus.

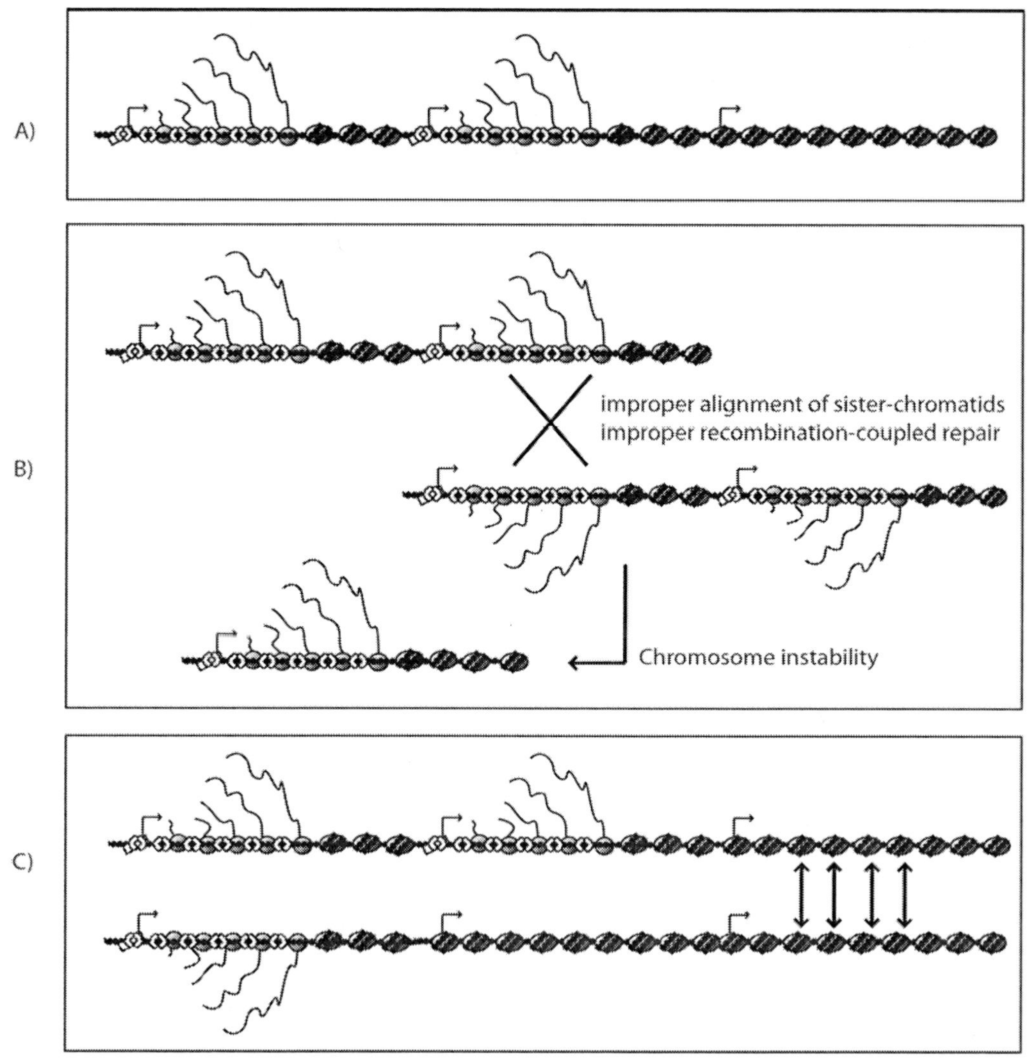

Figure 5. The role of rDNA silencing in promoting genome integrity. A) Representation of three rDNA units, with the first two being transcribed and the third one silenced. An increased need in rRNA will cause an increased production from the first two, but not the activation of the third one. B) POL I transcribes interferes with condensin association and with proper sister-chromatid cohesion [109,110]. This can in turn lead to improper replication-coupled repair and to chromosome instability. C) By maintaining silent copies, the cell ensures that proper alignment and repair can take place, thus maintaining a full array of undamaged rDNA copies.

Conclusion: Back to the Drawing Board

Our young student first drawing a nucleolus had little idea of all that was going on in this nuclear compartment, and we must humbly admit that there is still a lot that we, too, can learn about it. As more and more data comes in regarding the nature of the r-chromatin, it is certain that our understanding of it will evolve to take into account the link between its higher order organization and its regulation. To know more about this regulation, the reader is invited to refer to many excellent reviews recently published in the field [32,104-107].

Acknowledgments

My sincere thanks to Peter Moffet and Tom Moss for critical reading of this chapter.

References

[1] Wallace, H.&Birnstiel, M. L. (1966) Ribosomal cistrons and the nucleolar organizer. *Biochim Biophys Acta. 114*, 296-310.

[2] Schmickel, R. D. (1973) Quantitation of human ribosomal DNA: hybridization of human DNA with ribosomal RNA for quantitation and fractionation. *Pediatr Res. 7*, 5-12.

[3] Doudna, J. A.&Rath, V. L. (2002) Structure and function of the eukaryotic ribosome: the next frontier. *Cell. 109*, 153-156.

[4] Moss, T.&Stefanovsky, V. Y. (1995) Promotion and regulation of ribosomal transcription in eukaryotes by RNA polymerase I. *Prog Nucleic Acid Res Mol Biol. 50*, 25-66.

[5] Moss, T.&Stefanovsky, V. Y. (2002) At the center of eukaryotic life. *Cell. 109*, 545-548.

[6] Warner, J. R. (1999) The economics of ribosome biosynthesis in yeast. *Trends Biochem Sci. 24*, 437-440.

[7] Sorensen, P. D., Lomholt, B., Frederiksen, S.&Tommerup, N. (1991) Fine mapping of human 5S rRNA genes to chromosome 1q42.11----q42.13. *Cytogenet Cell Genet. 57*, 26-29.

[8] Birnstiel, M. L., Chipchase, M.&Speirs, J. (1971) The ribosomal RNA cistrons. *Prog Nucleic Acid Res Mol Biol. 11*, 351-389.

[9] Caburet, S., Conti, C., Schurra, C., Lebofsky, R., Edelstein, S. J.&Bensimon, A. (2005) Human ribosomal RNA gene arrays display a broad range of palindromic structures. *Genome Res. 15*, 1079-1085.

[10] Merz, K., Hondele, M., Goetze, H., Gmelch, K., Stoeckl, U.&Griesenbeck, J. (2008) Actively transcribed rRNA genes in S. cerevisiae are organized in a specialized chromatin associated with the high-mobility group protein Hmo1 and are largely devoid of histone molecules. *Genes Dev. 22*, 1190-1204.

[11] Gall, J. G. (1968) Differential synthesis of the genes for ribosomal RNA during amphibian oogenesis. *Proc Natl Acad Sci U S A. 60*, 553-560.

[12] Drouin, G., Hofman, J. D.&Doolittle, W. F. (1987) Unusual ribosomal RNA gene organization in copepods of the genus Calanus. *J Mol Biol. 196*, 943-946.

[13] Moss, T. (1983) A transcriptional function for the repetitive ribosomal spacer in Xenopus laevis. *Nature. 302*, 223-228.

[14] De Winter, R. F.&Moss, T. (1986) Spacer promoters are essential for efficient enhancement of X. laevis ribosomal transcription. *Cell. 44*, 313-318.

[15] Mayer, C., Schmitz, K. M., Li, J., Grummt, I.&Santoro, R. (2006) Intergenic transcripts regulate the epigenetic state of rRNA genes. *Mol Cell. 22*, 351-361.

[16] Miller, O. L., Jr.&Beatty, B. R. (1969) Visualization of nucleolar genes. *Science. 164*, 955-957.

[17] French, S. L., Osheim, Y. N., Cioci, F., Nomura, M.&Beyer, A. L. (2003) In exponentially growing Saccharomyces cerevisiae cells, rRNA synthesis is determined by the summed RNA polymerase I loading rate rather than by the number of active genes. *Mol Cell Biol. 23*, 1558-1568.

[18] Grummt, I.&Pikaard, C. S. (2003) Epigenetic silencing of RNA polymerase I transcription. *Nat Rev Mol Cell Biol. 4*, 641-649.

[19] Lawrence, R. J., Earley, K., Pontes, O., Silva, M., Chen, Z. J., Neves, N., Viegas, W., *et al.* (2004) A concerted DNA methylation/histone methylation switch regulates rRNA gene dosage control and nucleolar dominance. *Mol Cell. 13*, 599-609.

[20] Santoro, R., Li, J.&Grummt, I. (2002) The nucleolar remodeling complex NoRC mediates heterochromatin formation and silencing of ribosomal gene transcription. *Nat Genet. 32*, 393-396.

[21] Gagnon-Kugler, T., Langlois, F., Stefanovsky, V., Lessard, F.&Moss, T. (2009) Loss of human ribosomal gene CpG methylation enhances cryptic RNA polymerase II transcription and disrupts ribosomal RNA processing. *Mol Cell. 35*, 414-425.

[22] Kief, D. R.&Warner, J. R. (1981) Coordinate control of syntheses of ribosomal ribonucleic acid and ribosomal proteins during nutritional shift-up in Saccharomyces cerevisiae. *Mol Cell Biol. 1*, 1007-1015.

[23] Dammann, R., Lucchini, R., Koller, T.&Sogo, J. M. (1993) Chromatin structures and transcription of rDNA in yeast Saccharomyces cerevisiae. *Nucleic Acids Res. 21*, 2331-2338.

[24] Haaf, T., Hayman, D. L.&Schmid, M. (1991) Quantitative determination of rDNA transcription units in vertebrate cells. *Exp Cell Res. 193*, 78-86.

[25] Sanij, E., Poortinga, G., Sharkey, K., Hung, S., Holloway, T. P., Quin, J., Robb, E., *et al.* (2008) UBF levels determine the number of active ribosomal RNA genes in mammals. *J Cell Biol. 183*, 1259-1274.

[26] Sandmeier, J. J., French, S., Osheim, Y., Cheung, W. L., Gallo, C. M., Beyer, A. L.&Smith, J. S. (2002) RPD3 is required for the inactivation of yeast ribosomal DNA genes in stationary phase. *Embo J. 21*, 4959-4968.

[27] Banditt, M., Koller, T.&Sogo, J. (1999) Transcriptional activity and chromatin structure of enhancer-deleted rRNA genes in Saccharomyces cerevisiae. *Mol Cell Biol. 19*, 4953-4960.

[28] Stefanovsky, V.&Moss, T. (2006) Regulation of rRNA synthesis in human and mouse cells is not determined by changes in active gene count. *Cell Cycle. 5*, 735-739.

[29] Kobayashi, T., Heck, D. J., Nomura, M.&Horiuchi, T. (1998) Expansion and contraction of ribosomal DNA repeats in Saccharomyces cerevisiae: requirement of

replication fork blocking (Fob1) protein and the role of RNA polymerase I. *Genes Dev. 12*, 3821-3830.

[30] Stefanovsky, V., Langlois, F., Gagnon-Kugler, T., Rothblum, L. I.&Moss, T. (2006) Growth factor signaling regulates elongation of RNA polymerase I transcription in mammals via UBF phosphorylation and r-chromatin remodeling. *Mol Cell. 21*, 629-639.

[31] Conconi, A., Widmer, R., Koller, T.&Sogo, J. (1989) Two different chromatin structures coexist in ribosomal RNA genes throughout the cell cycle. *Cell. 57*, 753-761.

[32] Moss, T., Langlois, F., Gagnon-Kugler, T.&Stefanovsky, V. (2007) A housekeeper with power of attorney: the rRNA genes in ribosome biogenesis. *Cell Mol Life Sci. 64*, 29-49.

[33] Conconi, A., Sogo, J. M.&Ryan, C. A. (1992) Ribosomal gene clusters are uniquely proportioned between open and closed chromatin structures in both tomato leaf cells and exponentially growing suspension cultures. *Proc Natl Acad Sci U S A. 89*, 5256-5260.

[34] Stancheva, I., Lucchini, R., Koller, T.&Sogo, J. M. (1997) Chromatin structure and methylation of rat rRNA genes studied by formaldehyde fixation and psoralen cross-linking. *Nucleic Acids Res. 25*, 1727-1735.

[35] Dammann, R., Lucchini, R., Koller, T.&Sogo, J. M. (1995) Transcription in the yeast rRNA gene locus: distribution of the active gene copies and chromatin structure of their flanking regulatory sequences. *Mol Cell Biol. 15*, 5294-5303.

[36] Mutskov, V. J., Russanova, V. R., Dimitrov, S. I.&Pashev, I. G. (1996) Histones associated with non-nucleosomal rat ribosomal genes are acetylated while those bound to nucleosome-organized gene copies are not. *J Biol Chem. 271*, 11852-11857.

[37] Ness, P. J., Labhart, P., Banz, E., Koller, T.&Parish, R. W. (1983) Chromatin structure along the ribosomal DNA of Dictyostelium. Regional differences and changes accompanying cell differentiation. *J Mol Biol. 166*, 361-381.

[38] Prior, C. P., Cantor, C. R., Johnson, E. M., Littau, V. C.&Allfrey, V. G. (1983) Reversible changes in nucleosome structure and histone H3 accessibility in transcriptionally active and inactive states of rDNA chromatin. *Cell. 34*, 1033-1042.

[39] Schmid, M., Durussel, T.&Laemmli, U. K. (2004) ChIC and ChEC; genomic mapping of chromatin proteins. *Mol Cell. 16*, 147-157.

[40] Jones, H. S., Kawauchi, J., Braglia, P., Alen, C. M., Kent, N. A.&Proudfoot, N. J. (2007) RNA polymerase I in yeast transcribes dynamic nucleosomal rDNA. *Nat Struct Mol Biol. 14*, 123-130.

[41] Guillemette, B., Bataille, A. R., Gevry, N., Adam, M., Blanchette, M., Robert, F.&Gaudreau, L. (2005) Variant histone H2A.Z is globally localized to the promoters of inactive yeast genes and regulates nucleosome positioning. *PLoS Biol. 3*, e384.

[42] Zilberman, D., Coleman-Derr, D., Ballinger, T.&Henikoff, S. (2008) Histone H2A.Z and DNA methylation are mutually antagonistic chromatin marks. *Nature. 456*, 125-129.

[43] Santoro, R.&Grummt, I. (2001) Molecular mechanisms mediating methylation-dependent silencing of ribosomal gene transcription. *Mol Cell. 8*, 719-725.

[44] Jantzen, H. M., Admon, A., Bell, S. P.&Tjian, R. (1990) Nucleolar transcription factor hUBF contains a DNA-binding motif with homology to HMG proteins. *Nature. 344*, 830-836.

[45] Bell, S. P., Learned, R. M., Jantzen, H. M.&Tjian, R. (1988) Functional cooperativity between transcription factors UBF1 and SL1 mediates human ribosomal RNA synthesis. *Science. 241*, 1192-1197.

[46] Bell, S. P., Jantzen, H. M.&Tjian, R. (1990) Assembly of alternative multiprotein complexes directs rRNA promoter selectivity. *Genes Dev. 4*, 943-954.

[47] O'Sullivan, A. C., Sullivan, G. J.&McStay, B. (2002) UBF binding in vivo is not restricted to regulatory sequences within the vertebrate ribosomal DNA repeat. *Mol Cell Biol. 22*, 657-668.

[48] Denissov, S., Lessard, F., Mayer, C., Stefanovsky, V., van Driel, M., Grummt, I., Moss, T., *et al.* (2011) A model for the topology of active ribosomal RNA genes. *EMBO Rep. 12*, 231-237.

[49] Chen, D., Belmont, A. S.&Huang, S. (2004) Upstream binding factor association induces large-scale chromatin decondensation. *Proc Natl Acad Sci U S A. 101*, 15106-15111.

[50] Kermekchiev, M., Workman, J. L.&Pikaard, C. S. (1997) Nucleosome binding by the polymerase I transactivator upstream binding factor displaces linker histone H1. *Mol Cell Biol. 17*, 5833-5842.

[51] Dimitrov, S. I., Stefanovsky, V., Karagyozov, L., Angelov, D.&Pashev, I. G. (1990) The enhancers and promoters of the Xenopus laevis ribosomal spacer are associated with histones upon active transcription of the ribosomal genes. *Nucleic Acids Res. 18*, 6393-6397.

[52] Mais, C., Wright, J. E., Prieto, J. L., Raggett, S. L.&McStay, B. (2005) UBF-binding site arrays form pseudo-NORs and sequester the RNA polymerase I transcription machinery. *Genes Dev. 19*, 50-64.

[53] Moss, T., Stefanovsky, V., Langlois, F.&Gagnon Kugler, T. (2006) A new paradigm for the regulation of the mammalian ribosomal RNA genes. *Biochem Soc Trans. 34*, 1079-1081.

[54] Leblanc, B., Read, C.&Moss, T. (1993) Recognition of the Xenopus ribosomal core promoter by the transcription factor xUBF involves multiple HMG box domains and leads to an xUBF interdomain interaction. *Embo J. 12*, 513-525.

[55] Bazett-Jones, D. P., Leblanc, B., Herfort, M.&Moss, T. (1994) Short-range DNA looping by the Xenopus HMG-box transcription factor, xUBF. *Science. 264*, 1134-1137.

[56] Stefanovsky, V. Y., Pelletier, G., Bazett-Jones, D. P., Crane-Robinson, C.&Moss, T. (2001) DNA looping in the RNA polymerase I enhancesome is the result of non-cooperative in-phase bending by two UBF molecules. *Nucleic Acids Res. 29*, 3241-3247.

[57] Stefanovsky, V. Y., Langlois, F., Bazett-Jones, D., Pelletier, G.&Moss, T. (2006) ERK modulates DNA bending and enhancesome structure by phosphorylating HMG1-boxes 1 and 2 of the RNA polymerase I transcription factor UBF. *Biochemistry. 45*, 3626-3634.

[58] Stefanovsky, V. Y., Bazett-Jones, D. P., Pelletier, G.&Moss, T. (1996) The DNA supercoiling architecture induced by the transcription factor xUBF requires three of its five HMG-boxes. *Nucleic Acids Res. 24*, 3208-3215.

[59] O'Mahony, D. J.&Rothblum, L. I. (1991) Identification of two forms of the RNA polymerase I transcription factor UBF. *Proc Natl Acad Sci U S A. 88*, 3180-3184.

[60] Stefanovsky, V. Y.&Moss, T. (2008) The splice variants of UBF differentially regulate RNA polymerase I transcription elongation in response to ERK phosphorylation. *Nucleic Acids Res. 36*, 5093-5101.

[61] Sirri, V., Roussel, P.&Hernandez-Verdun, D. (2000) In vivo release of mitotic silencing of ribosomal gene transcription does not give rise to precursor ribosomal RNA processing. *J Cell Biol. 148*, 259-270.

[62] Sullivan, G. J., Bridger, J. M., Cuthbert, A. P., Newbold, R. F., Bickmore, W. A.&McStay, B. (2001) Human acrocentric chromosomes with transcriptionally silent nucleolar organizer regions associate with nucleoli. *Embo J. 20*, 2867-2874.

[63] Roussel, P., Andre, C., Masson, C., Geraud, G.&Hernandez-Verdun, D. (1993) Localization of the RNA polymerase I transcription factor hUBF during the cell cycle. *J Cell Sci. 104 (Pt 2)*, 327-337.

[64] Zatsepina, O. V., Voit, R., Grummt, I., Spring, H., Semenov, M. V.&Trendelenburg, M. F. (1993) The RNA polymerase I-specific transcription initiation factor UBF is associated with transcriptionally active and inactive ribosomal genes. *Chromosoma. 102*, 599-611.

[65] Junera, H. R., Masson, C., Geraud, G., Suja, J.&Hernandez-Verdun, D. (1997) Involvement of in situ conformation of ribosomal genes and selective distribution of upstream binding factor in rRNA transcription. *Mol Biol Cell. 8*, 145-156.

[66] Kamau, E., Bauerle, K. T.&Grove, A. (2004) The Saccharomyces cerevisiae high mobility group box protein HMO1 contains two functional DNA binding domains. *J Biol Chem. 279*, 55234-55240.

[67] Gadal, O., Labarre, S., Boschiero, C.&Thuriaux, P. (2002) Hmo1, an HMG-box protein, belongs to the yeast ribosomal DNA transcription system. *Embo J. 21*, 5498-5507.

[68] Hall, D. B., Wade, J. T.&Struhl, K. (2006) An HMG protein, Hmo1, associates with promoters of many ribosomal protein genes and throughout the rRNA gene locus in Saccharomyces cerevisiae. *Mol Cell Biol. 26*, 3672-3679.

[69] Berger, A. B., Decourty, L., Badis, G., Nehrbass, U., Jacquier, A.&Gadal, O. (2007) Hmo1 is required for TOR-dependent regulation of ribosomal protein gene transcription. *Mol Cell Biol. 27*, 8015-8026.

[70] Bradsher, J., Auriol, J., Proietti de Santis, L., Iben, S., Vonesch, J. L., Grummt, I.&Egly, J. M. (2002) CSB is a component of RNA pol I transcription. *Mol Cell. 10*, 819-829.

[71] Strohner, R., Nemeth, A., Jansa, P., Hofmann-Rohrer, U., Santoro, R., Langst, G.&Grummt, I. (2001) NoRC--a novel member of mammalian ISWI-containing chromatin remodeling machines. *Embo J. 20*, 4892-4900.

[72] Percipalle, P., Fomproix, N., Cavellan, E., Voit, R., Reimer, G., Kruger, T., Thyberg, J., *et al.* (2006) The chromatin remodelling complex WSTF-SNF2h interacts with nuclear myosin 1 and has a role in RNA polymerase I transcription. *EMBO Rep. 7*, 525-530.

[73] Citterio, E., Van Den Boom, V., Schnitzler, G., Kanaar, R., Bonte, E., Kingston, R. E., Hoeijmakers, J. H., *et al.* (2000) ATP-dependent chromatin remodeling by the Cockayne syndrome B DNA repair-transcription-coupling factor. *Mol Cell Biol. 20*, 7643-7653.

[74] Bartsch, I., Schoneberg, C.&Grummt, I. (1988) Purification and characterization of TTFI, a factor that mediates termination of mouse ribosomal DNA transcription. *Mol Cell Biol. 8*, 3891-3897.

[75] Morrow, B. E., Johnson, S. P.&Warner, J. R. (1989) Proteins that bind to the yeast rDNA enhancer. *J Biol Chem. 264*, 9061-9068.

[76] Venema, J., Mullenders, L. H., Natarajan, A. T., van Zeeland, A. A.&Mayne, L. V. (1990) The genetic defect in Cockayne syndrome is associated with a defect in repair of UV-induced DNA damage in transcriptionally active DNA. *Proc Natl Acad Sci U S A. 87*, 4707-4711.

[77] Yuan, X., Feng, W., Imhof, A., Grummt, I.&Zhou, Y. (2007) Activation of RNA polymerase I transcription by cockayne syndrome group B protein and histone methyltransferase G9a. *Mol Cell. 27*, 585-595.

[78] Zhou, Y., Santoro, R.&Grummt, I. (2002) The chromatin remodeling complex NoRC targets HDAC1 to the ribosomal gene promoter and represses RNA polymerase I transcription. *Embo J. 21*, 4632-4640.

[79] Strohner, R., Nemeth, A., Nightingale, K. P., Grummt, I., Becker, P. B.&Langst, G. (2004) Recruitment of the nucleolar remodeling complex NoRC establishes ribosomal DNA silencing in chromatin. *Mol Cell Biol. 24*, 1791-1798.

[80] Santoro, R.&Grummt, I. (2005) Epigenetic mechanism of rRNA gene silencing: temporal order of NoRC-mediated histone modification, chromatin remodeling, and DNA methylation. *Mol Cell Biol. 25*, 2539-2546.

[81] Langst, G., Becker, P. B.&Grummt, I. (1998) TTF-I determines the chromatin architecture of the active rDNA promoter. *Embo J. 17*, 3135-3145.

[82] Li, J., Langst, G.&Grummt, I. (2006) NoRC-dependent nucleosome positioning silences rRNA genes. *Embo J. 25*, 5735-5741.

[83] Moss, T.&Birnstiel, M. L. (1979) The putative promoter of a Xenopus laevis ribosomal gene is reduplicated. *Nucleic Acids Res. 6*, 3733-3743.

[84] Mayer, C., Neubert, M.&Grummt, I. (2008) The structure of NoRC-associated RNA is crucial for targeting the chromatin remodelling complex NoRC to the nucleolus. *EMBO Rep. 9*, 774-780.

[85] Schmitz, K. M., Mayer, C., Postepska, A.&Grummt, I. (2010) Interaction of noncoding RNA with the rDNA promoter mediates recruitment of DNMT3b and silencing of rRNA genes. *Genes Dev. 24*, 2264-2269.

[86] De Winter, R. F.&Moss, T. (1987) A complex array of sequences enhances ribosomal transcription in Xenopus laevis. *J Mol Biol. 196*, 813-827.

[87] Nemeth, A., Guibert, S., Tiwari, V. K., Ohlsson, R.&Langst, G. (2008) Epigenetic regulation of TTF-I-mediated promoter-terminator interactions of rRNA genes. *Embo J. 27*, 1255-1265.

[88] Bryk, M., Banerjee, M., Murphy, M., Knudsen, K. E., Garfinkel, D. J.&Curcio, M. J. (1997) Transcriptional silencing of Ty1 elements in the RDN1 locus of yeast. *Genes Dev. 11*, 255-269.

[89] Cockell, M.&Gasser, S. M. (1999) Nuclear compartments and gene regulation. *Curr Opin Genet Dev. 9*, 199-205.

[90] Smith, J. S.&Boeke, J. D. (1997) An unusual form of transcriptional silencing in yeast ribosomal DNA. *Genes Dev. 11*, 241-254.

[91] Straight, A. F., Shou, W., Dowd, G. J., Turck, C. W., Deshaies, R. J., Johnson, A. D.&Moazed, D. (1999) Net1, a Sir2-associated nucleolar protein required for rDNA silencing and nucleolar integrity. *Cell. 97*, 245-256.

[92] Peng, J. C.&Karpen, G. H. (2007) H3K9 methylation and RNA interference regulate nucleolar organization and repeated DNA stability. *Nat Cell Biol. 9*, 25-35.

[93] Gottlieb, S.&Esposito, R. E. (1989) A new role for a yeast transcriptional silencer gene, SIR2, in regulation of recombination in ribosomal DNA. *Cell. 56*, 771-776.

[94] Sinclair, D. A.&Guarente, L. (1997) Extrachromosomal rDNA circles--a cause of aging in yeast. *Cell. 91*, 1033-1042.

[95] Kaeberlein, M., McVey, M.&Guarente, L. (1999) The SIR2/3/4 complex and SIR2 alone promote longevity in Saccharomyces cerevisiae by two different mechanisms. *Genes Dev. 13*, 2570-2580.

[96] Kobayashi, T.&Ganley, A. R. (2005) Recombination regulation by transcription-induced cohesin dissociation in rDNA repeats. *Science. 309*, 1581-1584.

[97] Li, C., Mueller, J. E.&Bryk, M. (2006) Sir2 represses endogenous polymerase II transcription units in the ribosomal DNA nontranscribed spacer. *Mol Biol Cell. 17*, 3848-3859.

[98] Vasiljeva, L., Kim, M., Terzi, N., Soares, L. M.&Buratowski, S. (2008) Transcription termination and RNA degradation contribute to silencing of RNA polymerase II transcription within heterochromatin. *Mol Cell. 29*, 313-323.

[99] Shou, W., Sakamoto, K. M., Keener, J., Morimoto, K. W., Traverso, E. E., Azzam, R., Hoppe, G. J., *et al.* (2001) Net1 stimulates RNA polymerase I transcription and regulates nucleolar structure independently of controlling mitotic exit. *Mol Cell. 8*, 45-55.

[100] Fritze, C. E., Verschueren, K., Strich, R.&Easton Esposito, R. (1997) Direct evidence for SIR2 modulation of chromatin structure in yeast rDNA. *Embo J. 16*, 6495-6509.

[101] Ford, E., Voit, R., Liszt, G., Magin, C., Grummt, I.&Guarente, L. (2006) Mammalian Sir2 homolog SIRT7 is an activator of RNA polymerase I transcription. *Genes Dev. 20*, 1075-1080.

[102] Muth, V., Nadaud, S., Grummt, I.&Voit, R. (2001) Acetylation of TAF(I)68, a subunit of TIF-IB/SL1, activates RNA polymerase I transcription. *Embo J. 20*, 1353-1362.

[103] Ide, S., Miyazaki, T., Maki, H.&Kobayashi, T. (2010) Abundance of ribosomal RNA gene copies maintains genome integrity. *Science. 327*, 693-696.

[104] Birch, J. L.&Zomerdijk, J. C. (2008) Structure and function of ribosomal RNA gene chromatin. *Biochem Soc Trans. 36*, 619-624.

[105] McStay, B.&Grummt, I. (2008) The epigenetics of rRNA genes: from molecular to chromosome biology. *Annu Rev Cell Dev Biol. 24*, 131-157.

[106] Sanij, E.&Hannan, R. D. (2009) The role of UBF in regulating the structure and dynamics of transcriptionally active rDNA chromatin. *Epigenetics. 4*, 374-382.

[107] Moss, T. DNA methyltransferase inhibition may limit cancer cell growth by disrupting ribosome biogenesis. *Epigenetics. 6*, 128-133.

[108] Hong, E., Balakrishnan, R., Christie, K., Costanzo, M., Dwight, S., Engel, S., Fisk, D., *et al.* Saccharomyces cerevisiae database.

[109] Wang, B. D., Butylin, P.&Strunnikov, A. (2006) Condensin function in mitotic nucleolar segregation is regulated by rDNA transcription. *Cell Cycle. 5*, 2260-2267.

[110] Johzuka, K.&Horiuchi, T. (2007) RNA polymerase I transcription obstructs condensin association with 35S rRNA coding regions and can cause contraction of long repeat in Saccharomyces cerevisiae. *Genes Cells*. *12*, 759-771.

Reviewed by Dr. Tom Moss, Laval University, Quebec (QC) Canada.

In: Histones: Class, Structure and Function
Editor: Chang-Hui Shen

ISBN: 978-1-62100-274-1

Chapter 12

THE EPIGENOME AND ITS THERAPEUTIC POTENTIAL IN BREAST CANCER

***Virginia A. Spencer*[1,*] *and Jimmie E. Fata*[2,†]**
[1] Cell Culture Essentials, Life Technologies, Frederick, MD, US
[2]Department of Biology, College of Staten Island, City University of New York, NY, US

ABSTRACT

Cancer development is a multi-factorial event, involving nuclear processes that promote the expression of genes linked to cell survival and proliferation. A clear connection has been established between epigenetic events and cancer biology. However, the nature of this relationship as it relates to disease progression and treatment remains unclear. With a particular emphasis on breast cancer, here we review advancements in the fields of DNA methylation, histone acetylation and histone methylation as they relate to tumor development, progression, and treatment.

1. INTRODUCTION

Residing in the nucleus of a eukaryotic cell is a dynamic yet highly organized network containing approximately 2 meters of DNA surrounded by protein and RNA. DNA is packaged into discrete filaments referred to as chromatin through its association with histone and non-histone chromosomal proteins. In order for chromatin to fit within the confines of the nucleus, it must be compacted approximately 10,000 fold into a higher order structure by a sequence of intra- and inter-chromatin fiber interactions [reviewed in [1,2]]. This high level of compaction is thought to structurally organize DNA, making some regions of DNA accessible while other regions relatively inaccessible to their respective trans-acting factors [3]. Epigenetic events, otherwise known as heritable changes in gene expression which occur

* E-mail: ginspencer@hotmail.com.
† E-mail: jimmie.fata@csi.cuny.edu.

independently of alterations in the primary DNA sequence [4], promote changes in gene expression either by inducing changes in chromatin structure which affect gene accessibility to transcription factors or by flagging the histones associated with specific DNA sequences for recognition by specific transcription factors [5]. In the past few decades, two epigenetic mechanisms have emerged as the most critical players of transcriptional regulation: DNA methylation and covalent post-translational histone modification. Substantial evidence exists to suggest that these events play an important role in cancer development and treatment. In this chapter, we discuss the role of DNA methylation and histone modifications in events related to tumor development, progression and treatment, with a particular emphasis on breast cancer.

2. DNA Methylation

DNA methylation is characterized by the addition of a methyl group typically to the 5th position of the pyrimidine ring of the cytosine residue in a cytosine-guanine dinucleotide sequence [6]. This DNA modification is catalyzed by the cooperative action of *de novo* DNA cytosine-5-methyltransferases 3A and 3B (DNMT3A, DNMT3B). Additional DNA methyltransferases include DNMT1 which is thought to copy the methylation pattern from one strand to another during DNA replication, and DNMT2, an enzyme shown to have weak activity [reviewed in [7]]. DNA methylation is linked with the transcriptional silencing of associated genes and is thought to repress transcription by either inhibiting the association of transcription factors with their cognate DNA sequences [8] or by binding to methyl-CpG-binding proteins (MBPs) which recruit co-repressors such as histone deacetylases [9].

The methylation status of a CpG dinucleotide appears to depend on its location within the genome. CpG dinucleotides that are distributed as individual units throughout the genome are generally methylated and it has been proposed that the modification of these residues may act to prevent their transcription, resulting in a reduction of overall transcriptional noise [7]. In contrast, the majority of CpG dinucleotide clusters (otherwise referred to as CpG islands) that are localized to the 5' end of more than half of all human gene promoters are generally unmethylated during normal cell development [10].

To date, DNA methylation is generally thought to be a mitotically stable event. However, the promoter region of multiple genes in an estrogen-dependent human breast cancer cell line has been shown to fluctuate between a state of DNA hyper- and hypo-methylation through the dual DNA-methylating and -unmethylating actions of DNMTs [11]. Thus, it is possible that the state of DNA methylation within a particular CpG island is a net result of the balance between methylating and demethylating activities which are localized within close proximity of this region.

3. DNA Methylation in Cancer

CpG island methylation patterns have been shown to vary between normal and tumorigenic cells and to have a consequential effect on the promoter activity of several tumor suppressor genes. At least 100 genes have been reported to be hypermethylated in breast tumors or breast cancer cell lines [12] and DNA hypermethylation has been shown to be associated with genomic instability through the silencing of DNA repair genes such as BRCA1 and MLH1 [13,14]. In an extensive analysis of 215 primary breast and ovarian tumor specimens, 13% of the analyzed tumors displayed DNA hypermethylation along the BRCA1 gene promoter. However, a strong association was observed between DNA hypermethylation of BRCA1 and specific tumors subtypes that are known to develop frequently in BRCA1 families [13]. Similarly, the MLH1 gene has been shown to be hypermethylated and repressed in a significant fraction of colon carcinomas [14].

In addition to targeting specific DNA repair genes, DNA hypermethylation also occurs on the promoters of different tumor-suppressor genes and transcription factors. In an analysis of primary neoplasms testing negative for p16 expression, nearly 20% of tumor samples exhibited high levels of DNA methylation along the p16 gene promoter when compared to normal patient tissue [15]. Similarly, approximately 20% of renal tumors displayed high levels of DNA methylation along the promoter region of the von Hippel-Lindau (VHL) tumor suppressor gene [16]. In both studies, treatment of cell lines with 5-deoxyazacytidine reversed either p16 or VHL promoter hypermethylation and rescued the expression of these genes. In colorectal and gastric cancers, DNA hypermethyation has been shown to silence the expression of GATA-4 and GATA-5 transcription factors and their downstream antitumor target genes [17]. MALDI-TOF mass array analysis of 22 genes from 48 breast cancer and 48 paired normal parraffin-embedded tissues identified CpG hypermethylation near the consensus sequences of binding sites for BRCA1, P16 and additional transcription factors including APC, BIN1, BMP6, CST6, ESRb, GSTP1, P21 and TIMP3 [18].

Even though DNA hypermethylation has been shown to be localized to specific gene promoters, overall steady state levels of DNA methylation tend to be lower in various cancer cell types compared to normal cells [19-22]. An analysis of DNA from human colon tissue revealed substantial and equivalent DNA hypomethylation in both benign colon polyps and malignant carcinomas compared to levels in adjacent normal tissue [20]. Similarly, in an analysis of 103 human tumor specimens, DNA methylation was shown to be significantly lower in metastatic neoplasms compared to benign neoplasms or normal tissue [19]. Global decreases in DNA methylation have been reported to promote tumorigenesis by inducing chromosomal instability. DNMT1-deficient mice display high levels of global DNA hypomethylation, develop an aggressive form of T cell lymphoma and display a high frequency of chromosomal instability [23]. As well, DNA hypomethylation activates retrotransposons and promotes their translocation to other regions of the genome [24]. DNA hypomethylation has also been shown to occur on and promote the activation of growth-promoting genes in gastric, colon and prostate cancers [reviewed in [4]].

4. Histone Modifications

At its most basic level of structural organization, the chromatin fiber resembles a pearl necklace with its strand of DNA wrapping around and interlacing circular protein complexes referred to as nucleosomes [25]. The segment of DNA that extends between two neighboring nucleosomes is referred to as linker DNA. A nucleosome contains 147 bp of DNA wrapped 1.65 times around a histone octamer composed of two H2A-H2B dimers and a histone H3-H4 tetramer [2]. Each of the four core histones has a basic N terminal tail, a central globular domain and a C terminal tail. The histone N terminal tails emanate radially from the nucleosome where they are able to interact with linker DNA, other nucleoprotein complexes and neighboring chromatin fibers [3].

Histones undergo several types of post-translational modifications at residues in response to cellular and environmental cues [26]. The nature and extent of histone modifications plays an important role in gene regulation by influencing both chromatin structure and transcription factor recruitment. Examples of enzymes which mediate these modifications are histone acetyltransferases and methyltransferases which respectively acetylate and methylate specific residues in the histone N terminal tails and globular domains. With the exception of arginine methylation, these modifications can be reversed by the activity of histone deacetylases and demethylases.

5. Histone Acetylation and Deacetylation

Histone acetylation is a dynamic post-translational process involving the addition of acetyl groups from acetyl-coA to the ε-amino group on a lysine residue within the histone N terminal tails [27,28]. Depending on the histone type, up to four acetyl groups can be transferred onto the N terminal tail lysine residues. This event is catalyzed by histone acetyltransferases which are assembled into multi-subunit complexes that influence their activity and substrate specificity [29]. To date, at least 20 histone acetyltransferases (HATs) have been identified and the majority of these have been classified into three major families: the GNAT (Gcn5-related N-acetyl transferase) family, the MYST (MOZ, YBF2/SAS3, SAS2, TIP60) family, and the p300/CBP family [30] Acting in opposition to the HATs are four classes of histone deacetylases (HDACs): class 1 (HDACs 1-3 and 8), class II (HDACs 4-7 and 9), class III (sirtuins 1-7), and class IV (HDAC 11) [reviewed in [31]]. By themselves, HDACs are able to deacetylate the four core histones. However, like HATs, HDACs exist in multi-protein complexes that contain subunits which control their substrate preference [3].

In general, histone acetylation promotes transcription by causing specific DNA sequences within compact chromatin to become accessible to transcription factors. To achieve this effect, histone acetylation disturbs either histone-DNA contacts [32] possibly through charge neutralization [26] and/or chromatin fiber-fiber interactions [33]. Histone acetylation also flags specific lysine residues for recognition by nuclear factors which, in turn, recruit transcription complexes [5]. In addition to transcription, histone acetylation plays a role in DNA repair and replication [26].

6. Histone Lysine Methylation and Demethylation

Histone lysine methylation is characterized by the transfer of one to three methyl groups from S-adenosylmethionine to specific lysine residues [34]. Lysine methyltransferases catalyze this event and are classified into some of the following families according to the sequence of their SET catalytic domain: SET1, SET2, SUV39, EZH, SYMD and PRDM [31].

The reversal of this modification is mediated by histone demethylases which are classified into two groups based on their distinct catalytic domains: LSD1, and the JHDMs (jumonji-domain-containing demethylases) [29]. Both HMTs and LSDs display a high degree of specificity for particular histone lysine residues [26]. The specificity of HMTs combined with the fact that histone methylation does not alter the charge of the N terminal tails [31] points to a role for histone methylation in the presentation of specific histone codes which, in turn, function as recognition sites for nuclear factor recruitment. The type and degree of histone methylation varies with gene activity, and specific methylated forms of histones have been shown to recruit various protein factors involved in transcriptional initiation, transcriptional elongation, and gene silencing [26]. In addition, HMTs such as G9a and SUV39H1 have been shown to recruit DNMTs to specific DNA sequences [35,36]. As was stated previously, DNMTs interact with HDACs and MBPs. Thus, DNA methylation, histone methylation and histone acetylation are processes which can overlap with one another.

7. Histone Acetylation and Methylation in Cancer

Histone acetyltransferases and deacetylases exist in a dynamic equilibrium. The same is true for histone methyltransferases and demethylases. Therefore, it is likely that cellular events which offset the balance between these enzymes can incur detrimental effects on gene expression. The misregulation of histone modifications has been proposed to be a major contributor to cancer development and progression [37]. In particular, strong correlations have been observed between tumor phenotype and the acetylation and methylation of specific histone residues. In a panel of cancer cell lines and primary lymphoma and colon tumor samples, significant losses in total monoacetylated H4 (H4K16) and tri-methylated H4 (H4K20) were observed predominantly at hypomethylated repetitive DNA sequences [38]. Furthermore, an analysis of a mouse multistage tumor progression model revealed that the extent of these losses was positively correlated with advanced stages in tumor progression [38]. Following this trend, low levels of lysine acetylation (H3K9ac, H3K18ac, and H4K12ac), lysine methylation (H3K4me2 and H4K20me3) and arginine methylation (H4R3me2) have been correlated with breast carcinomas of poor prognostic subtypes [39].

A large body of evidence indicates that the expression levels of histone modifiers can differ quite significantly between normal and cancer cells. The levels of many HMTs have been reported to be elevated in breast, prostate, liver, endometrial and skin cancer [4]. Moreover, a strong association has been established between high levels of the EZH2 methyltransferase and advanced breast cancers displaying an aggressive angiogenic phenotype [40]. Overexpression of the Hbo1 acetyltransferase has been identified in the estrogen-dependent human breast cancer cell line, MCF-7, as well as in primary carcinomas of the testis, ovary, breast, stomach, and bladder [41]. Similarly, AIB1 has been identified as a HAT which is overexpressed in different types of carcinomas, including breast cancers that are resistant to anti-estrogen therapy [42]. HDACS have also been reported to be over-expressed in gastric and prostate cancer [43,44]. As well, specific HDACs involved in the downregulation of the ARH1 tumor suppressor gene have been identified as being upregulated in breast cancer cells [31]. In a recent study, the transition of a normal breast epithelium to ductal carcinoma *in situ* was accompanied by a decrease in H4 acetylation but also a decrease in HDAC1, HDAC2 and HDAC6 expression [45]. Although these two observations contradict one another, observations that the activity of specific HDAC subtypes is influenced by different mechanisms [46], suggest that this progressive decline in histone acetylation may be an event mediated by other HDACs. Lastly, overexpression of the histone demethylase LSD1 has been identified in prostate cancers bearing a poor prognostic phenotype [47], as well as in poorly differentiated neuroblastoma [48].

The function of several histone modifiers in different cancer types has also been shown to be altered or misdirected through gene mutation or chromosomal translocation events. Simultaneous mutations of CBP and p300 have been found in colorectal, gastric, hepatocellular and breast cancers, and CBP- or p300-deficient mice develop hematological malignancies [49]. Chromosomal translocation events leading to the production of CBP-MOZ and CBP-MLL fusion proteins have been identified in a subset of acute myeloid leukemia patients [50,51]. In addition, the mistargeting of DOT1L histone methyltransferase through its chromosomal translocation to MLL leads to leukemic transformation and the expression of a number of leukemia-relevant genes [52]. Collectively, the evidence discussed in this section clearly shows that cancer development does indeed involve an offset in the balance or specificity of histone modifying activities.

8. Current Breast Cancer Therapies

Despite many advances in detection and treatment, breast cancer remains a leading cause of female deaths and affects more than 1 million individuals every year. This statement is the major force behind a constant need to develop novel therapies or evaluate combinations of existing therapies. Clinically, there are many therapeutic modalities when considering the treatment or management of breast cancer. These include surgery, radiotherapy, chemotherapy, hormonal therapy, and targeted therapy. These therapies are most often dictated by the stage, grade and subtype of the cancer. Subtype has been defined by either gene signatures (luminal A, luminal B, Her-2-positive or basal-like breast cancer) or by immunohistochemistry (estrogen and progesterone receptor-positive, Her-2-positive, or triple negative which lacks all three). As with therapies, the stage and subtype also predict prognostic outcome. For instance, triple negative breast cancer, which accounts for ~15% of all breast cancers, has an aggressive clinical course [53], and cannot be treated with endocrine therapies or Her-2-targeting therapies.

Cytotoxic chemotherapy is a predominant modality of breast cancer treatment; however toxicity and non-specific effects limit the duration and often the effectiveness of this approach. Examples of commonly used chemotherapies for breast cancer include paclitaxel and doxorubicin. Targeted therapy, which is not considered cytotoxic, aims to attenuate pathways involved in growth and progression of breast cancer. Estrogen receptor (ER) therapies, such as Tamoxifen, aromatase inhibitors such as Anastrazole, and anti-Her-2 inhibitors such as Herceptin are all commonly used targeted therapies. Combinations of chemotherapy with targeted adjuvant (post-operative) or neoadjuvant (pre-operative) therapies have proven to be effective in breast cancer [54,55]. In the last 10 years, the development of new regimens and new therapies in the treatment of breast cancer has had a direct and positive impact on overall patient survival, but during this time there has been a constant concern over long-term toxicity effects or unnecessary treatment of individuals and the need to find effective targeted-treatment modalities for triple negative breast cancers [56].

Existing targeted therapies are predominated with endocrine-targeted and receptor-targeted approaches. Aside from these two large groups, there are a number of alternative targeted therapies that involve inhibiting other cellular and extracellular factors that are known to promote growth and progression of cancer. These alternative targeted therapies that are in current use or in clinical trials for breast cancer include those targeting microRNAs [57], poly(ADP) ribose polymerase-1 (PARP-1)[58], angiogenesis [59], non-receptor tyrosine kinases [60,61], mTOR [62], and the epigenome [63-65]. Breast cancer therapies directed against epigenetic events have been primarily grouped into either those targeting DNA methylation [66] or those targeting histone acetylation [31].

9. THE CLINICAL RELEVANCE OF THE EPIGENOME

Drugs that target DNA methylation (demethylating agents) or histone deacetylation are designed to restore normal DNA-dependent processes that are known to be deregulated in cancer such as gene transcription, DNA replication and repair, and chromosome condensation and segregation. These drugs provide a unique promise since, unlike genetic alterations, epigenetic alterations are reversible. A number of existing epigenome therapies are being evaluated through the clinical trial process and primarily include two types; DNA methyltransferase inhibitors (DNMTi) and histone deacetylase inhibitors (HDACi). A search on ClinicalTrials.gov, which compiles all clinical trials initiated since Sept. 27, 2007, reveals 14 trials completed, ongoing or recruiting since this time (Table I).

The epigenome signature of cancer cells, similar to the stage and subtype of the tumor, can provide important information in determining treatment and prognosis [64]. For instance the relevance of DNA hypermethylation in the silencing of tumor suppressor genes has been well established [64,67,68] and there is considerable evidence that specific alterations in DNA methylation are linked to specific aberrant patterns of histone modifications [69,70]. Since altered DNA methylation and histone modifications are invariably associated with cancer cells, there is a continued effort to develop and refine techniques that evaluate epigenome modifications as predictors of cancer type, prognosis and treatment. Currently, three clinical trials are investigating the predictive value in DNA methylation for breast cancer outcome. One study aims to evaluate the association between gene promoter methylation and breast tumor cytology in woman at increased risk for breast cancer (clinical trial identifier: NCT00323908; Table 1). The second trial is designed to determine whether the DNA methylation patterns from individuals with metastatic breast cancer can be used to predict response after chemotherapy. The third trial involves determining whether chemotherapy alters DNA methylation patterns in breast tumors and if these alterations correlate with clinical response outcome (NCT00698477; Table 1). The results of these three important trials will begin to shed light on whether genome wide DNA methylation or gene-specific methylation can serve as predictive markers for breast cancer prognosis and treatment.

Table 1. Current DNMTi and HDACi Clinical Trials

Drug	Identifier	Status	Phase	Brief Title
HDAC inhibitors				
SAHA	NCT00365599	Ongoing	Phase II	Phase II Trial of SAHA & Tamoxifen for Patients With Breast Cancer
SAHA	NCT00616967	Recruiting	Phase II	Carboplatin and Paclitaxel Albumin-Stabilized Nanoparticle Formulation With or Without Vorinostat in Treating Women With Breast Cancer That Can Be Removed by Surgery
SAHA	NCT00788112	Recruiting	N/A	Vorinostat in Treating Women With Ductal Carcinoma in Situ of the Breast
SAHA	NCT00262834	Recruiting	Phase II	Vorinostat in Treating Women Who Are Undergoing Surgery For Newly Diagnosed Stage I, Stage II, or Stage III Breast Cancer
SAHA	NCT00416130	Recruiting	Phase I/II	Clinical Trial of SAHA in Patients With Breast Cancer
SAHA	NCT00719875	Recruiting	Phase I/II	HDAC Inhibitor Vorinostat (SAHA) With Capecitabine (Xeloda) Using a New Weekly Dose Regimen for Advanced Breast Cancer
CUDC-101	NCT01171924	Recruiting	Phase I	A Phase Ib Expansion Study Investigating the Safety, Efficacy, and Pharmacokinetics of Intravenous CUDC-101 in Subjects With Advanced Head and Neck, Gastric, Breast, Liver and Non-small Cell Lung Cancer Tumors
MS-275	NCT00020579	Completed	Phase I	MS-275 in Treating Patients With Advanced Solid Tumors or Lymphoma
Entinostat	NCT01234532	Not Yet Open	Phase II	Entinostat and Anastrozole in Treating Postmenopausal Women With Triple-Negative Breast Cancer That Can Be Removed by Surgery
DNMT inhibitors				
Decitabine	NCT0119490	Recruiting	Phase I/II	Re-expression of ER in Triple Negative Breast Cancers
5-AzaC	NCT00748553	Recruiting	Phase I/II	A Phase I/II Clinical Trial of Vidaza With Abraxane in the Treatment of Patients With Advanced or Metastatic Solid Tumors and Breast Cancer
Exploratory				

Drug	Identifier	Status	Phase	Brief Title
None	NCT00323908	Ongoing	N/A	Protocol to Examine Methylation of Tumor Suppression Genes in Women at High Risk of Developing Breast Cancer
None	NCT00698477	Recruiting	N/A	Chemotherapy on Methylation Patterns in Breast Tumor Tissue Correlating With Clinical Response and Outcomes
None	NCT00899548	Ongoing	N/A	DNA in Predicting Response After Systemic Therapy in Women With Metastatic Breast Cancer

10. Demethylating Agents in Clinical Trials for the Treatment of Breast Cancer

A number of preclinical investigations into how breast cancer cells respond to demethylating agents have provided a strong foundation for further investigating these drugs as possible therapies against this disease. These findings have found that inhibitors of DNA methylation can suppress proliferation [71-73], induce apoptosis [74], activate the expression of silenced genes [72,74-76] and increase drug sensitivity [76]. Inhibitors of DNA methylation have been found to restore the expression levels of specific hypermethylated genes in breast cancer cells. Because of this, these inhibitors may show promise in restoring the expression of hypermethylated genes that are correlated with a good prognostic outcome in breast cancer such as BRCA1, estrogen receptor, progesterone receptor, E-cadherin, and p16 expression [77].

Two cytosine analogue drugs that reduce DNA methylation are currently approved by the Federal Drug Administration (FDA). Low doses of these two drugs, 5-azacytidine (5-AzaC) and 5-aza-2'-deoxycytidine (decitabine), have been shown to attenuate tumor growth attenuation. As a result, these drugs are approved as elective treatments for a pre-leukemic disease, myelodisplastic syndrome [78]. Once incorporated into DNA, these drugs covalently bind to DNA methyltransferase and sequester it from further activity [79].

Clinically, a number of previously ran trials using 5-AzaC as a monotherapy have unfortunately not fared well against breast cancer. The responses have been generally low but with some promise. In one Phase I trial with 30 patients displaying solid breast cancer tumors, treatment with 5-AzaC (1 to 24 mg/kg per day for at least 8 days) reduced tumor growth in 7 of 11 patients [80], whereas only a 25% tumor response was observed in another study where 5-AzaC was administered by subcutaneous injection [81]. In another trial which included 58 patients with metastatic breast cancer, only 2 of 27 (7%) patients receiving 5-AzaC responded to treatment [82]. In an important Phase II clinical trial where 170 patients were treated with 5-AzaC (1.6 mg/kg per day for 10 days), antitumor activity was reported in 17% of patients with breast carcinoma; however the response was transient and rapid resistance to treatment became evident [83]. This Phase II trial confirms that, unlike the effectiveness of 5-AzaC in myelodisplastic syndrome, treatment with this drug alone does not have a significant positive impact on breast cancer progression.

Although 5-AzaC does not appear to be a good frontline drug against breast cancer, two trials have been designed to revisit demethylating agents as therapies. One trial will address whether decitabine treatment can lead to re-expression of ER in triple negative breast cancer (NCT0119490; Table 1), while the other is aimed at evaluating the combinatorial treatment of 5-AzaC with Paclitaxel in individuals with advanced metastatic breast cancer (NCT00748553; Table 1). Revisiting the effectiveness of demethylating agents as breast cancer therapies for HER-2 positive aggressive cancer cells should also be considered in light of three associated hypermethylated genes which are associated with this phenotype: PgR, HSD17B4 (a hydroxysteroid hydrogenase) and H-cadherin [84]. Both the completion of these two clinical trials and the design for future similar trials is needed to determine whether DNA demethylating agents are indeed promising drugs in the fight against breast cancer.

11. Histone Deacetylase Inhibitors in Clinical Trials for the Treatment of Breast Cancer

HDAC inhibitors which are mostly non-selective maintain or increase the amount of acetylated histones in the genome. One large potential of HDAC inhibitors is their ability to possibly remove transcriptional repression of dormant tumor suppressor genes [85], which may then lead to the growth arrest, apoptosis, or differentiation of cancer cells. In preclinical studies, a large number of HDAC inhibitors have been examined against breast cancer cells and breast tumors. The data from these studies indicate that these inhibitors prevent proliferation [73,86-91], prevent tumor growth [87,92,93], prevent metastasis [94], reactivate silenced genes [88,95-97], induce differentiation [98], induce apoptosis [99-101], increase drug sensitivity [102-104], induce protein degradation [105,106], and inhibit lipid raft remodeling [107,108] . Although HDAC inhibitors, like demethylating agents, produce pleiotropic outcomes in cancer cells, their promise as potential therapies that target epigenetic events has materialized into a large number of clinical trials and the drug suberoylanilide hydroxamic acid (SAHA; Vorinostat) which has been approved for the treatment of cutaneous T-cell lymphoma [109].

Preclinical data indicates that HDAC inhibitors can increase the effectiveness of topoisomerases on cancer cells [110,111]. On the basis of this evidence, two clinical trials were performed to evaluate the combination of epirubicin, a drug which triggers DNA cleavage by topoisomerase II, with the epileptic drug, valproic acid, a drug which inhibits HDAC activity [112,113]. Although the results of these two trials were promising, the neurotoxicity and low potency of the valproic acid limited future studies. In an effort to alleviate the toxicity of this HDAC inhibitor, another Phase 1 trial was initiated and completed, this time examining the toxicity profile of SAHA, another HDAC inhibitor, when used in combination with the topoisomerase inhibitor, doxorubicin [114]. In individuals harboring solid tumors, including breast tumors, there was minimal antitumor activity found but it was determined that SAHA at 800 mg/day can be tolerated with doxorubicin [114]. Currently nine clinical trials are investigating the use of HDAC inhibitors as therapies against breast cancer (Table 1). All of these trials are in Phase I or II with the majority still in the recruiting stage. Of the 9 trials, 6 are investigating SAHA, while the others are evaluating other HDAC inhibitors such as CUDC-101, MS-275, and Entinostat. There is a strong sense that HDAC inhibitors are likely to move forward in clinical trials and in time may eventually become important therapies used in combination with other drugs against breast cancer.

CONCLUSION

A role for DNA methylation and post-translational histone modifications in normal cell physiology has been well established and a relatively large body of evidence shows an important connection between these epigenetic events and cancer development. However, the fact that the field of histone epigenetics appears to be expanding at a rapid and unyielding pace, suggests that both our understanding of chromatin structure and the relationship between epigenetic events as they relate to normal and cancer cell development are in their infancy. Advancements in our knowledge of these areas are essential if more effective treatments are to be discovered. At this time, a number of Phase I and II clinical trials that aim to evaluate the toxicity and effectiveness of HDAC inhibitors and DNA methyltransferase inhibitors as new therapies for breast cancer treatment are ongoing. Successful development of a breast cancer drug that targets epigenetic modifications of either DNA or histones may therefore occur in the near future and could benefit a large number of patients that show limited response to current therapies.

REFERENCES

[1] Belmont, A.S. (2006). Mitotic Chromosome Structure And Condensation. *Curr Opin Cell Biol 18*, 632-638.

[2] Luger, K. (2006). Dynamic Nucleosomes. *Chromosome Res 14*, 5-16.

[3] Spencer, V.A., And Davie, J.R. (1999). Role Of Covalent Modifications Of Histones In Regulating Gene Expression. *Gene 240*, 1-12.

[4] Sharma, S., Kelly, T.K., And Jones, P.A. (2010). Epigenetics In Cancer. *Carcinogenesis 31*, 27-36.

[5] Jenuwein, T., And Allis, C.D. (2001). Translating The Histone Code. *Science 293*, 1074-1080.

[6] Bird, A. (2002). Dna Methylation Patterns And Epigenetic Memory. *Genes Dev 16*, 6-21.

[7] Klose, R.J., And Bird, A.P. (2006). Genomic Dna Methylation: The Mark And Its Mediators. *Trends Biochem Sci 31*, 89-97.

[8] Watt, F., And Molloy, P.L. (1988). Cytosine Methylation Prevents Binding To Dna Of A Hela Cell Transcription Factor Required For Optimal Expression Of The Adenovirus Major Late Promoter. *Genes Dev 2*, 1136-1143.

[9] Jones, P.L., Veenstra, G.J., Wade, P.A., Vermaak, D., Kass, S.U., Landsberger, N., Strouboulis, J., And Wolffe, A.P. (1998). Methylated Dna And Mecp2 Recruit Histone Deacetylase To Repress Transcription. *Nat Genet 19*, 187-191.

[10] Suzuki, M.M., And Bird, A. (2008). Dna Methylation Landscapes: Provocative Insights From Epigenomics. *Nat Rev Genet 9*, 465-476.

[11] Kangaspeska, S., Stride, B., Metivier, R., Polycarpou-Schwarz, M., Ibberson, D., Carmouche, R.P., Benes, V., Gannon, F., And Reid, G. (2008). Transient Cyclical Methylation Of Promoter Dna. *Nature 452*, 112-115.

[12] Hinshelwood, R.A., And Clark, S.J. (2008). Breast Cancer Epigenetics: Normal Human Mammary Epithelial Cells As A Model System. *J Mol Med 86*, 1315-1328.

[13] Esteller, M., Silva, J.M., Dominguez, G., Bonilla, F., Matias-Guiu, X., Lerma, E., Bussaglia, E., Prat, J., Harkes, I.C., Repasky, E.A., *Et Al.* (2000). Promoter Hypermethylation And Brca1 Inactivation In Sporadic Breast And Ovarian Tumors. *J Natl Cancer Inst 92*, 564-569.

[14] Deng, G., Bell, I., Crawley, S., Gum, J., Terdiman, J.P., Allen, B.A., Truta, B., Sleisenger, M.H., And Kim, Y.S. (2004). Braf Mutation Is Frequently Present In Sporadic Colorectal Cancer With Methylated Hmlh1, But Not In Hereditary Nonpolyposis Colorectal Cancer. *Clin Cancer Res 10*, 191-195.

[15] Merlo, A., Herman, J.G., Mao, L., Lee, D.J., Gabrielson, E., Burger, P.C., Baylin, S.B., And Sidransky, D. (1995). 5' Cpg Island Methylation Is Associated With Transcriptional Silencing Of The Tumour Suppressor P16/Cdkn2/Mts1 In Human Cancers. *Nat Med 1*, 686-692.

[16] Herman, J.G., Latif, F., Weng, Y., Lerman, M.I., Zbar, B., Liu, S., Samid, D., Duan, D.S., Gnarra, J.R., Linehan, W.M., *Et Al.* (1994). Silencing Of The Vhl Tumor-Suppressor Gene By Dna Methylation In Renal Carcinoma. *Proc Natl Acad Sci U S A 91*, 9700-9704.

[17] Akiyama, Y., Watkins, N., Suzuki, H., Jair, K.W., Van Engeland, M., Esteller, M., Sakai, H., Ren, C.Y., Yuasa, Y., Herman, J.G., *Et Al.* (2003). Gata-4 And Gata-5 Transcription Factor Genes And Potential Downstream Antitumor Target Genes Are Epigenetically Silenced In Colorectal And Gastric Cancer. *Mol Cell Biol 23*, 8429-8439.

[18] Radpour, R., Kohler, C., Haghighi, M.M., Fan, A.X., Holzgreve, W., And Zhong, X.Y. (2009). Methylation Profiles Of 22 Candidate Genes In Breast Cancer Using High-Throughput Maldi-Tof Mass Array. *Oncogene 28*, 2969-2978.

[19] Gama-Sosa, M.A., Slagel, V.A., Trewyn, R.W., Oxenhandler, R., Kuo, K.C., Gehrke, C.W., And Ehrlich, M. (1983). The 5-Methylcytosine Content Of Dna From Human Tumors. *Nucleic Acids Res 11*, 6883-6894.

[20] Goelz, S.E., Vogelstein, B., Hamilton, S.R., And Feinberg, A.P. (1985). Hypomethylation Of Dna From Benign And Malignant Human Colon Neoplasms. *Science 228*, 187-190.

[21] Wilson, A.S., Power, B.E., And Molloy, P.L. (2007). Dna Hypomethylation And Human Diseases. *Biochim Biophys Acta 1775*, 138-162.

[22] Piotrowski, A., Benetkiewicz, M., Menzel, U., Diaz De Stahl, T., Mantripragada, K., Grigelionis, G., Buckley, P.G., Jankowski, M., Hoffman, J., Bala, D., *Et Al.* (2006). Microarray-Based Survey Of Cpg Islands Identifies Concurrent Hyper- And Hypomethylation Patterns In Tissues Derived From Patients With Breast Cancer. *Genes Chromosomes Cancer 45*, 656-667.

[23] Eden, A., Gaudet, F., Waghmare, A., And Jaenisch, R. (2003). Chromosomal Instability And Tumors Promoted By Dna Hypomethylation. *Science 300*, 455.

[24] Howard, G., Eiges, R., Gaudet, F., Jaenisch, R., And Eden, A. (2008). Activation And Transposition Of Endogenous Retroviral Elements In Hypomethylation Induced Tumors In Mice. *Oncogene 27*, 404-408.

[25] Fussner, E., Ching, R.W., And Bazett-Jones, D.P. (2011). Living Without 30nm Chromatin Fibers. *Trends Biochem Sci 36*, 1-6.

[26] Kouzarides, T. (2007). Chromatin Modifications And Their Function. *Cell 128*, 693-705.

[27] Kuo, M.H., And Allis, C.D. (1998). Roles Of Histone Acetyltransferases And Deacetylases In Gene Regulation. *Bioessays 20*, 615-626.

[28] Spencer, V.A., And Davie, J.R. (2001). Dynamically Acetylated Histone Association With Transcriptionally Active And Competent Genes In The Avian Adult Beta-Globin Gene Domain. *J Biol Chem 276*, 34810-34815.

[29] Sawan, C., And Herceg, Z. (2010). Histone Modifications And Cancer. *Adv Genet 70*, 57-85.

[30] Berdasco, M., And Esteller, M. (2010). Aberrant Epigenetic Landscape In Cancer: How Cellular Identity Goes Awry. *Dev Cell 19*, 698-711.

[31] Dalvai, M., And Bystricky, K. (2010). The Role Of Histone Modifications And Variants In Regulating Gene Expression In Breast Cancer. *J Mammary Gland Biol Neoplasia 15*, 19-33.

[32] Norton, V.G., Imai, B.S., Yau, P., And Bradbury, E.M. (1989). Histone Acetylation Reduces Nucleosome Core Particle Linking Number Change. *Cell 57*, 449-457.

[33] Tse, C., Sera, T., Wolffe, A.P., And Hansen, J.C. (1998). Disruption Of Higher-Order Folding By Core Histone Acetylation Dramatically Enhances Transcription Of Nucleosomal Arrays By Rna Polymerase Iii. *Mol Cell Biol 18*, 4629-4638.

[34] Lachner, M., And Jenuwein, T. (2002). The Many Faces Of Histone Lysine Methylation. *Curr Opin Cell Biol 14*, 286-298.

[35] Lehnertz, B., Ueda, Y., Derijck, A.A., Braunschweig, U., Perez-Burgos, L., Kubicek, S., Chen, T., Li, E., Jenuwein, T., And Peters, A.H. (2003). Suv39h-Mediated Histone H3 Lysine 9 Methylation Directs Dna Methylation To Major Satellite Repeats At Pericentric Heterochromatin. *Curr Biol 13*, 1192-1200.

[36] Tachibana, M., Matsumura, Y., Fukuda, M., Kimura, H., And Shinkai, Y. (2008). G9a/Glp Complexes Independently Mediate H3k9 And Dna Methylation To Silence Transcription. *Embo J 27*, 2681-2690.

[37] Chi, P., Allis, C.D., And Wang, G.G. (2010). Covalent Histone Modifications--Miswritten, Misinterpreted And Mis-Erased In Human Cancers. *Nat Rev Cancer 10*, 457-469.

[38] Fraga, M.F., Ballestar, E., Villar-Garea, A., Boix-Chornet, M., Espada, J., Schotta, G., Bonaldi, T., Haydon, C., Ropero, S., Petrie, K., *Et Al.* (2005). Loss Of Acetylation At Lys16 And Trimethylation At Lys20 Of Histone H4 Is A Common Hallmark Of Human Cancer. *Nat Genet 37*, 391-400.

[39] Elsheikh, S.E., Green, A.R., Rakha, E.A., Powe, D.G., Ahmed, R.A., Collins, H.M., Soria, D., Garibaldi, J.M., Paish, C.E., Ammar, A.A., *Et Al.* (2009). Global Histone Modifications In Breast Cancer Correlate With Tumor Phenotypes, Prognostic Factors, And Patient Outcome. *Cancer Res 69*, 3802-3809.

[40] Collett, K., Eide, G.E., Arnes, J., Stefansson, I.M., Eide, J., Braaten, A., Aas, T., Otte, A.P., And Akslen, L.A. (2006). Expression Of Enhancer Of Zeste Homologue 2 Is

Significantly Associated With Increased Tumor Cell Proliferation And Is A Marker Of Aggressive Breast Cancer. *Clin Cancer Res 12*, 1168-1174.

[41] Iizuka, M., Takahashi, Y., Mizzen, C.A., Cook, R.G., Fujita, M., Allis, C.D., Frierson, H.F., Jr., Fukusato, T., And Smith, M.M. (2009). Histone Acetyltransferase Hbo1: Catalytic Activity, Cellular Abundance, And Links To Primary Cancers. *Gene 436*, 108-114.

[42] Lahusen, T., Henke, R.T., Kagan, B.L., Wellstein, A., And Riegel, A.T. (2009). The Role And Regulation Of The Nuclear Receptor Co-Activator Aib1 In Breast Cancer. *Breast Cancer Res Treat 116*, 225-237.

[43] Halkidou, K., Gaughan, L., Cook, S., Leung, H.Y., Neal, D.E., And Robson, C.N. (2004). Upregulation And Nuclear Recruitment Of Hdac1 In Hormone Refractory Prostate Cancer. *Prostate 59*, 177-189.

[44] Song, J., Noh, J.H., Lee, J.H., Eun, J.W., Ahn, Y.M., Kim, S.Y., Lee, S.H., Park, W.S., Yoo, N.J., Lee, J.Y., *Et Al.* (2005). Increased Expression Of Histone Deacetylase 2 Is Found In Human Gastric Cancer. *Apmis 113*, 264-268.

[45] Suzuki, J., Chen, Y.Y., Scott, G.K., Devries, S., Chin, K., Benz, C.C., Waldman, F.M., And Hwang, E.S. (2009). Protein Acetylation And Histone Deacetylase Expression Associated With Malignant Breast Cancer Progression. *Clin Cancer Res 15*, 3163-3171.

[46] Duong, V., Bret, C., Altucci, L., Mai, A., Duraffourd, C., Loubersac, J., Harmand, P.O., Bonnet, S., Valente, S., Maudelonde, T., *Et Al.* (2008). Specific Activity Of Class Ii Histone Deacetylases In Human Breast Cancer Cells. *Mol Cancer Res 6*, 1908-1919.

[47] Kahl, P., Gullotti, L., Heukamp, L.C., Wolf, S., Friedrichs, N., Vorreuther, R., Solleder, G., Bastian, P.J., Ellinger, J., Metzger, E., *Et Al.* (2006). Androgen Receptor Coactivators Lysine-Specific Histone Demethylase 1 And Four And A Half Lim Domain Protein 2 Predict Risk Of Prostate Cancer Recurrence. *Cancer Res 66*, 11341-11347.

[48] Schulte, J.H., Lim, S., Schramm, A., Friedrichs, N., Koster, J., Versteeg, R., Ora, I., Pajtler, K., Klein-Hitpass, L., Kuhfittig-Kulle, S., *Et Al.* (2009). Lysine-Specific Demethylase 1 Is Strongly Expressed In Poorly Differentiated Neuroblastoma: Implications For Therapy. *Cancer Res 69*, 2065-2071.

[49] Iyer, N.G., Ozdag, H., And Caldas, C. (2004). P300/Cbp And Cancer. *Oncogene 23*, 4225-4231.

[50] Borrow, J., Stanton, V.P., Jr., Andresen, J.M., Becher, R., Behm, F.G., Chaganti, R.S., Civin, C.I., Disteche, C., Dube, I., Frischauf, A.M., *Et Al.* (1996). The Translocation T(8;16)(P11;P13) Of Acute Myeloid Leukaemia Fuses A Putative Acetyltransferase To The Creb-Binding Protein. *Nat Genet 14*, 33-41.

[51] Sobulo, O.M., Borrow, J., Tomek, R., Reshmi, S., Harden, A., Schlegelberger, B., Housman, D., Doggett, N.A., Rowley, J.D., And Zeleznik-Le, N.J. (1997). Mll Is Fused To Cbp, A Histone Acetyltransferase, In Therapy-Related Acute Myeloid Leukemia With A T(11;16)(Q23;P13.3). *Proc Natl Acad Sci U S A 94*, 8732-8737.

[52] Okada, Y., Feng, Q., Lin, Y., Jiang, Q., Li, Y., Coffield, V.M., Su, L., Xu, G., And Zhang, Y. (2005). Hdot1l Links Histone Methylation To Leukemogenesis. *Cell 121*, 167-178.

[53] Gluz, O., Liedtke, C., Gottschalk, N., Pusztai, L., Nitz, U., And Harbeck, N. (2009). Triple-Negative Breast Cancer--Current Status And Future Directions. *Ann Oncol 20*, 1913-1927.

[54] Guarneri, V., Barbieri, E., Dieci, M.V., Piacentini, F., And Conte, P. (2010). Anti-Her2 Neoadjuvant And Adjuvant Therapies In Her2 Positive Breast Cancer. *Cancer Treat Rev 36 Suppl 3*, S62-66.

[55] Orlando, L., Schiavone, P., Fedele, P., Calvani, N., Nacci, A., Rizzo, P., Marino, A., D'amico, M., Sponziello, F., Mazzoni, E., *Et Al.* (2010). Molecularly Targeted Endocrine Therapies For Breast Cancer. *Cancer Treat Rev 36 Suppl 3*, S67-71.

[56] Foulkes, W.D., Smith, I.E., And Reis-Filho, J.S. (2010). Triple-Negative Breast Cancer. *N Engl J Med 363*, 1938-1948.

[57] Corcoran, C., Friel, A.M., Duffy, M.J., Crown, J., And O'driscoll, L. (2010). Intracellular And Extracellular Micrornas In Breast Cancer. *Clin Chem 57*, 18-32.

[58] Rouleau, M., Patel, A., Hendzel, M.J., Kaufmann, S.H., And Poirier, G.G. (2010). Parp Inhibition: Parp1 And Beyond. *Nat Rev Cancer 10*, 293-301.

[59] Bossung, V., And Harbeck, N. (2010). Angiogenesis Inhibitors In The Management Of Breast Cancer. *Curr Opin Obstet Gynecol 22*, 79-86.

[60] Carvalho, S., And Schmitt, F. (2010). Potential Role Of Pi3k Inhibitors In The Treatment Of Breast Cancer. *Future Oncol 6*, 1251-1263.

[61] Saad, F., And Lipton, A. (2010). Src Kinase Inhibition: Targeting Bone Metastases And Tumor Growth In Prostate And Breast Cancer. *Cancer Treat Rev 36*, 177-184.

[62] Ghayad, S.E., And Cohen, P.A. (2010). Inhibitors Of The Pi3k/Akt/Mtor Pathway: New Hope For Breast Cancer Patients. *Recent Pat Anticancer Drug Discov 5*, 29-57.

[63] Wang, G.G., Allis, C.D., And Chi, P. (2007). Chromatin Remodeling And Cancer, Part I: Covalent Histone Modifications. *Trends Mol Med 13*, 363-372.

[64] Esteller, M. (2007). Cancer Epigenomics: Dna Methylomes And Histone-Modification Maps. *Nat Rev Genet 8*, 286-298.

[65] Jones, P.A., And Baylin, S.B. (2007). The Epigenomics Of Cancer. *Cell 128*, 683-692.

[66] Szyf, M., Pakneshan, P., And Rabbani, S.A. (2004). Dna Methylation And Breast Cancer. *Biochem Pharmacol 68*, 1187-1197.

[67] Costello, J.F., Fruhwald, M.C., Smiraglia, D.J., Rush, L.J., Robertson, G.P., Gao, X., Wright, F.A., Feramisco, J.D., Peltomaki, P., Lang, J.C., *Et Al.* (2000). Aberrant Cpg-Island Methylation Has Non-Random And Tumour-Type-Specific Patterns. *Nat Genet 24*, 132-138.

[68] Esteller, M., Corn, P.G., Baylin, S.B., And Herman, J.G. (2001). A Gene Hypermethylation Profile Of Human Cancer. *Cancer Res 61*, 3225-3229.

[69] Fahrner, J.A., Eguchi, S., Herman, J.G., And Baylin, S.B. (2002). Dependence Of Histone Modifications And Gene Expression On Dna Hypermethylation In Cancer. *Cancer Res 62*, 7213-7218.

[70] Vire, E., Brenner, C., Deplus, R., Blanchon, L., Fraga, M., Didelot, C., Morey, L., Van Eynde, A., Bernard, D., Vanderwinden, J.M., *Et Al.* (2006). The Polycomb Group Protein Ezh2 Directly Controls Dna Methylation. *Nature 439*, 871-874.

[71] Yang, Q., Shan, L., Yoshimura, G., Nakamura, M., Nakamura, Y., Suzuma, T., Umemura, T., Mori, I., Sakurai, T., And Kakudo, K. (2002). 5-Aza-2'-Deoxycytidine Induces Retinoic Acid Receptor Beta 2 Demethylation, Cell Cycle Arrest And Growth Inhibition In Breast Carcinoma Cells. *Anticancer Res 22*, 2753-2756.

[72] Parker, B.S., Cutts, S.M., Nudelman, A., Rephaeli, A., Phillips, D.R., And Sukumar, S. (2003). Mitoxantrone Mediates Demethylation And Reexpression Of Cyclin D2,

Estrogen Receptor And 14.3.3sigma In Breast Cancer Cells. *Cancer Biol Ther 2*, 259-263.

[73] Hurtubise, A., And Momparler, R.L. (2006). Effect Of Histone Deacetylase Inhibitor Laq824 On Antineoplastic Action Of 5-Aza-2'-Deoxycytidine (Decitabine) On Human Breast Carcinoma Cells. *Cancer Chemother Pharmacol 58*, 618-625.

[74] Ferguson, A.T., Vertino, P.M., Spitzner, J.R., Baylin, S.B., Muller, M.T., And Davidson, N.E. (1997). Role Of Estrogen Receptor Gene Demethylation And Dna Methyltransferase.Dna Adduct Formation In 5-Aza-2'deoxycytidine-Induced Cytotoxicity In Human Breast Cancer Cells. *J Biol Chem 272*, 32260-32266.

[75] Ferguson, A.T., Lapidus, R.G., Baylin, S.B., And Davidson, N.E. (1995). Demethylation Of The Estrogen Receptor Gene In Estrogen Receptor-Negative Breast Cancer Cells Can Reactivate Estrogen Receptor Gene Expression. *Cancer Res 55*, 2279-2283.

[76] Kastl, L., Brown, I., And Schofield, A.C. (2010). Effects Of Decitabine On The Expression Of Selected Endogenous Control Genes In Human Breast Cancer Cells. *Mol Cell Probes 24*, 87-92.

[77] Santini, V., Kantarjian, H.M., And Issa, J.P. (2001). Changes In Dna Methylation In Neoplasia: Pathophysiology And Therapeutic Implications. *Ann Intern Med 134*, 573-586.

[78] Mack, G.S. (2006). Epigenetic Cancer Therapy Makes Headway. *J Natl Cancer Inst 98*, 1443-1444.

[79] Santi, D.V., Garrett, C.E., And Barr, P.J. (1983). On The Mechanism Of Inhibition Of Dna-Cytosine Methyltransferases By Cytosine Analogs. *Cell 33*, 9-10.

[80] Weis, J. (2003). [Current Aspects Of Myopathology]. *Ther Umsch 60*, 413-417.

[81] Bellet, R.E., Mastrangelo, M.J., Engstrom, P.F., Strawitz, J.G., Weiss, A.J., And Yarbro, J.W. (1974). Clinical Trial With Subcutaneously Administered 5-Azacytidine (Nsc-102816). *Cancer Chemother Rep 58*, 217-222.

[82] Nemoto, T.K., Fukuma, Y., Itoh, H., Takagi, T., And Ono, T. (2006). A Disulfide Bridge Mediated By Cysteine 574 Is Formed In The Dimer Of The 70-Kda Heat Shock Protein. *J Biochem 139*, 677-687.

[83] Weiss, A.J., Metter, G.E., Nealon, T.F., Keanan, J.P., Ramirez, G., Swaiminathan, A., Fletcher, W.S., Moss, S.E., And Manthei, R.W. (1977). Phase Ii Study Of 5-Azacytidine In Solid Tumors. *Cancer Treat Rep 61*, 55-58.

[84] Fiegl, H., Millinger, S., Goebel, G., Muller-Holzner, E., Marth, C., Laird, P.W., And Widschwendter, M. (2006). Breast Cancer Dna Methylation Profiles In Cancer Cells And Tumor Stroma: Association With Her-2/Neu Status In Primary Breast Cancer. *Cancer Res 66*, 29-33.

[85] Richon, V.M., Sandhoff, T.W., Rifkind, R.A., And Marks, P.A. (2000). Histone Deacetylase Inhibitor Selectively Induces P21waf1 Expression And Gene-Associated Histone Acetylation. *Proc Natl Acad Sci U S A 97*, 10014-10019.

[86] Lee, B.I., Park, S.H., Kim, J.W., Sausville, E.A., Kim, H.T., Nakanishi, O., Trepel, J.B., And Kim, S.J. (2001). Ms-275, A Histone Deacetylase Inhibitor, Selectively Induces Transforming Growth Factor Beta Type Ii Receptor Expression In Human Breast Cancer Cells. *Cancer Res 61*, 931-934.

[87] Vigushin, D.M., Ali, S., Pace, P.E., Mirsaidi, N., Ito, K., Adcock, I., And Coombes, R.C. (2001). Trichostatin A Is A Histone Deacetylase Inhibitor With Potent Antitumor Activity Against Breast Cancer In Vivo. *Clin Cancer Res 7*, 971-976.

[88] Bovenzi, V., And Momparler, R.L. (2001). Antineoplastic Action Of 5-Aza-2'-Deoxycytidine And Histone Deacetylase Inhibitor And Their Effect On The Expression Of Retinoic Acid Receptor Beta And Estrogen Receptor Alpha Genes In Breast Carcinoma Cells. *Cancer Chemother Pharmacol 48*, 71-76.

[89] Lee, K.W., Kim, J.H., Park, J.H., Kim, H.P., Song, S.H., Kim, S.G., Kim, T.Y., Jong, H.S., Jung, K.H., Im, S.A., *Et Al.* (2006). Antitumor Activity Of Sk-7041, A Novel Histone Deacetylase Inhibitor, In Human Lung And Breast Cancer Cells. *Anticancer Res 26*, 3429-3438.

[90] Kwon, H.K., Ahn, S.H., Park, S.H., Park, J.H., Park, J.W., Kim, H.M., Park, S.K., Lee, K., Lee, C.W., Choi, E., *Et Al.* (2009). A Novel Gamma-Lactam-Based Histone Deacetylase Inhibitor Potently Inhibits The Growth Of Human Breast And Renal Cancer Cells. *Biol Pharm Bull 32*, 1723-1727.

[91] Shi, Y.K., Li, Z.H., Han, X.Q., Yi, J.H., Wang, Z.H., Hou, J.L., Feng, C.R., Fang, Q.H., Wang, H.H., Zhang, P.F., *Et Al.* (2010). The Histone Deacetylase Inhibitor Suberoylanilide Hydroxamic Acid Induces Growth Inhibition And Enhances Taxol-Induced Cell Death In Breast Cancer. *Cancer Chemother Pharmacol 66*, 1131-1140.

[92] Baschnagel, A., Russo, A., Burgan, W.E., Carter, D., Beam, K., Palmieri, D., Steeg, P.S., Tofilon, P., And Camphausen, K. (2009). Vorinostat Enhances The Radiosensitivity Of A Breast Cancer Brain Metastatic Cell Line Grown In Vitro And As Intracranial Xenografts. *Mol Cancer Ther 8*, 1589-1595.

[93] Shankar, S., Davis, R., Singh, K.P., Kurzrock, R., Ross, D.D., And Srivastava, R.K. (2009). Suberoylanilide Hydroxamic Acid (Zolinza/Vorinostat) Sensitizes Trail-Resistant Breast Cancer Cells Orthotopically Implanted In Balb/C Nude Mice. *Mol Cancer Ther 8*, 1596-1605.

[94] Palmieri, D., Lockman, P.R., Thomas, F.C., Hua, E., Herring, J., Hargrave, E., Johnson, M., Flores, N., Qian, Y., Vega-Valle, E., *Et Al.* (2009). Vorinostat Inhibits Brain Metastatic Colonization In A Model Of Triple-Negative Breast Cancer And Induces Dna Double-Strand Breaks. *Clin Cancer Res 15*, 6148-6157.

[95] Keen, J.C., Yan, L., Mack, K.M., Pettit, C., Smith, D., Sharma, D., And Davidson, N.E. (2003). A Novel Histone Deacetylase Inhibitor, Scriptaid, Enhances Expression Of Functional Estrogen Receptor Alpha (Er) In Er Negative Human Breast Cancer Cells In Combination With 5-Aza 2'-Deoxycytidine. *Breast Cancer Res Treat 81*, 177-186.

[96] Jang, E.R., Lim, S.J., Lee, E.S., Jeong, G., Kim, T.Y., Bang, Y.J., And Lee, J.S. (2004). The Histone Deacetylase Inhibitor Trichostatin A Sensitizes Estrogen Receptor Alpha-Negative Breast Cancer Cells To Tamoxifen. *Oncogene 23*, 1724-1736.

[97] Li, Y., Yuan, Y.Y., Meeran, S.M., And Tollefsbol, T.O. (2010). Synergistic Epigenetic Reactivation Of Estrogen Receptor-Alpha (Eralpha) By Combined Green Tea Polyphenol And Histone Deacetylase Inhibitor In Eralpha-Negative Breast Cancer Cells. *Mol Cancer 9*, 274.

[98] Munster, P.N., Troso-Sandoval, T., Rosen, N., Rifkind, R., Marks, P.A., And Richon, V.M. (2001). The Histone Deacetylase Inhibitor Suberoylanilide Hydroxamic Acid Induces Differentiation Of Human Breast Cancer Cells. *Cancer Res 61*, 8492-8497.

[99] Louis, M., Rosato, R.R., Brault, L., Osbild, S., Battaglia, E., Yang, X.H., Grant, S., And Bagrel, D. (2004). The Histone Deacetylase Inhibitor Sodium Butyrate Induces Breast Cancer Cell Apoptosis Through Diverse Cytotoxic Actions Including Glutathione Depletion And Oxidative Stress. *Int J Oncol 25*, 1701-1711.

[100] Park, H., Im, J.Y., Kim, J., Choi, W.S., And Kim, H.S. (2008). Effects Of Apicidin, A Histone Deacetylase Inhibitor, On The Regulation Of Apoptosis In H-Ras-Transformed Breast Epithelial Cells. *Int J Mol Med 21*, 325-333.

[101] Huang, X., Gao, L., Wang, S., Lee, C.K., Ordentlich, P., And Liu, B. (2009). Hdac Inhibitor Sndx-275 Induces Apoptosis In Erbb2-Overexpressing Breast Cancer Cells Via Down-Regulation Of Erbb3 Expression. *Cancer Res 69*, 8403-8411.

[102] Fuino, L., Bali, P., Wittmann, S., Donapaty, S., Guo, F., Yamaguchi, H., Wang, H.G., Atadja, P., And Bhalla, K. (2003). Histone Deacetylase Inhibitor Laq824 Down-Regulates Her-2 And Sensitizes Human Breast Cancer Cells To Trastuzumab, Taxotere, Gemcitabine, And Epothilone B. *Mol Cancer Ther 2*, 971-984.

[103] Butler, L.M., Liapis, V., Bouralexis, S., Welldon, K., Hay, S., Thai Le, M., Labrinidis, A., Tilley, W.D., Findlay, D.M., And Evdokiou, A. (2006). The Histone Deacetylase Inhibitor, Suberoylanilide Hydroxamic Acid, Overcomes Resistance Of Human Breast Cancer Cells To Apo2l/Trail. *Int J Cancer 119*, 944-954.

[104] Restall, C., Doherty, J., Liu, H.B., Genovese, R., Paiman, L., Byron, K.A., Anderson, R.L., And Dear, A.E. (2009). A Novel Histone Deacetylase Inhibitor Augments Tamoxifen-Mediated Attenuation Of Breast Carcinoma Growth. *Int J Cancer 125*, 483-487.

[105] Alao, J.P., Stavropoulou, A.V., Lam, E.W., Coombes, R.C., And Vigushin, D.M. (2006). Histone Deacetylase Inhibitor, Trichostatin A Induces Ubiquitin-Dependent Cyclin D1 Degradation In Mcf-7 Breast Cancer Cells. *Mol Cancer 5*, 8.

[106] Yi, X., Wei, W., Wang, S.Y., Du, Z.Y., Xu, Y.J., And Yu, X.D. (2008). Histone Deacetylase Inhibitor Saha Induces Eralpha Degradation In Breast Cancer Mcf-7 Cells By Chip-Mediated Ubiquitin Pathway And Inhibits Survival Signaling. *Biochem Pharmacol 75*, 1697-1705.

[107] Ostapkowicz, A., Inai, K., Smith, L., Kreda, S., And Spychala, J. (2006). Lipid Rafts Remodeling In Estrogen Receptor-Negative Breast Cancer Is Reversed By Histone Deacetylase Inhibitor. *Mol Cancer Ther 5*, 238-245.

[108] Fan, J., Yin, W.J., Lu, J.S., Wang, L., Wu, J., Wu, F.Y., Di, G.H., Shen, Z.Z., And Shao, Z.M. (2008). Er Alpha Negative Breast Cancer Cells Restore Response To Endocrine Therapy By Combination Treatment With Both Hdac Inhibitor And Dnmt Inhibitor. *J Cancer Res Clin Oncol 134*, 883-890.

[109] Thompson, C.A. (2006). Vorinostat Approved For Rare Lymphoma. *Am J Health Syst Pharm 63*, 2168.

[110] Tsai, S.C., Valkov, N., Yang, W.M., Gump, J., Sullivan, D., And Seto, E. (2000). Histone Deacetylase Interacts Directly With Dna Topoisomerase Ii. *Nat Genet 26*, 349-353.

[111] Kurz, E.U., Wilson, S.E., Leader, K.B., Sampey, B.P., Allan, W.P., Yalowich, J.C., And Kroll, D.J. (2001). The Histone Deacetylase Inhibitor Sodium Butyrate Induces Dna Topoisomerase Ii Alpha Expression And Confers Hypersensitivity To Etoposide In Human Leukemic Cell Lines. *Mol Cancer Ther 1*, 121-131.

[112] Munster, P., Marchion, D., Bicaku, E., Lacevic, M., Kim, J., Centeno, B., Daud, A., Neuger, A., Minton, S., And Sullivan, D. (2009). Clinical And Biological Effects Of Valproic Acid As A Histone Deacetylase Inhibitor On Tumor And Surrogate Tissues: Phase I/Ii Trial Of Valproic Acid And Epirubicin/Fec. *Clin Cancer Res 15*, 2488-2496.

[113] Munster, P., Marchion, D., Bicaku, E., Schmitt, M., Lee, J.H., Deconti, R., Simon, G., Fishman, M., Minton, S., Garrett, C., *Et Al.* (2007). Phase I Trial Of Histone Deacetylase Inhibition By Valproic Acid Followed By The Topoisomerase Ii Inhibitor Epirubicin In Advanced Solid Tumors: A Clinical And Translational Study. *J Clin Oncol 25*, 1979-1985.

[114] Munster, P.N., Marchion, D., Thomas, S., Egorin, M., Minton, S., Springett, G., Lee, J.H., Simon, G., Chiappori, A., Sullivan, D., *Et Al.* (2009). Phase I Trial Of Vorinostat And Doxorubicin In Solid Tumours: Histone Deacetylase 2 Expression As A Predictive Marker. *Br J Cancer 101*, 1044-1050.

INDEX

D

E

F

G

H

I

N

O

P

Q

R

S

T

U

V

W

X

Z